Microplastics

Ajay Kumar Mishra · Pankaj Raizada ·
Elsayed T. Helmy · Santhiagu Arockiasamy ·
Rangabhashiyam Selvasembian
Editors

Microplastics

Environmental Pollution and Degradation Process

Springer

Editors
Ajay Kumar Mishra
Department of Chemistry
University of the Western Cape
Cape Town, South Africa

Elsayed T. Helmy
National Institute of Oceanography
and Fisheries
Alexandria, Egypt

Rangabhashiyam Selvasembian
Department of Environmental Science
and Engineering
SRM University
Amaravati, Andhra Pradesh, India

Pankaj Raizada
School of Advanced Chemical Sciences
Shoolini University of Biotechnology
and Management Sciences
Solan, Himachal Pradesh, India

Santhiagu Arockiasamy
Department of Bioscience and Engineering
National Institute of Technology Calicut
Kozhikode, Kerala, India

ISBN 978-981-97-6460-0 ISBN 978-981-97-6461-7 (eBook)
https://doi.org/10.1007/978-981-97-6461-7

This Springer imprint is published by the registered company Springer Nature Singapore Pte Ltd.
The registered company address is: 152 Beach Road, #21-01/04 Gateway East, Singapore 189721, Singapore

Contents

Chapter 1
Microplastics Pollution: Global Challenge and Future Potential Solution

Annu Sheokand, Sarita Sindhu, Mamta Bulla, Raman Devi, Payal Chaudhary, Ajay Kumar Mishra, and Vinay Kumar

Abstract The global issue of plastic pollution is intensifying day by day due to the growing demand for plastics in every sector and the poor plastic waste management practices. The extensive use and improper disposal of plastics lead to the inevitable introduction of microplastics into the water, soil, and environment. The mechanical and physiochemical characteristics of plastics change due to environmental exposures that lead to the fragmentation of plastics and formation of microplastics (size < 5 mm). Microplastics have the ability to alter the soil natural biogeochemical processes that will affect the functioning of soil microorganisms, directly affecting soil fertility and human life. According to reports, about 4 and 12 million tons of plastic enter the aquatic ecosystems through the seas and oceans, causing the deaths of many aquatic animals. Microplastic plastic pollution is a central issue due to its high stability, long residence time, fragment rapidly and can adsorb other toxic substances, and slow degradation in water and soil. This chapter primarily focuses on the accumulation of microplastics in global environment. This chapter addresses the global scenario of microplastic pollution, highlighting data and source contributions from various sources while discussing, ecosystem impacts, and removal methods. This chapter extensively explores the worldwide issue of microplastic pollution, highlighting various contributing sources and investigating its significant effects on ecosystems, as well as effective removal strategies.

A. Sheokand · S. Sindhu · M. Bulla · R. Devi · V. Kumar (✉)
Department of Physics, COBS&H, CCS Haryana Agricultural University, Hisar 125004, Haryana, India
e-mail: vinay23@hau.ac.in

P. Chaudhary
Department of Microbiology, CCS University, Meerut 250004, UP, India

A. K. Mishra
Department of Chemistry, Durban University of Technology, Steve Biko Road, Durban 4001, South Africa

Department of Chemistry, University of the Western Cape, Robert Sobukwe Road, Bellville 7535, South Africa

A. K. Mishra et al. (eds.), *Microplastics*, https://doi.org/10.1007/978-981-97-6461-7_1

Keywords Microplastic · Pollution · Degradation · Environmental health · Waste management

1 Introduction

Plastic pollution is a major environmental problem that we are facing nowadays due to ever increasing demand for plastic products. Plastic is a low-cost material that is long-lasting, inert, easy to mold, and resistant to disintegration. As a result, it has emerged as an essential material for daily use (Bajt, 2021). Since 1950s, the rate of production of plastics has increased tremendously surpassing any other material. By 2050, it is anticipated that global production of plastic would have increased by 1600 million tonnes. Today, the idea of a plastics-free future appears unrealistic (Rafey & Siddiqui, 2021). A huge amount of plastic garbage is released into the environment as a result of the continuing growth in plastics generation and application. More than 6000 million tons of plastic waste had been produced by the year 2015, with around 80% of that waste being dumped into landfills or released into the environment (Wang et al., 2020a). When exposed to the environment, it undergoes repeated fragmentation due to ultraviolet radiation and mechanical abrasion. This process partially decomposes the plastic trash and transforms it into microplastics which have dimensions of less than 5 mm (Wang et al., 2020b). Microplastics have been categorized into two classes based upon their origin, i.e., primary microplastics and secondary microplastics. Primary microplastics are produced at microscopic sizes for a particular purpose like cosmetic and medical items, raw material for plastic manufacturing units, plastic fibers for synthetic clothing, etc. Secondary microplastics originate as the result of breakdown of bulk plastic materials caused by exposure to sunlight, heat, wind or various other mechanical stress. Out of all degradation mechanisms, photodegradation is the leading one which involves the exposure of light and absorption of photons to transform a compound. When plastic material degrades, it develops obvious flaws like fractures, color transformation, changes in physical and surface characteristics, and breaks up into smaller pieces to form microplastics.

Microplastic pollution has recently gained global attention. Various types of morphologies of microplastics are found in environmental samples, including granules, fibers, and pellets, due to their various sources (Akdogan & Guven, 2019). Primary microplastics enter the marine systems via domestic sewage disposal and leakage of air blasting-related plastic resin particles. Microfibers enter the environment through textile industry wastage and washing machine affluent due to continuous erosion of synthetic clothes which are made of polyester, acrylic, and polyamide. Approximately 1900 fibers per clothing item are produced after each washing which end up in the water bodies. Another big source of microplastics is wastewater treatment plants where large particles are removed from water but microplastics generally penetrate the filtration system and enter the ocean and other water habitats (Padervand et al., 2020). Because of the increased specific surface area and stronger adsorption capabilities of microplastics, pollutants (heavy metals and carcinogenic substances) and pathogens ingested over their surfaces pose a major threat to aquatic species. A large amount of microfibers has been observed in the deeper layers of agricultural soil fertilized with sewage sludge. Microplastics have the ability to alter the physical, chemical, and microbial properties of soils which in turn affects the soil fertility. Also, pollutant availability could be dangerous for agricultural yield and plant health (Campanale et al., 2020). According to the reports, microplastics are found in various food products such as sugar, bottled water, canned fish, mussels, table salt, etc., are directly consumed by people (Lin et al., 2022; Zhang et al., 2020a). Research findings suggest that sea salt contains a higher concentration of microplastics compared to rock salt, which may be attributed to the greater microplastics pollution near coastal areas.

Approximately, 39,000–52,000 particles of microplastics per person are ingested through food each year. Particles can enter the digestive system by contaminated food or mucociliary clearance during inhalation and lead to chronic inflammation which can be the reason for cancer. In addition to ingestion through food, exposure through skin contact and inhalation can also result in the entry of microplastics into the human body. Due to their tiny size and low density, microplastics derived from rubber tires, urban dust, and synthetic clothing fibers can be easily dispersed through the air by wind and enter a person's respiratory system through inhalation. Therefore, a better awareness of the environmental MPs' concentration, source, and risks is required. The interior décor and cleaning practices constantly produce microplastics and the indoor air renovation rates may not be as high, which might result in significant amounts of microplastics in indoor air. On the other hand, the amount of MP is significantly lessened by the outdoor air's dilution. While microplastics and nanoplastics cannot travel through the epidermal membrane, these can penetrate through cuts, sweat glands, or hair follicles (Chen et al., 2020; Prata et al., 2020). After ingestion of microplastics, the body experiences various physical and physiological complications such as blockage, swelling, poor growth, oxidative stress, and hormonal imbalances. Due to their small size and low concentration, microplastics can permeate peripheral tissues and the blood circulation system, exposing the whole body to them. Due to the increased tissue permeability brought on by inflammatory illnesses, patients with gastrointestinal infections will have much higher transmission efficiency of microplastics (Shen et al., 2020). Therefore, it is crucial to get microplastics out of the ecosystem. There are several techniques that are utilized to eliminate the microplastics from both drinking water and effluent of wastewater treatment plants such as coagulation method, photocatalysis, adsorption, degradation by microorganisms and various oxidation processes, for example, Fenton/Fenton-like reactions (Xu et al., 2021). Out of these, photocatalysis and coagulation methods are the recent research hotspots because of their simple and low-cost operation. The sludge from wastewater treatment plants contains a large amount of microplastics and hence must be treated before using as fertilizer. High density solutions like $ZnCl_2$ and NaI are used to remove microplastics from sludge because these have higher density than water. Microplastics float on the saline solution and are captured in the filter where they can be removed from sludge. Aluminum oxide, glass microfiber, polycarbonate, and nitrocellulose are the most often utilized filter materials. The filter must meet a number of specific parameters, including water resistance, an adequate pore size so that water can penetrate through it and negligible spectroscopic interference to minimize errors during the detection and characterization process of microplastics (Sol et al., 2020). The present chapter describes the ongoing initiatives to develop potential methods for elimination, stimulated degradation, and reprocessing of microplastics. In order to better comprehend and manage microplastics pollution, it is critical to properly comprehend all the resources of microplastics and corresponding fate in environment.

2 Origin of Microplastics and Classification

Plastic pollution is currently a major problem on a global scale due to its frequent production and accumulation in natural ecosystems that adversely affects wildlife, humans, and their habitats. Plastic is chemically inert material and hence it does not degrade completely once disposed of. However, when it is exposed to various environmental factors, mechanical abrasions and ultraviolet (UV) light from sunlight, it breaks down into secondary granules. The particles with size less than 5 mm are termed as microplastics whereas the particles with size less than 0.1 μm are called nanoplastics (Yee et al., 2021). Depending on where they come from, microplastics can be broadly classified as either primary or secondary.

2.1 Primary Microplastics

Primary microplastics refer to the small size particles of plastic that are intentionally produced for certain specific applications and dispersed to the environment during washing and cleaning activities. Primary microplastics can enter the environment through several pathways, some are by synthetic fibers produced during laundry, generation of plastic dust during the production of bulk plastic and their transportation, accumulation of microplastics to aquatic environments during marine coating, etc. Most of the primary microplastics are formed as a consequence of using plastic items on a regular basis. For instance, while painting a wall, plastic paint particles fall to the ground and are released into the air inside the house. Despite the fact that the emission mechanism is typically undetectable, there could be serious ecological consequences (Wang et al., 2019). Another major source of primary microplastics is microbeads which are present in various industrial and daily use products such as face wash, scrubbers used for deep cleaning purposes, cosmetics, and medicines. It has been claimed that microbeads like polyethylene come in a variety of shapes and concentrations in facial scrubbers and other personal care items (Karbalaei et al., 2018). Primary microplastics are also produced as byproducts during the manufacturing of other materials. There are mainly two sinks for these particles originating from various sources, i.e. land and water. Microplastics from daily use products enter into the aquatic environment through sewage system. It can even pass the waste treatment mechanisms due to its smaller size and lower density. These particles, like polystyrene, accumulate in the bodies of aquatic animals and can have very dangerous effects due to their being resistant to biodegradation (Li et al., 2018). The sources of primary microplastics are quite predictable and it is relatively easier to control their contamination.

2.2 *Secondary Microplastics*

The production of secondary microplastics is due to degradation or disintegration of larger plastic particles to the smaller ones (Efimova et al., 2018). It is very well known that plastic materials have a high chemical and physical stability and these can survive for thousands of years. But climatic weathering still disintegrates plastics, albeit extremely slowly, and modifies the polymer characteristic via biological or non-biological mechanisms. Abiotic or non-biological mechanisms include the degradation of plastic due to abiotic elements like temperature, water, light, mechanical abrasions, etc. The main cause of secondary microplastics production is poor plastics management (Pourebrahimi & Pirooz, 2023). The main sources of microplastics are shown in Fig. 1. Microplastics have become ubiquitous across the Earth's ecosystems, infiltrating the air, land, and aquatic environments. Their pervasive presence raises concerns about the potential consequences for biodiversity, human health, and ecological balance. The distribution and transport of microplastics vary significantly in each ecosystem, highlighting the diverse sources and pathways through which these tiny pollutants enter and circulate within the environment.

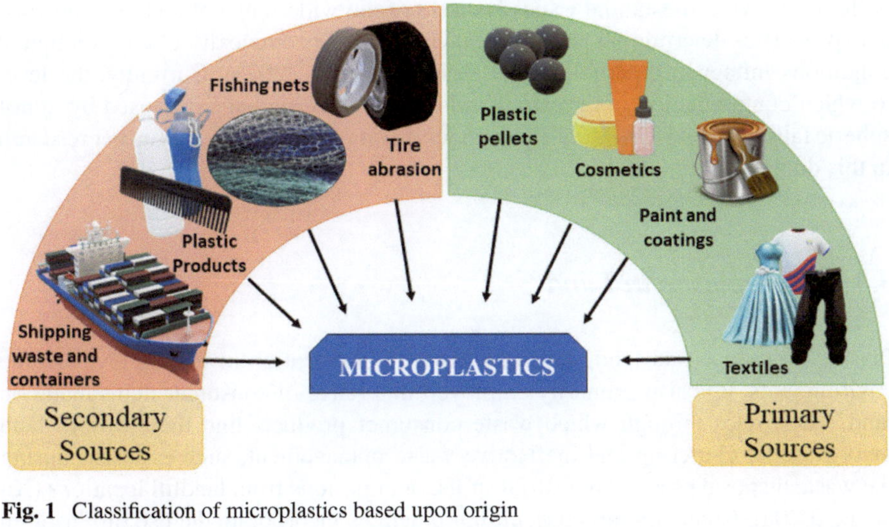

Fig. 1 Classification of microplastics based upon origin

3 Sources of Microplastics Within the Environment

Microplastics can enter the environment through various channels, either in the form of primary microplastics or larger plastic items that eventually break down into secondary microplastics. Microplastics stemming from household items, like microbeads, may be found in wastewater and subsequently discharged into rivers. During production processes, nurdles can be unintentionally released into freshwater. The sources of secondary microplastic include deliberate actions like illegal dumping, mismanaged waste (such as litter), and accidental losses like fishing gear or shipping cargo (Boucher & Friot, 2017). The extent and routes of microplastic release differ across terrestrial, freshwater, and marine environments.

3.1 Microplastic in Air

Recently, it has been acknowledged that numerous microplastic particles, owing to their lightweight nature, have the potential to be suspended and carried in the air, constituting what is referred to as 'urban dust' (Dehghani et al., 2017). These particles commonly stem from sources like road dust (such as tire and paint particles) and fibers originating from synthetic textiles, particularly soft furnishings (Dris et al., 2017). This airborne microplastic matter can result in the deposition of microplastics in both terrestrial and aquatic environments. While urban dust predominantly originates in cities and densely populated areas, the movement of air streams and wind can transport these particles considerable distances from their source (Dris et al., 2016). Events related to the environment, including prolonged rain, favor the settling of particles on land. The tracking and prediction of airborne microplastic pollution provide substantial issues because of the wide range of sources, the variable properties determining particle behavior, and the complexity of environmental conditions influencing particle travel (Sridharan et al., 2021). Currently, the level to which contamination in terrestrial and aquatic environments is caused by atmospheric fallout is not well-established, underscoring the necessity for further research in this domain.

3.2 Microplastic in Land

Plastic is produced on land, and aside from its application in fishing or maritime environments, it is also primarily employed in a range of consumer items made on land. The routes through which waste consumer products find their way to land involve direct cluttering and ineffective waste management, such as losses during the waste disposal process, industrial spills, or emissions from landfill locations (Xu et al., 2020a). Contemporary agricultural practices incorporate the use of plastic in diverse ways, including mulches that can degrade in situ, as well as bale twine and packaging that may be disposed of improperly, contributing to environmental degradation. Additionally, these items can break down, forming secondary microplastics within the environment (Siegfried et al., 2017). Microplastics can also be directly introduced to the land through the use of sewage sludge as a fertilizer in agriculture. While wastewater treatment plants are relatively more efficient at eliminating microplastic particles from the wastewater stream, repeatedly achieving approximately 99% removal, a significant portion of these particles settles into the sludge (Talvitie et al., 2017a). It is estimated that in Europe, an annual addition of 125–850 tons of microplastics per million people occurs in agricultural soils due to sewage sludge application (Nizzetto et al., 2016). According to Horton and his team an estimated 473,000–910,000 metric tonnes of plastic waste, including microplastics from sewage sludge, along with anticipated inputs from litter and ineffectively managed waste, stays within European continental environments (both land and freshwater) each year (Horton et al., 2017). This encompasses instances where plastics remain on land without being transported to rivers or seas, potentially leading to substantial accumulation. However, there is a paucity of studies exploring the abundance of microplastics in terrestrial environments or establishing connections between excess and specific input routes, making it presently unfeasible to directly correlate accumulation with distinct environmental characteristics or human actions.

3.3 Microplastic in Freshwater

Freshwater systems pose the most intricate challenges in terms of microplastic transport and retention. They serve as recipients of microplastics from terrestrial environments, act as routes for transporting microplastics to the marine environment, function as sites for microplastic production through the fragmentation of larger items, and act as repositories collecting microplastics within sediments (Li et al., 2020). The category of "freshwater" encompasses various environments such as rivers, streams, ditches, lakes, and ponds, each exhibiting distinct characteristics. Sizeable plastic items can infiltrate freshwater ecosystems due to improper waste discarding, either through cluttering or leakage from landfills, and can be carried from land by airstream or surface runoff (Yang et al., 2021a). Microplastics have a significant direct impact on freshwater systems in addition to macroplastics. Farmland runoff and drainage can introduce microbeads and fibers made of sewage sludge and agricultural plastics. Microplastics arising from deteriorated road paint and vehicle wear may be carried by storm drainage and urban runoff, which are frequently untreated and unfiltered (Fahrenfeld et al., 2019). Even while wastewater treatment facilities are effective at removing microplastics, these particles might still be present in direct effluent flow. Furthermore, combined sewage overflows (CSOs) discharge raw sewage into adjacent rivers during times of exceptionally high flow, discharging both micro- and macroplastic waste. While studies indicate that microplastic hotspots may be concentrated near urban areas, the primary entry of microplastics into water bodies is likely through drainage systems (Li et al., 2018). Consequently, it is imperative to take into account inputs such CSOs, storm drains, and effluent outfalls, which must be located distant from areas with a high population density (Wang et al., 2021). While the majority of research on microplastics in freshwater environments focuses on rivers, it is acknowledged that microplastics can also be found in ponds and lakes. Similar to rivers, these standing water bodies receive inputs from land runoff and debris from airflow (Kowalski et al., 2016). But, because of the enclosed nature of lakes, the accumulation of microplastics in standing water bodies is expected over time.

3.4 Microplastic in Marine Water

The amount and distribution of microplastics in oceans have been the subject of extensive research. Given that oceans are typically thought of as the ultimate repository for all plastic in the environment, there are a variety of microplastic sources in marine ecosystems (Rocha et al., 2021). Apart from contributions from rivers, plastics can directly enter oceans through improperly managed maritime or fishing waste, encompassing discarded fishing gear, unintentional cargo loss, and illicit dumping (Mai et al., 2023). Typically, this involves macroplastic waste that degrades into microplastics within the marine setting (Thushari & Senevirathna, 2020). Microplastics have been discovered ubiquitously across various sites and within aquatic organisms globally, and ocean streams contribute to specific gathering zones (Alfaro-Núñez et al., 2021), like the well-documented 'Great Pacific Garbage Patch. Various models have been designed to explore the transport processes and how microplastics end up in oceans, potentially enhancing our comprehension of the factors influencing microplastic entry in freshwater environments (Cai et al., 2023).

4 Transport Process

The prevailing belief is that the ocean serves as a repository for a substantial portion of microplastics, while terrestrial and freshwater environments act as crucial sources and routes for these particles to reach the sea (Fig. 2). As microplastics are lighter in weight and potential for extensive dispersal, it is probable that air currents contribute to the transportation of microplastics, thereby contributing to contamination in both terrestrial and aquatic ecosystems (Horton et al., 2017; Vaid et al., 2021).

Numerous studies indicate that rivers transfer macro and microplastic litter to oceans, originating primarily from land-based sources. However, there is a growing acknowledgment that freshwaters and soils are not just conduits for plastic waste but can also function as repositories, retaining a large amount of the microplastic pollution they receive (Strokal et al., 2023). Contrary to being mere conduits for plastic waste, certain terrestrial and freshwater areas may gather microplastics at concentrations higher than those observed in the ocean, owing to their closeness

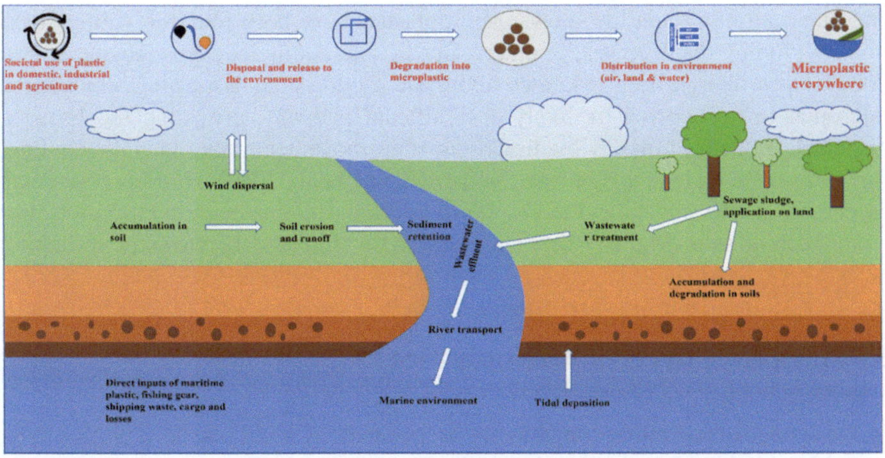

Fig. 2 Conceptual model for the transport of microplastic into the environment

and the scale of plastic involvements. To comprehensively understand microplastic pollution, it is crucial to establish connections between sources, particle behaviors, and transport mechanisms to discern where and how microplastics will accumulate (Su et al., 2022). Agricultural soils may serve as substantial sources of microplastics entering rivers, particularly through the use of sewage slush as fertilizer. However, a considerable proportion is likely retained within the soil. Studies have shown that synthetic fibers from sewage sludge persist up to 15 years after the treatment of agricultural land with sewage sludge, with accumulation hotspots even at significant depths (Tian et al., 2022).

Soil retention is further eased by methods like bioturbation, which draws particles into deeper soil layers. Agricultural and forest soils are more adept at retaining particles compared to urban land due to their permeable nature and lower rates of overland flow. When microplastics enter rivers, they undergo transport processes similar to other sediments, such as sand and silt (Yang et al., 2021b). However, most rivers are likely to be supply-limited, meaning they can transport all plastics delivered to them. In slower-moving sections, it is more probable that microplastics settle along with sediment particles that are sinking. In lakes with high sediment accumulation rates, microplastics may be retained and incorporated into sediments, leading to long-term preservation. The density and shape of microplastic particles play critical roles in their transport and retention in sediments. Size, shape, and density variations add complexity to predicting microplastic fate and transport them in rivers. Sediment transport and deposition in rivers exhibit temporal and spatial variability, influenced by factors such as turbulence, floods, and changes in river channel morphology (Román-Sánchez et al., 2019; Swarnkar et al., 2020).

Once in the oceans, microplastics can disperse widely due to water streams, and winds, in the vast areas enclosed. Vertical transportation of microplastic within the oceans is facilitated by processes like biofouling, egestion in faecal pellets, and integration into marine snows. Microplastics have been found in various locations, including the deep sea, Southern Ocean, and Arctic ice cores (Haque & Fan, 2022). Even though little is known about the progressions governing atmospheric microplastic transport, it is likely a significant pathway, providing the broadest dispersal, influenced mainly by directions of air movement rather than the confined flows observed on land and in water bodies (Zhang et al., 2020b). Further research is essential to better understand atmospheric microplastic transport and its connection to aquatic and terrestrial contamination due to the limited data currently available.

5 Harmful Effects of Microplastics

In biological organization, negative consequences of microplastic pollution have been identified, at different levels and it currently seems to present a threat to the environment. It appears that microplastic invasion happens in all ecosystems, from terrestrial to marine settings at various trophic levels (Ogonowski et al., 2018).

5.1 Effect on Marine Life

Consumption of microplastics is a hazard for marine life due to their microscopic size and universal presence throughout the aquatic ecosystem. The size, density, abundance, and color of microplastics influence their bioavailability to a particular organism (Courtene-Jones et al., 2022). Microplastics possess a large specific surface area and are smaller in size, which makes them accessible to a variety of aquatic creatures (Xu et al., 2020b). Many researchers worldwide have studied the accumulation of microplastic in various marine living organisms, including sea cucumbers, copepods, amphipods, fishers, turtles, and birds that eat these sea creatures. Their findings suggest that microplastic, once consumed, can move within the tissues of different organs, undergo excretion via pseudo-faeces, or accumulate in a specific body tissue. When microplastics build up in internal tissue, they may cause internal or external damage, growth retardation, infertility, blockage of body passages, and other harmful consequences on an organism's health (Gola et al., 2021). According to studies, fish tissues contain the same compounds that are used to make plastics. Even though the organisms released a portion of the accumulated microplastic particles, some remained inside the intestinal villi and epithelial cells. Additionally, the study found that the organic pollutant benzo [a] pyrene (BaP) was transferred through the microplastic particles from nauplii to zebrafish and that the material was retained in the intestines. These findings demonstrate the ability of microplastics and related harmful substances to travel along the food chains across different trophic levels (Auta et al., 2017). Numerous investigations on aquatic animals have shown the harmful, if not fatal, effects of microplastics on the molecular, cellular, and individual levels. Exposure to microplastics causes a variety of negative consequences and stress reactions in cells. Microplastics change the behavior of genes linked to a wide range of functions at the molecular level (Xu et al., 2020b).

5.2 *Effects of Microplastic on Soil*

Microplastics have the ability to change the bulk density of soil which is a critical characteristic for predicting soil carbon storage as plastics typically have a much lower density than soil particles. Polymer sheet Microplastics in soils have the potential to alter the porosity of the soil, which would accelerate soil water evaporation and lead to soil cracking (Ya et al., 2021). The biophysical properties of soil can be influenced by microplastic by changing the pH of soil, soil structure, soil microorganisms, soil fertility and nutrients, and water-stable aggregates. As microplastics in soil frequently contain other chemical pollutants such as dioxins, heavy metals, and persistent organic pollutants, they may be considered a composite environmental contaminant (Yang et al., 2021b). This may have a substantial influence on the structural and functional variability of the microbial communities in the soil, potentially leading to major environmental issues with the soil. This comprises the range of functions that soil bacteria and enzymes perform. Due to a decrease in microbial community growth, this alteration in the structure and composition of the soil may also have an impact on the soil's fertility (Khalid et al., 2020). Plant roots find it more difficult to absorb nutrients and water when MPs are present. In addition, mechanical obstruction slows down the development of roots. MPs in the form of aggregates can be absorbed by or adsorbed by plants ("Uptake & accumulation of nano, microplastics in plants: a critical review", 2021). The most plastic-polluted places are agricultural ecosystems that are close to landfills or metropolitan areas and enter the food chain. In order to ensure safe food production techniques, it is imperative to give the effects of MPs in agroecosystems immediate attention (Ullah et al., 2021).

5.3 *Effect on Human Health*

Numerous investigations have demonstrated that there are easily inhaled MPs in the atmosphere. It is well known that breathing in microplastic particles exposes the lungs, which is why respiratory discomfort is more common (Blackburn & Green, 2022). One of the main ways that microplastics enter the human body is through the consumption of contaminated food. Particles of size smaller than 2.5 μm have the ability to penetrate the circulatory system through the gastrointestinal tract after being consumed through M cells' endocytosis. Due to the persistent nature of microplastics and their special qualities, such as their chemical composition and hydrophobic nature, the resulting toxicity manifests as inflammation. This toxicity is thought to have a dose-dependent effect (Campanale et al., 2020). A significant portion of the human diet consists of sea food, and the high concentration of microplastics in this food poses a major risk to human health. Numerous investigations have revealed the existence of microplastics in consumable fish, and prawns; thus, the microplastics are able to reach human systems through biomagnification (Ajith et al., 2020). Possible microplastics have also been linked to other foods, such as honey, sugar, sea salt, and others, in addition to seafood. Airborne microplastics have the potential to land on plants like flowers and foliage, where they may mix with pollen and be carried by bees back to their hive. When it comes to salt, the most probable cause of contamination is probably the coastal waters used to make sea salt, however microplastics might also have been deposited there by the atmosphere as well (Wright & Kelly, 2017). However, using personal care products, such as body washes, face scrubs, makeup products might result in dermal exposure to microplastics. Synthetic clothing and textiles, which can shed microfibers after washing and wear, can also contain microplastics. Although effects of microplastic exposure on human health are still being investigated, there is evidence that microplastic exposure has a number of adverse effects on human health. Numerous studies demonstrate that microplastics may cause genotoxicity, cellular damage, and inflammation in human cells. To safeguard both human health and ecosystems, it is crucial to limit acquaintance with microplastics and their discharge into the environment (Ghosh et al., 2023).

6 Treatment of Microplastics Pollution

Microplastic pollution has emerged as a global environmental concern due to its pervasive presence in various ecosystems. Microplastics are found virtually everywhere on Earth, indicating their widespread distribution. They have infiltrated terrestrial ecosystems, including soils and sediments, aquatic environments such as oceans, rivers, and lakes, and even the Earth's atmosphere (Issac & Kandasubramanian, 2021). Microplastics do not readily decompose or break down into harmless components over time. Their durable nature allows them to remain in the environment for extended periods, contributing to long-term contamination and potential harm to ecosystems (Bandow et al., 2017). Microplastics have high mobility due to their small size in aquatic environments, they can be transported by currents, tides, and water flow (Zhang, 2017). Wind can carry microplastics through the atmosphere, contributing to their dispersion over large distances. Additionally, human activities, such as plastic waste mismanagement and runoff, play a role in the mobility of microplastics within ecosystems. The presence of microplastics in the environment poses potential ecological threats. In aquatic ecosystems, marine life may ingest microplastics, leading to physical harm, reduced feeding efficiency, and potential bioaccumulation up the food chain (Roy et al., 2022). The interaction of microplastics with soil may affect nutrient cycling and soil health. Moreover, the inhalation of airborne microplastics and their deposition in terrestrial ecosystems could have implications for both human and environmental health (Prata et al., 2021). Therefore, there is a pressing need for effective treatment approaches. These approaches aim to mitigate the adverse effects of microplastic pollution, including strategies for the removal, degradation, and prevention of further microplastic contamination in various environmental compartments.

Although microplastics are present in soil, air, and water more strategies have been implemented for the removal of microplastics in water systems because these particles pose a direct threat to aquatic ecosystems, as water bodies are primary repositories for plastic pollution. Also, microplastics can enter the food chain through aquatic organisms, possibly leading to human ingestion. Furthermore, water bodies serve as the sink of microplastics coming from various pathways (Schell et al., 2020). This has raised public health concerns, contributing to a focus on water-based strategies.

6.1 Physical Methods

Physical methods for removing microplastics include flotation, sedimentation, filtration, adsorption on algae, membrane bioreactors, etc. The primary step of the water treatment is sedimentation and most of the heavy particles are removed by this process. At the end of this step. approximately, 41% of microplastics can be removed from water (Liu et al., 2019). The next step is removal of microplastics by using dissolved air floatation method. In this method, air is introduced into the polluted water at high pressure which causes the formation of air bubbles on the surfaces of the suspended solids, oils, etc. These bubbles lead to the movement of these insoluble particles to the surface of water where these can be removed by skimming process. This technique is a highly efficient method to remove microplastics from water (Talvitie et al., 2017b).

(a) **Filtration Technology**

Filtration technology is the most common and widely used method for water treatment. Most commonly used filters are shown in Fig. 3. Sand filters use a bed of sand as the filtering medium to trap the particles on the surface of the sand grains and remove them, providing a relatively simple and effective method for water filtration. Sand filters are commonly employed in various applications, including municipal water treatment plants, swimming pools, and industrial water treatment (Bui et al., 2020). In rapid sand filters multilayer media is used for the filtration of particles depending upon their size. Most widely used filters use the three-layer medium with anthracite grains, silica sand, and gravel particles. The sand is carefully graded to create a porous medium with specific particle sizes that allow for effective filtration (Bayo et al., 2020). Recently activated carbons have gained significant attention in filtration technology due to their appreciable porosity, high surface area, and adsorption capacity (Bui et al., 2020). Another type of filter are disc filters that have specific pore size to remove the microplastics from wastewater. Disc filters use a filtration

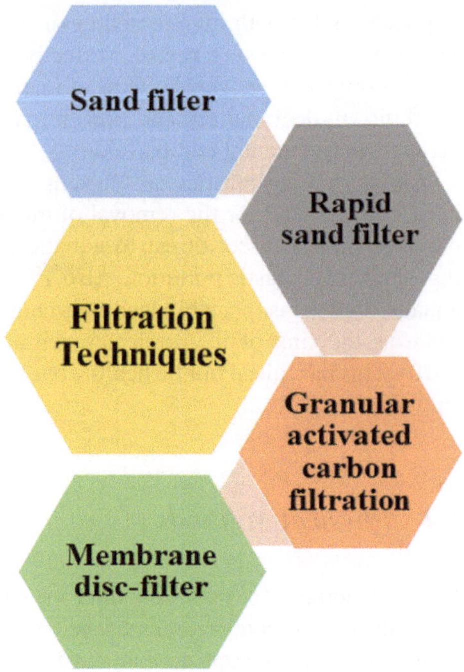

Fig. 3 Various types of filtration techniques

mechanism to trap particles, including microplastics, as water passes through the fine filter media. Many disc filters have a backwashing capability. Backwashing entails reversing the water flow through the filter that displaces and eliminates gathered particles from the filter media, thereby prolonging the operational lifespan of the filter (Poerio et al., 2019).

(b) **Dynamic Membrane Technology**

In this approach, a secondary membrane, or dynamic membrane, is generated on a supporting layer made up of a filter cloth, mesh, or membrane. This occurs when solid particles such as microbial cells and aggregates are suspended in the solution. Figure 4 demonstrates the schematics of the formation of the cake layer. This layer on the surface of membrane is crucial in determining the rejection properties of the system. It acts as a "secondary" membrane preceding the "real" membrane or the supporting membrane. This method is cost-effective as compared to advanced ultrafiltration membranes as economical materials such as mesh, non-woven and woven filter cloth can serve as supporting layers to form a dynamic membrane layer. The particles comprising the secondary membrane are larger than the pores of the supporting membrane, enabling them to capture microplastics. However, a notable drawback of this approach is that the membrane got clogged after sometime (Coolen et al., International Innovation, Engineering & Entrepreneurship).

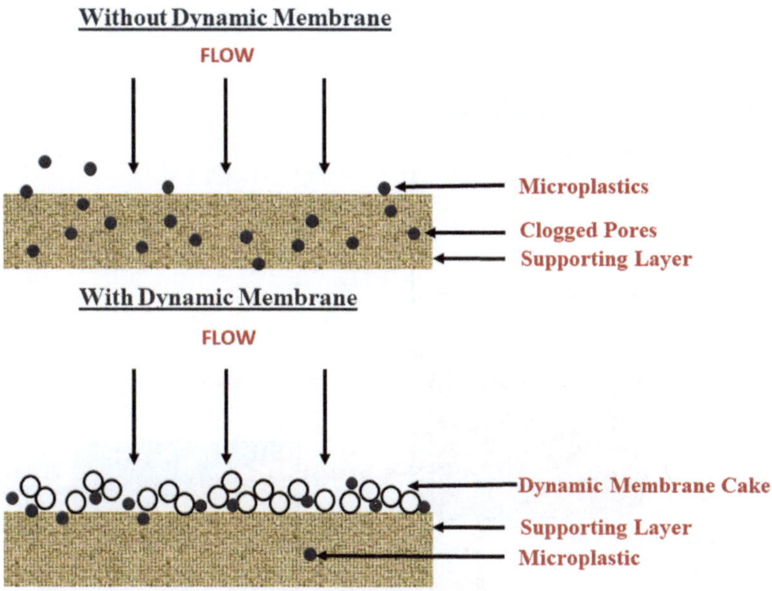

Fig. 4 Schematics of formation of the dynamic cake layer

(c) **Activated Sludge Process**

The Activated Sludge Process (ASP) is a well-established and commonly used wastewater treatment approach in which activated sludge, that is a mixture of microorganisms, decomposes organic contaminants in the wastewater. Subsequently, the processed water is recovered from the activated sludge via sedimentation process in a tank. The microplastics in ASP may either attach to solids suspended in the solution and be removed during the settling process or act as carriers for microbial attachment, contributing to the overall treatment of wastewater. About 42–77 % of microplastics can be removed from water by using this method (Hidayaturrahman & Lee, 2019).

(d) **Membrane Bioreactors**

Membrane Bioreactors (MBRs) are systems in which biological catalysis by microorganisms, enzymes, bacteria, and membrane filtration are coupled together for removal of microplastics. The biodegradation of organic material reduces the complexity of the contaminated water and facilitates the removal of microplastics. Firstly, the pre-treated water streams enter the bioreactor, where the biodegradation process of organic matter takes place. Then the resultant mixture is conveyed to a semi-crossflow filtration scheme for the separation process, concentrating the MP in the retentate stream (Judd, 2016; Poerio et al., 2019) as shown in Fig. 5.

Compared to simple dynamic membranes, membrane bioreactors exhibit enhanced capacities for the separation of microplastics. This method is used to treat industrial wastewater which is a complex mixture of various contaminants that

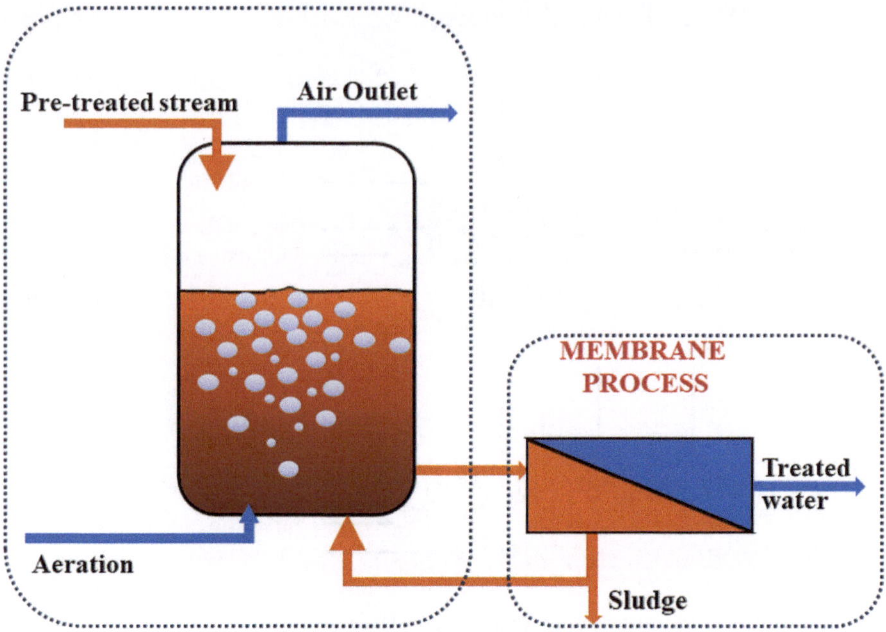

Fig. 5 Schematics of the membrane reactor system

confirms its applicability for treating potent contaminants like polymeric debris and microplastics. The successful treatment of various complex industrial wastewaters by membrane bioreactors confirms the applicability of this technology for treating high-strength pollutants like polymeric debris and microplastics. The removal efficiency of the membranes is particularly dependent on factors such as durability, incoming flow, concentration, and size of the microplastics. Integration of the porous membranes and biological processes has the potential to significantly improve the removal efficiency of the system, reaching up to 99.9% (Gurung et al., 2016; Lares et al., 2018; Talvitie et al., 2017b).

6.2 Chemical Methods

In this methodology, a chemical agent is introduced into contaminated water to facilitate the agglomeration of waste particles into larger entities, thereby enabling their facile separation. Typically, coagulants such as Fe^{3+} and Al^{3+} based salts are employed, operating through ligand exchange mechanisms to establish robust bonds between waste particles. For microplastics with dimensions below 0.5 mm, polyacrylamide has demonstrated notable efficacy as a coagulating agent, achieving a removal efficiency of approximately 61.19% at a concentration of 15 mg/L. The efficacy of removal is contingent upon variables such as pH, coagulant concentration, and solution composition (Padervand et al., 2020). Another approach is the agglomeration of microplastics due to the formation of alkoxy-silyl bond via sol–gel reactions. As a result of the reaction, there is the formation of large-size 3D agglomerated particles that that can be eliminated from the solution easily (Herbort et al., 2018). The photocatalytic degradation of microplastics by a semiconducting photocatalyst in the presence of light is the most recent technology that is being explored nowadays. A protein-based N-TiO_2 photocatalyst can be employed to decompose the polyethylene microplastics in both solid and aqueous phases. Although it has low efficiency (1.1% after 20 h of treatment) it depends upon surface area and band gap of the photocatalyst which can be controlled to get higher removal efficiency (Ariza-Tarazona et al., 2019).

6.3 Biodegradation of Microplastics

Microorganisms such as fungi, and bacteria can disintegrate the microplastic waste materials at different physical conditions such as moisture, temperature, and pressure called biodegradation (Pramila & Ramesh, 2011; Usha et al., 2011). These microorganisms have the capacity to excrete extracellular enzymes, leading to the disintegration of the polymers into monomers. Some of the microbial species can form a substantial biofilm on the weathered surface of microplastics, eliciting modifications in surface properties. In the course of bacterial growth, various metabolic cycles may ensue, utilizing these plastics as carbon source to derive the energy. Some bacteria such as *Bacillus amyloliquefaciens, Bacillus cereus, Alcaligenes faecalis, Paenibacillus* sp., *Pseudomonas* sp., and *Streptomyces* sp. can degrade natural and synthetic plastics (Kumar et al., 2013). Microplastics may be classified into high-density polyethylene (HDPE) and low-density polyethylene (LDPE). The observed variations in the biodegradation rates among LDPE, HDPE, and polyester could be attributed to the presence of different types of enzymes and their corresponding concentration which are specific to the degradation of different types of microplastics (Park & Kim, 2019).

Microbial enzyme-mediated degradation comprises two sequential stages: extracellular and intracellular processes. During the extracellular phase, LDPE undergoes break down into shorter chains, followed by the subsequent intracellular mineralization, transforming the polymer into carbon dioxide (CO_2) and water (H_2O).

Many extracellular hydrolytic enzymes such as lipase, xylanase, keratinase CMCase, protease, and chitinase are secreted by the bacteria that play a charismatic role in microplastic degradation. Bio-plastics such as polylactic acid, polyhydroxy butyrate, and polybutylene succinate can be readily decomposed by microorganisms. Synthetic polymers, including polyethylene, polystyrene, and polyamide cannot be easily degraded by environmental conditions. Various studies have indicated that the degradation rate of polyethylene ranges from 1.5 to 13%, depending on the specific assays and types of microbes employed (Das & Kumar, 2015).

7 Conclusions

In our technologically advanced era, plastics play an integral role in various aspects of daily life. Despite extensive efforts to identify alternatives, a suitable substitute for plastics remains elusive, resulting in a continuous accumulation of a substantial volume of plastic waste on Earth. Despite possessing high resistance to degradation, plastics undergo partial degradation due to mechanical stress and environmental factors such as wind, temperature, and sunlight. This degradation process contributes to the formation of microplastics. Furthermore, micro-sized plastic particles are intentionally created for specific applications, and upon disposal, they enter various ecosystems. Microplastics, whether originating from primary or secondary sources, disrupt the natural processes within ecosystems, posing a significant threat to aquatic environments, soil health, air quality, and ultimately, human well-being. Consequently, there is an urgent need to develop effective techniques to address the pervasive issue of microplastic pollution. This chapter extensively examines the sources, impacts, and treatment technologies associated with mitigating microplastic pollution. While considerable efforts have been made to comprehend the potential harms of microplastics and assess the efficiency of removal technologies, there is a notable gap in research. As a result, more research and advancements in this area are crucial to gain precise insights into the potential threats posed by microplastics and to develop effective solutions. Moreover, the global community must unite in implementing responsible waste management practices to curb the influx of microplastics into our oceans, rivers, and soils.

References

Ajith, N., Arumugam, S., Parthasarathy, S., Manupoori, S., & Janakiraman, S. (2020). Global distribution of microplastics and its impact on marine environment—A review. *Environmental Science and Pollution Research, 27*, 25970–25986.

Akdogan, Z., & Guven, B. (2019). Microplastics in the environment: A critical review of current understanding and identification of future research needs. *Environmental Pollution, 254*, 113011.

Alfaro-Núñez, A., Astorga, D., Cáceres-Farías, L., Bastidas, L., Soto Villegas, C., Macay, K., & Christensen, J. H. (2021). Microplastic pollution in seawater and marine organisms across the Tropical Eastern Pacific and Galápagos. *Scientific Reports, 11*(1), 6424.

Ariza-Tarazona, M. C., Villarreal-Chiu, J. F., Barbieri, V., Siligardi, C., & Cedillo-González, E. I. (2019). New strategy for microplastic degradation: Green photocatalysis using a protein-based porous N-TiO2 semiconductor. *Ceramics International, 45*(7), 9618–9624.

Auta, H. S., Emenike, C. U., & Fauziah, S. H. (2017). Distribution and importance of microplastics in the marine environment: A review of the sources, fate, effects, and potential solutions. *Environment International, 102*, 165–176.

Azeem, I., Adeel, M., Ahmad, M. A., Shakoor, N., Jiangcuo, G. D., Azeem, K., & Ishfaq, M., et al. (2021). Uptake and accumulation of nano/microplastics in plants: a critical review. *Nanomaterials, 11*, 2935.

Bajt, O. (2021). From plastics to microplastics and organisms. *FEBS Open Bio, 11*(4), 954–966.

Bandow, N., Will, V., Wachtendorf, V., & Simon, F. G. (2017). Contaminant release from aged microplastic. *Environmental Chemistry, 14*(6), 394–405.

Bayo, J., López-Castellanos, J., & Olmos, S. (2020). Membrane bioreactor and rapid sand filtration for the removal of microplastics in an urban wastewater treatment plant. *Marine Pollution Bulletin, 156*, 111211.

Blackburn, K., & Green, D. (2022). The potential effects of microplastics on human health: What is known and what is unknown. *Ambio, 51*(3), 518–530.

Boucher, J., & Friot, D. (2017). *Primary microplastics in the oceans: A global evaluation of sources* (Vol. 10). Gland, Switzerland: IUCN.

Bui, X. T., Nguyen, P. T., Nguyen, V. T., Dao, T. S., & Nguyen, P. D. (2020). Microplastics pollution in wastewater: Characteristics, occurrence and removal technologies. *Environmental Technology & Innovation, 19*, 101013.

Cai, C., Zhu, L., & Hong, B. (2023). A review of methods for modeling microplastic transport in the marine environments. *Marine Pollution Bulletin, 193*, 115136.

Campanale, C., Massarelli, C., Savino, I., Locaputo, V., & Uricchio, V. F. (2020). A detailed review study on potential effects of microplastics and additives of concern on human health. *International Journal of Environmental Research and Public Health, 17*(4), 1212.

Chen, G., Feng, Q., & Wang, J. (2020). Mini-review of microplastics in the atmosphere and their risks to humans. *Science of the Total Environment, 703*, 135504.

Coolen, M., van Gerwen, M., Schot, H., Szyłak, M., & Tax, E. Dynamic microplastic filter. *International Innovation, Engineering & Entrepreneurship*, 57.

Courtene-Jones, W., Clark, N. J., Fischer, A. C., Smith, N. S., & Thompson, R. C. (2022). Ingestion of microplastics by marine animals. *Plastics and the Ocean: Origin, Characterization, Fate, and Impacts*, 349–366.

Das, M. P., & Kumar, S. (2015). An approach to low-density polyethylene biodegradation by Bacillus amyloliquefaciens. *3 Biotech, 5*(1), 81–86.

Dehghani, S., Moore, F., & Akhbarizadeh, R. (2017). Microplastic pollution in deposited urban dust, Tehran metropolis. *Iran. Environmental Science and Pollution Research, 24*, 20360–20371.

Dris, R., Gasperi, J., Saad, M., Mirande, C., & Tassin, B. (2016). Synthetic fibers in atmospheric fallout: A source of microplastics in the environment? *Marine Pollution Bulletin, 104*(1–2), 290–293.

Dris, R., Gasperi, J., Mirande, C., Mandin, C., Guerrouache, M., Langlois, V., & Tassin, B. (2017). A first overview of textile fibers, including microplastics, in indoor and outdoor environments. *Environmental Pollution, 221*, 453–458.

Efimova, I., Bagaeva, M., Bagaev, A., Kileso, A., & Chubarenko, I. P. (2018). Secondary microplastics generation in the sea swash zone with coarse bottom sediments: Laboratory experiments. *Frontiers in Marine Science, 5*, 313.

Fahrenfeld, N. L., Arbuckle-Keil, G., Beni, N. N., & Bartelt-Hunt, S. L. (2019). Source tracking microplastics in the freshwater environment. *TrAC Trends in Analytical Chemistry, 112*, 248–254.

Ghosh, S., Sinha, J. K., Ghosh, S., Vashisth, K., Han, S., & Bhaskar, R. (2023). Microplastics as an emerging threat to the global environment and human health. *Sustainability, 15*(14), 10821.

Gola, D., Tyagi, P. K., Arya, A., Chauhan, N., Agarwal, M., Singh, S. K., & Gola, S. (2021). The impact of microplastics on marine environment: A review. *Environmental Nanotechnology, Monitoring & Management, 16*, 100552.

Gurung, K., Ncibi, M. C., Fontmorin, J. M., Särkkä, H., & Sillanpää, M. J. J. M. S. T. (2016). Incorporating submerged MBR in conventional activated sludge process for municipal wastewater treatment: A feasibility and performance assessment. *Journal of Membrane Science & Technology, 6*(3).

Haque, F., & Fan, C. (2022). Microplastics in the marine environment: A review of their sources, formation, fate, and ecotoxicological impact.

Herbort, A. F., Sturm, M. T., Fiedler, S., Abkai, G., & Schuhen, K. (2018). Alkoxy-silyl induced agglomeration: A new approach for the sustainable removal of microplastic from aquatic systems. *Journal of Polymers and the Environment, 26*, 4258–4270.

Hidayaturrahman, H., & Lee, T. G. (2019). A study on characteristics of microplastic in wastewater of South Korea: Identification, quantification, and fate of microplastics during treatment process. *Marine Pollution Bulletin, 146*, 696–702.

Horton, A. A., Walton, A., Spurgeon, D. J., Lahive, E., & Svendsen, C. (2017). Microplastics in freshwater and terrestrial environments: Evaluating the current understanding to identify the knowledge gaps and future research priorities. *Science of the Total Environment, 586*, 127–141.

Issac, M. N., & Kandasubramanian, B. (2021). Effect of microplastics in water and aquatic systems. *Environmental Science and Pollution Research, 28*, 19544–19562.

Judd, S. J. (2016). The status of industrial and municipal effluent treatment with membrane bioreactor technology. *Chemical Engineering Journal, 305*, 37–45.

Karbalaei, S., Hanachi, P., Walker, T. R., & Cole, M. (2018). Occurrence, sources, human health impacts and mitigation of microplastic pollution. *Environmental Science and Pollution Research, 25*, 36046–36063.

Khalid, N., Aqeel, M., & Noman, A. (2020). Microplastics could be a threat to plants in terrestrial systems directly or indirectly. *Environmental Pollution, 267*, 115653.

Kowalski, N., Reichardt, A. M., & Waniek, J. J. (2016). Sinking rates of microplastics and potential implications of their alteration by physical, biological, and chemical factors. *Marine Pollution Bulletin, 109*(1), 310–319.

Kumar, S., Das, M. P., Rebecca, L. J., & Sharmila, S. (2013). Isolation and identification of LDPE degrading fungi from municipal solid waste. *Journal of Chemical and Pharmaceutical Research, 5*(3), 78–81.

Lares, M., Ncibi, M. C., Sillanpää, M., & Sillanpää, M. (2018). Occurrence, identification and removal of microplastic particles and fibers in conventional activated sludge process and advanced MBR technology. *Water Research, 133*, 236–246.

Li, J., Liu, H., & Chen, J. P. (2018). Microplastics in freshwater systems: A review on occurrence, environmental effects, and methods for microplastics detection. *Water Research, 137*, 362–374.

Li, C., Busquets, R., & Campos, L. C. (2020). Assessment of microplastics in freshwater systems: A review. *Science of the Total Environment, 707*, 135578.

Lin, Q., Zhao, S., Pang, L., Sun, C., Chen, L., & Li, F. (2022). Potential risk of microplastics in processed foods: Preliminary risk assessment concerning polymer types, abundance, and human exposure of microplastics. *Ecotoxicology and Environmental Safety, 247*, 114260.

Liu, X., Yuan, W., Di, M., Li, Z., & Wang, J. (2019). Transfer and fate of microplastics during the conventional activated sludge process in one wastewater treatment plant of China. *Chemical Engineering Journal, 362*, 176–182.

Mai, L., Sun, X., & Zeng, E. Y. (2023). Country-specific riverine contributions to marine plastic pollution. *Science of the Total Environment, 874*, 162552.

Nizzetto, L., Futter, M., & Langaas, S. (2016). Are agricultural soils dumps for microplastics of urban origin?

Ogonowski, M., Gerdes, Z., & Gorokhova, E. (2018). What we know and what we think we know about microplastic effects–A critical perspective. *Current Opinion in Environmental Science & Health, 1*, 41–46.

Padervand, M., Lichtfouse, E., Robert, D., & Wang, C. (2020). Removal of microplastics from the environment. A review. *Environmental Chemistry Letters, 18*, 807–828.

Park, S. Y., & Kim, C. G. (2019). Biodegradation of micro-polyethylene particles by bacterial colonization of a mixed microbial consortium isolated from a landfill site. *Chemosphere, 222*, 527–533.

Poerio, T., Piacentini, E., & Mazzei, R. (2019). Membrane processes for microplastic removal. *Molecules, 24*(22), 4148.

Pourebrahimi, S., & Pirooz, M. (2023). Microplastic pollution in the marine environment: A review. *Journal of Hazardous Materials Advances*, 100327.

Pramila, R., & Ramesh, K. V. (2011). Biodegradation of low density polyethylene (LDPE) by fungi isolated from marine water a SEM analysis. *African Journal of Microbiology Research, 5*(28), 5013–5018.

Prata, J. C., da Costa, J. P., Lopes, I., Duarte, A. C., & Rocha-Santos, T. (2020). Environmental exposure to microplastics: An overview on possible human health effects. *Science of the Total Environment, 702*, 134455.

Prata, J. C., da Costa, J. P., Lopes, I., Andrady, A. L., Duarte, A. C., & Rocha-Santos, T. (2021). A One Health perspective of the impacts of microplastics on animal, human and environmental health. *Science of the Total Environment, 777*, 146094.

Rafey, A., & Siddiqui, F. Z. (2021). A review of plastic waste management in India–challenges and opportunities. *International Journal of Environmental Analytical Chemistry*, 1–17.

da Rocha, F. O. C., Martinez, S. T., Campos, V. P., da Rocha, G. O., & de Andrade, J. B. (2021). Microplastic pollution in Southern Atlantic marine waters: review of current trends, sources, and perspectives. *Science of the Total Environment, 782*, 146541.

Román-Sánchez, A., Reimann, T., Wallinga, J., & Vanwalleghem, T. (2019). Bioturbation and erosion rates along the soil-hillslope conveyor belt, part 1: Insights from single-grain feldspar luminescence. *Earth Surface Processes and Landforms, 44*(10), 2051–2065.

Roy, P., Mohanty, A. K., & Misra, M. (2022). Microplastics in ecosystems: Their implications and mitigation pathways. *Environmental Science: Advances, 1*(1), 9–29.

Schell, T., Rico, A., & Vighi, M. (2020). Occurrence, fate and fluxes of plastics and microplastics in terrestrial and freshwater ecosystems. *Reviews of Environmental Contamination and Toxicology, 250*, 1–43.

Shen, M., Song, B., Zhu, Y., Zeng, G., Zhang, Y., Yang, Y., ... & Yi, H. (2020). Removal of microplastics via drinking water treatment: current knowledge and future directions. *Chemosphere, 251*, 126612.

Siegfried, M., Koelmans, A. A., Besseling, E., & Kroeze, C. (2017). Export of microplastics from land to sea. A modelling approach. *Water Research, 127*, 249–257.

Sol, D., Laca, A., Laca, A., & Díaz, M. (2020). Approaching the environmental problem of microplastics: Importance of WWTP treatments. *Science of the Total Environment, 740*, 140016.

Sridharan, S., Kumar, M., Singh, L., Bolan, N. S., & Saha, M. (2021). Microplastics as an emerging source of particulate air pollution: A critical review. *Journal of Hazardous Materials, 418*, 126245.

Strokal, M., Vriend, P., Bak, M. P., Kroeze, C., van Wijnen, J., & van Emmerik, T. (2023). River export of macro-and microplastics to seas by sources worldwide. *Nature Communications, 14*(1), 4842.

Su, L., Xiong, X., Zhang, Y., Wu, C., Xu, X., Sun, C., & Shi, H. (2022). Global transportation of plastics and microplastics: A critical review of pathways and influences. *Science of the Total Environment, 831*, 154884.

Swarnkar, S., Sinha, R., & Tripathi, S. (2020). Morphometric diversity of supply-limited and transport-limited river systems in the Himalayan foreland. *Geomorphology, 348*, 106882.

Talvitie, J., Mikola, A., Setälä, O., Heinonen, M., & Koistinen, A. (2017a). How well is microlitter purified from wastewater?–A detailed study on the stepwise removal of microlitter in a tertiary level wastewater treatment plant. *Water Research, 109*, 164–172.

Talvitie, J., Mikola, A., Koistinen, A., & Setälä, O. (2017b). Solutions to microplastic pollution–Removal of microplastics from wastewater effluent with advanced wastewater treatment technologies. *Water Research, 123*, 401–407.

Thushari, G. G. N., & Senevirathna, J. D. M. (2020). Plastic pollution in the marine environment. *Heliyon, 6*(8).

Tian, L., Jinjin, C., Ji, R., Ma, Y., & Yu, X. (2022). Microplastics in agricultural soils: Sources, effects, and their fate. *Current Opinion in Environmental Science & Health, 25*, 100311.

Ullah, R., Tsui, M. T. K., Chen, H., Chow, A., Williams, C., & Ligaba-Osena, A. (2021). Microplastics interaction with terrestrial plants and their impacts on agriculture. *Journal of Environmental Quality, 50*(5), 1024–1041.

Usha, R., Sangeetha, T., & Palaniswamy, M. (2011). Screening of polyethylene degrading microorganisms from garbage soil. *Libyan Agriculture Research Center Journal International, 2*(4), 200–204.

Vaid, M., Mehra, K., & Gupta, A. (2021). Microplastics as contaminants in Indian environment: A review. *Environmental Science and Pollution Research*, 1–28.

Wang, T., Li, B., Zou, X., Wang, Y., Li, Y., Xu, Y., ... & Yu, W. (2019). Emission of primary microplastics in mainland China: Invisible but not negligible. *Water Research, 162*, 214-224.

Wang, W., Ge, J., Yu, X., & Li, H. (2020a). Environmental fate and impacts of microplastics in soil ecosystems: Progress and perspective. *Science of the Total Environment, 708*, 134841.

Wang, W., Ge, J., & Yu, X. (2020b). Bioavailability and toxicity of microplastics to fish species: A review. *Ecotoxicology and Environmental Safety, 189*, 109913.

Wang, Z., Zhang, Y., Kang, S., Yang, L., Shi, H., Tripathee, L., & Gao, T. (2021). Research progresses of microplastic pollution in freshwater systems. *Science of the Total Environment, 795*, 148888.

Wright, S. L., & Kelly, F. J. (2017). Plastic and human health: A micro issue? *Environmental Science & Technology, 51*(12), 6634–6647.

Xu, C., Zhang, B., Gu, C., Shen, C., Yin, S., Aamir, M., & Li, F. (2020a). Are we underestimating the sources of microplastic pollution in terrestrial environment? *Journal of Hazardous Materials, 400*, 123228.

Xu, S., Ma, J., Ji, R., Pan, K., & Miao, A. J. (2020b). Microplastics in aquatic environments: Occurrence, accumulation, and biological effects. *Science of the Total Environment, 703*, 134699.

Xu, Q., Huang, Q. S., Luo, T. Y., Wu, R. L., Wei, W., & Ni, B. J. (2021). Coagulation removal and photocatalytic degradation of microplastics in urban waters. *Chemical Engineering Journal, 416*, 129123.

Ya, H., Jiang, B., Xing, Y., Zhang, T., Lv, M., & Wang, X. (2021). Recent advances on ecological effects of microplastics on soil environment. *Science of the Total Environment, 798*, 149338.

Yang, L., Zhang, Y., Kang, S., Wang, Z., & Wu, C. (2021a). Microplastics in freshwater sediment: A review on methods, occurrence, and sources. *Science of the Total Environment, 754*, 141948.

Yang, L., Zhang, Y., Kang, S., Wang, Z., & Wu, C. (2021b). Microplastics in soil: A review on methods, occurrence, sources, and potential risk. *Science of the Total Environment, 780*, 146546.

Yee, M. S. L., Hii, L. W., Looi, C. K., Lim, W. M., Wong, S. F., Kok, Y. Y., ... & Leong, C. O. (2021). Impact of microplastics and nanoplastics on human health. *Nanomaterials, 11*(2), 496.

Zhang, H. (2017). Transport of microplastics in coastal seas. *Estuarine, Coastal and Shelf Science, 199*, 74–86.

Zhang, Q., Xu, E. G., Li, J., Chen, Q., Ma, L., Zeng, E. Y., & Shi, H. (2020a). A review of microplastics in table salt, drinking water, and air: Direct human exposure. *Environmental Science & Technology, 54*(7), 3740–3751.

Zhang, Y., Kang, S., Allen, S., Allen, D., Gao, T., & Sillanpää, M. (2020b). Atmospheric microplastics: A review on current status and perspectives. *Earth-Science Reviews, 203*, 103118.

Chapter 2
Sources, Types, and Occurrences of Microplastics in Soil, Water, and Air

Pranav Prashant Dagwar, Parth Saole, Amulya Jeevanasai, M. Seetha Rama Raju, Hemant sharma, and Kanchan Deoli Bahukhandi

Abstract The presence of microplastics in the world's water bodies is an urgent and pressing issue that must be addressed immediately. The increasing amounts of microplastics found in water bodies and sediments highlight the crucial need for comprehensive waste management strategies that can reduce plastic consumption while also improving recycling and waste disposal systems. The prevalence of microplastics in water raises serious concerns about their potential influence on both aquatic ecosystems and human health. According to research, a wide variety of aquatic animals, including fish, shellfish, and plankton, can swallow microplastics. Ingestion of microplastics is thought to increase the risk of physical injury, disrupting feeding and digestive processes, and transferring hazardous compounds. However, the specific implications of microplastic consumption are still being investigated. The major goal of this chapter is to provide a complete overview of the origins, classification, and environmental prevalence of microplastics in India and around the world. This chapter aims to shed light on the complex character of the microplastics problem by investigating the numerous types and sources of microplastics, as well as their distribution and abundance in various environmental contexts. It also seeks to understand the intricate relationships between environmental dynamics, human activities, and the spread of microplastics. It also aims to catalyze future research into microplastic mitigation and management by identifying significant research gaps and emphasizing critical research needs. Finally, this chapter aims to serve as a catalyst for educated decision-making, policy formulation, and collective action to address the growing worldwide challenge of microplastic contamination.

P. Saole · A. Jeevanasai · M. Seetha Rama Raju · H. sharma · K. D. Bahukhandi (✉)
Sustainability Cluster, School of Engineering, UPES, Dehradun 248001, India
e-mail: kanchan@ddn.upes.ac.in

P. P. Dagwar
Department of Environmental Science and Engineering, School of Engineering and Sciences, SRM University AP, Amravathi 522240, Andhra Pradesh, India

A. K. Mishra et al. (eds.), *Microplastics*, https://doi.org/10.1007/978-981-97-6461-7_2

1 Introduction

Plastic is defined as a polymeric substance that consists of long chains of carbon, and it has also been observed that due to its commendable physiochemical properties and economic viability, it is being used by many industrial and household sectors (Geyer et al., 2017; Shams et al., 2021; Silva et al., 2020). Plastics are generally adapted from binders, polymers, fillers, pigments, plasticizers, and some other additives. In the total of 100% plastic production throughout the world 90% of them are found consisting of synthetic polymers such as polyethylene (PE), polypropylene (PP), polystyrene (PS), polyethylene terephthalate (PET), polyvinyl chloride (PVC), and polyurethane (PUR) (Abbas et al., 2020; Evode et al., 2021). Plastic waste generated affects the environment on a global scale, it is figured out that human-caused setback which harms the ecosystem and potentially threatens human health as well as animal health (Dijkstra et al., 2020; Singh & Ordoñez, 2016). It has been estimated that about 13.6 billion plastic waste would reach landfills and environment by 2050 if the current plastic waste management practices have no improvements (Shams et al., 2021; Silva et al., 2020). Only around 9% of plastic waste generated is recycled out of the total plastic waste produced and it has been reported that more than 80% of the total generated plastic accumulates in the landfills or in the natural environment. It's been studied that about 8–12 million tons of plastic waste which has been generated on land have ended up in water bodies like ocean; as per reports every year a figure between 4.2 and 12.7 million metric tons of mishandled plastic ends up in aquatic streams like ocean, river, seas as a result of the anthropogenic activities (Akdogan & Guven, 2019; Jambeck et al., 2015). This plastic once drifted to the environment is more susceptible to degradation process involving abiotic and biotic forces (Atugoda et al., 2021; Shams et al., 2021). With the effect of different conditions, plastic decomposes under UV light into its monomeric constituents which also include "Microplastics" which are the most difficult and complicated structures and also most impossible to recover and they tend to affect the food chain and disturb human and environmental health (Evode et al., 2021; Li et al., 2021; Zhao et al., 2021). The plastics are in sizes which includes Macro (>25 mm) and Meso (5–25 mm) break down and degrade into Micro (<5 mm) and nanoscale (<100 mm) plastic overtimes (Boyle & Örmeci, 2020). Microplastics are termed as substances that appear in the environment by the process of parent degradation into smaller fragments that can protract in the environment for centuries without any further breakdown. These particles are by design produced by the consumer products in size of >5 mm by physical activities that result in the decomposition of larger factions to highly persistent contaminants that possess harmful effects on organisms and ecosystem (Silori et al., 2023; Yang et al., 2021; De Bhowmick et al., 2021; Brooks et al., 2018; Kosuth et al., 2018a; Cole et al., 2011; Thompson et al., 2004). The pervasive presence of microplastics in the soil and aquatic environments like seas, rivers, lakes, and estuaries has embarked a potential threat to the health of ecosystems due to improper anthropogenic activities and incompetent waste management (Golwala et al., 2021;

Yang et al., 2021; Klingelhöfer et al., 2020; Yukioka et al., 2020). The microplastics presence in the environment has its concerning negative impacts on the health of humans and animals (Carbery et al., 2018; Waring et al., 2018; Wright & Kelly, 2017a, b). Microplastics should be learnt as a threat to human, animal, and aquatic health. In this chapter, our objective is to provide a thorough comprehension of the origins, categorization, and prevalence of microplastics in both India and globally. Drawing from recent research, the objective is to tackle this pressing environmental concern. The emphasis is on the necessity for concerted action to counter the detrimental effects of microplastics on the Earth's aquatic ecosystems and the dependent organisms therein.

2 Microplastics

Microplastics are small plastic particles with a diameter of less than 5 mm. They are generated by the degradation and fragmentation of larger plastic debris, as well as by the release of plastic microbeads from personal care products and synthetic fibers from clothing (Silori et al., 2023). Microplastics are ubiquitous in the environment, found in freshwater and marine systems, soil, and even the atmosphere and these microplastics have shown to affect soil biota and plant growth (Ding et al., 2019). There is growing concern that microplastics may also have negative impacts on human health. Microplastics have been found in food, drinking water, and even in the air we breathe (Atugoda et al., 2021; Chang et al., 2020). While the health risks associated with consuming microplastics are still uncertain, studies have shown that microplastics can cross the gut barrier and accumulate in human tissues, potentially leading to inflammation and other adverse health effects (Danso et al., 2019; Di & Wang, 2018). The ubiquity of microplastics in the environment and their potential negative impacts on both environmental and human health highlight the need for greater understanding of their sources, distribution, and fate. Addressing the issue of microplastics requires concerted efforts from policymakers, industry, and the public to reduce plastic consumption and increase recycling and waste management practices. Microplastics are a major environmental concern due to their potential impacts on ecosystems and human health (Side et al., 2021; He et al., 2019). They can be ingested by a wide range of organisms, from plankton to larger marine animals, and can accumulate in the food chain, leading to unknown long-term effects on the ecosystem. Additionally, the presence of microplastics in water and air has been linked to health issues such as respiratory problems. The presence of microplastics in water bodies has escalated dramatically over the past few decades, largely driven by the exponential increase in plastic production and consumption. According to Lebreton et al. (2018) it has been estimated that approximately 8.3 billion metric tons of plastic have been produced since the 1950s, with approximately 60% of this plastic waste ending up in landfills or the natural environment. Therefore, microplastics have become a significant component of marine litter, and their abundance has been observed even in remote and pristine environments.

Microplastics enter aquatic ecosystems through various pathways, with primary sources being urban runoff, industrial discharges, and inadequate waste management practices (Jha et al., 2019; Gallo et al., 2018). Additionally, plastic debris can persist in the environment for extended periods, undergoing physical and chemical degradation processes that lead to the formation of microplastics (Cózar et al., 2014). These small particles can then be transported by ocean currents and atmospheric deposition to distant regions, contributing to the global dispersion of microplastics in marine environments (Bergmann et al., 2017). The occurrence of microplastics in water bodies has significant ecological implications. As per Rochman et al. (2016) it proved that a wide range of marine organisms, including fish, shellfish, and plankton, can ingest microplastics and this ingested microplastics can lead to physical damage, intestinal blockages, and hinder feeding and digestion processes, ultimately affecting the growth, reproduction, and survival of marine organisms. Furthermore, microplastics have the potential to function as vectors for toxic chemicals, as they can absorb and transport persistent organic pollutants and heavy metals in the aquatic environment (Lambert et al., 2024). Moreover, the potential transfer of microplastics and associated chemicals through the food chain raises concerns about their impact on human health. Humans may be exposed to microplastics through the consumption of seafood, drinking water, and even inhalation of airborne particles (Eerkes-Medrano et al., 2019). Although research on the direct health effects of microplastics on humans is still in its infancy, as there is a growing body of evidence suggesting that microplastics may pose health risks, such as inflammation and oxidative stress, as well as the potential for these particles to translocate across biological barriers (Hartmann et al., 2019). Considering these concerns, it is an urgent need for effective waste management strategies and policies aimed at reducing plastic pollution. This includes promoting the use of biodegradable alternatives, improving recycling practices, and implementing better waste disposal systems to prevent further microplastic contamination of water bodies. Additionally, checking and surveillance programs must be enhanced to assess the extent of microplastic pollution and its impact on aquatic ecosystems and human health. Plastics are weather sensitive to varying degrees. The chemical oxidation of polymers due to exposure to sunlight is often the most impactful (John et al., 2022; Andrady, 2015). There is a possible intervention that the additives can reduce the degradation whereas photo oxidation is fastest on water surfaces, beaches, and exposed land scenes, and is negligible if protected in aquatic sediments, soils, or landfills. According to Weinstein et al. (2016) it has been studied that biofilm formation on plastic surfaces can reduce UV penetration by up to 99%. Interestingly, Khalid et al. (2021) reported that incorporating brominated flame-retardant additives into polystyrene films increases UV absorption and later photooxidation of the polymer. it also provided by-products from the flame retardant itself and from the reaction of polystyrene with bromine radicals. The amount of degradation products leached into the water was 14% of the weight of the original polymer film. Weathering of plastic by UV oxidation also increases vulnerability to further biodegradation. The production of plastics typically involves the extraction and processing of fossil fuels, primarily oil and natural gas. These processes release carbon dioxide (CO2) and other greenhouse gases into the

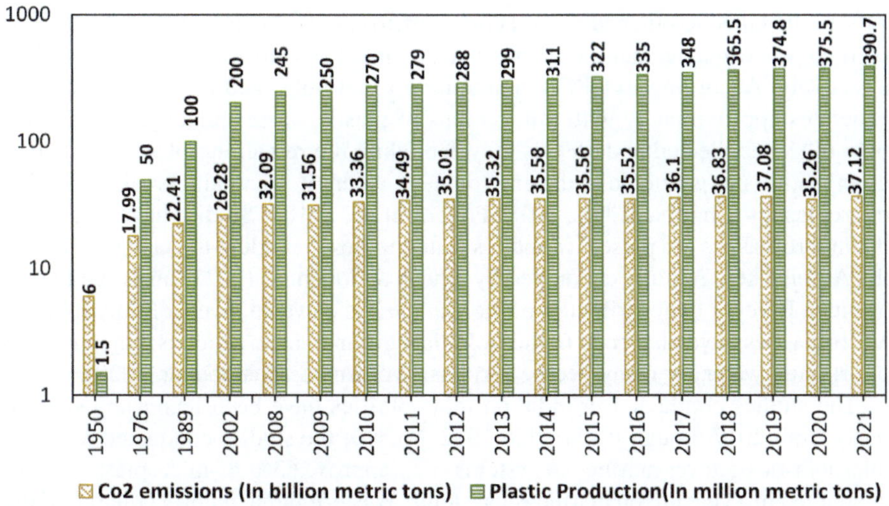

Fig. 1 Amount of carbon emissions per year in billion metric tons and plastic production per year in million metric tons as per Statista

atmosphere, contributing to global carbon emissions. Additionally, plastic production and disposal processes emit other pollutants, further exacerbating environmental concerns. Figure 1 shows the amount of carbon emissions per year in billion metric tons and plastic production per year in million metric tons as per statista. These plastics being produced every year significantly contribute to the overall generation of plastic waste that eventually ends in environment and then release the microplastics as it disassociate from the parent particle (Silori et al., 2023).

2.1 *Indian Scenario of Microplastics*

Microplastics are a growing concern in India due to its impact on the environment and human health which are found in various forms such as microbeads, fibers, and fragments. They are released into the environment through various sources such as plastic waste, synthetic textiles, and personal care products.

In India, as per the data given by the Annual report on plastic waste by CPCB (Central Pollution Control Board) states like Maharashtra, Tamil Nadu, Gujarat, West Bengal, and Karnataka produced 55% which is nearly half of the plastic produced from the country. Plastic waste is a major contributor to microplastic pollution and as Fig. 2 shows plastic waste generation across Indian states and union territories varies significantly. Starting with Maharashtra it leads with 13%, followed closely by Delhi and Telangana at 7% each, whereas Karnataka and West Bengal at 9% each, Uttar Pradesh, Kerala, and Haryana contribute 5, 4, and 4% respectively, while Punjab and Madhya Pradesh each account for 3% (Kumar & Agrawal, 2020). The remaining

11% is attributed to other states and union territories. This data provides the need for tailored waste management strategies at the regional level to address plastic pollution effectively. According to CPCB annual report that published in year 19–20, India generates approximately 9.46 million tons of plastic waste annually, out of which only 60% is collected and 15% is recycled while the remaining plastic waste ends up in landfills or is littered in the environment, where it eventually breaks down into microplastics (Sharma & Jain, 2019; Prateek et al., 2018) . Studies have also shown that microplastics are present in India's water bodies, including the Ganga River and the Arabian sea. A study conducted by Ghosh and Kumari (2021) analyzed that the Yamuna River, a major tributary of the Ganga, has elevated levels of microplastics. Microplastics have also been found in India's groundwater, which is a major source of drinking water for many people (Sri Sasi Jyothsna & Chakradhar, 2020).

The studies conducted in India on microplastics have been increasing continuously from 2017 though it started in 1982. Most of the studies conducted related to microplastics are on marine ecosystems, i.e., approx. 63% of microplastic studies conducted in a year are on marine ecosystems. Western India majorly Gujarat, Maharashtra, Madhya Pradesh, Daman and Diu, Chhattisgarh, and Dadra and Nagar Haveli use large numbers of plastics compared to other parts of India therefore a greater number of microplastics have been recorded (Fig. 3).

The dots in the map represent the states where the studies on microplastics have been carried out. This map includes various states of the Indian subcontinent with coastal areas as well as riverine areas. Various studies conducted on the western coastal region include states like Gujarat, Goa, Karnataka, and Kerala (Vaid et al.,

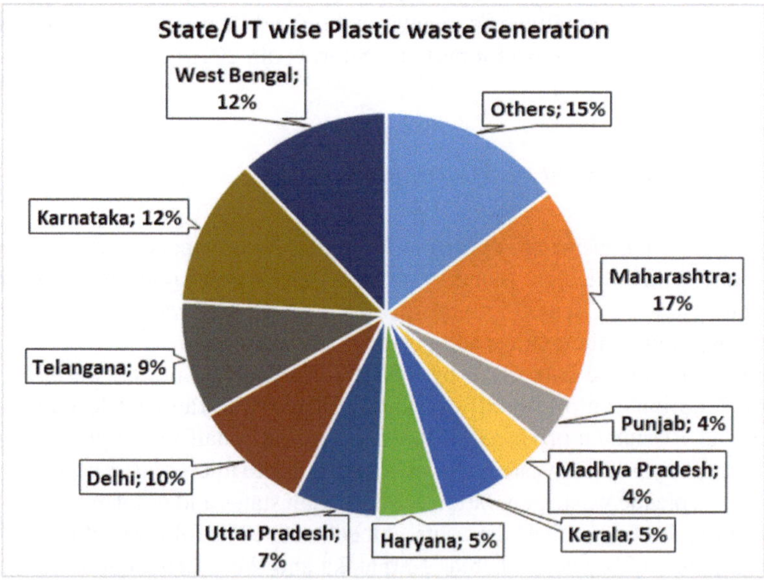

Fig. 2 Various states indicating percentage of plastic waste generated during 2019–20

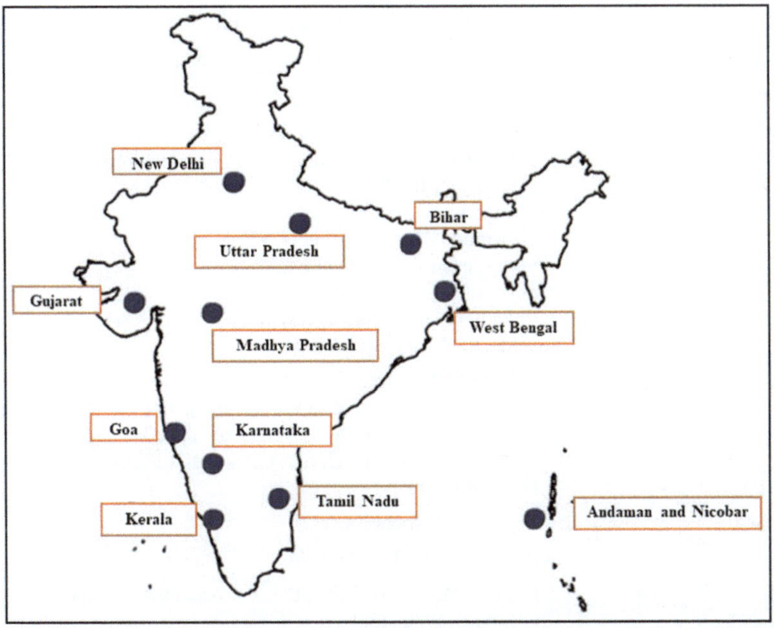

Fig. 3 Indian subcontinent map showing various locations with microplastic studies conducted

2022; Raha et al., 2021). In Table 1 it is signified that in the various sources of water different polymers of microplastics are detected and all these are found in the different states of the country.

The fragments of microplastics are more abundant along 10 sandy beaches which are spread across the states of Maharashtra, Goa, and Karnataka consisting of fibers, films, and pellets (Jayasiri et al., 2014). Different morphotypes of microplastics have been detected and found in the beach sands near Marine Drive Mumbai, Maharashtra state of different shape and size most of them being fibers (51%) followed by 40% granules and 9% plastic films. The presence of microplastics varying in color, shape, and size has been reported from nine stations along the coast of Odisha with high abundance in Swargadwara, followed by Chandipur, Chandrabhaga, Golden beach, Beleswar, Gopalpur, Paradeep, Sunapur and Abhaya Chandpur (Patchaiyappan et al., 2021; Yaranal et al., 2021; Tiwari et al., 2019). Also, the southern part of Vembanad lake, Kochi district, and in the Nattika beach, Thrissur district of Kerala state detected fragment, film, foam, fiber, and pellet-shaped microplastics present in the waters. In the west coast of India compared to different morphotypes of microplastics it has been shown that fragment is in higher percentages. Comparing polymers of west coast Indian states such as Kerala Maharashtra, Goa, and Karnataka polyethylene concentration is more than polypropylene (Mashirin & Chitra, 2022).

Due to El Niño phenomenon South India experienced heavy rains in 2015. During which floods in Chennai had increased the presence of microplastic content in the city's water according to a study conducted before and after the floods in Chennai.

Table 1 Status of microplastics detected in various states of India

States	Types of MP polymer detected	References
In marine waters		
Andaman and Nicobar Islands	Surlyn ionomer, PEI, acrylic, PPS, ethylene vinyl alcohol, NY, EVA, PIP, PU, PVC, Poly (dimer acid-co-alkyl polyamine), PP, melamine, PVF, poly (perfluoro ethylene oxide), polysulfide, polybutadiene, poly (butadiene-acrylonitrile acrylic acid), PVB, nylon 6, epoxy epichlorohydrin, ABS	Mohan et al. (2022); Krishnakumar et al. (2020)
Tamil Nadu	PE, PVC, PP, PS, PET, NY, PET, PA, CP, PU, PEST, PVA, PVC, alkyd resin, PEU, NY, CP, blended PE-PP	Sathish et al. (2020); Sundar et al. (2021); Yogeshwaran et al. (2020)
Karnataka	PE, PP and others	Maharana et al. (2020)
Goa	PE, PEST, PA, PS, PP, PVA	Maharana et al. (2020)
Kerala	PE, PP, PA, PS, PET, RY, PU, CE, ABS, PVC, PVF	Anandhan et al. (2022)
Maharashtra	PE, PET, PS, PP, PVC and others	Maharana et al. (2020)
Pondicherry	PVC, EAA copolymer, CA, PVK, polypropylene acrylic acid, polymer resin, poly (ester urethane), polyvinyl behenate, acrylonitrile, PP, HDPE, LDPE, PS	Dowarah and Devipriya (2019)
Gujarat	PU, NY, PEST, PS	Rabari et al. (2022)
In river waters		
Tamil Nadu	HDPE, LDPE, PP, PS, PET	Sathish et al. (2020)
Karnataka	PE, PET, PP, PVC	Maharana et al. (2020)
Kerala	PE, PP, PVC, PS, NY, HDPE, LDPE	Robin et al. (2020)
Uttar Pradesh, Prayagraj, Sangam (Ganga & Yamuna River)	EVOH, polyacetylene, PIP, polyimide, PVAL, PP, styrene/ isoprene, polyvinyl pyrrolidone, lanoline, PVC, polyacrylamide, polyacetal, CE	Gupta et al. (2024)

As per the studies conducted in the western parts of India for the states of Gujrat, Maharashtra, and Kerala, in the state of Maharashtra, 41.85% of microplastics were found after degradation from parent plastic materials. In the Juhu, Bandra, and Navi Mumbai beaches, microplastic content was found in the form of fragments and fibers. The concentration of microplastics in the state was highest in Central Kerala than

Southern Kerala and Northern Kerala. Among the Coromandel states, the State of Tamil Nadu recorded the highest amount of microplastic pollution. Microplastics found in the stomach and intestine of Scomber *japonicus* are present in the northern part of the Indian ocean. This is the fish that has the highest mean of microplastic ingestion as per the study given by Karuppasamy et al. (2020). The main importance must be given to the fish Indian markel (*Rastrelliger kanagurta*), since microplastic content was also found in this fish which is an important source of proteins for humans as shown in Fig. 4.

Therefore, caution should be taken in selecting the fish for the meal. The coral reefs of Tamil Nadu are also affected by the microplastics. The Black tailed godwit, a bird in Indian coastal, eats *Scrobicularia plana* shell type species, this shell type species has accumulated microplastics. This species is to be tested since they are near threatened (likely to become endangered in near future). As per the study on hard corals has showed nematocyst discharge and ingestion of different microplastics, suggesting the presence of phagostimulants which is potentially related to toxic compounds in it (Vikas Madhav et al., 2020). As it is a growing concern in India it is also important to take immediate action to address this issue. The government, industries, and individuals should work together to reduce plastic waste and promote

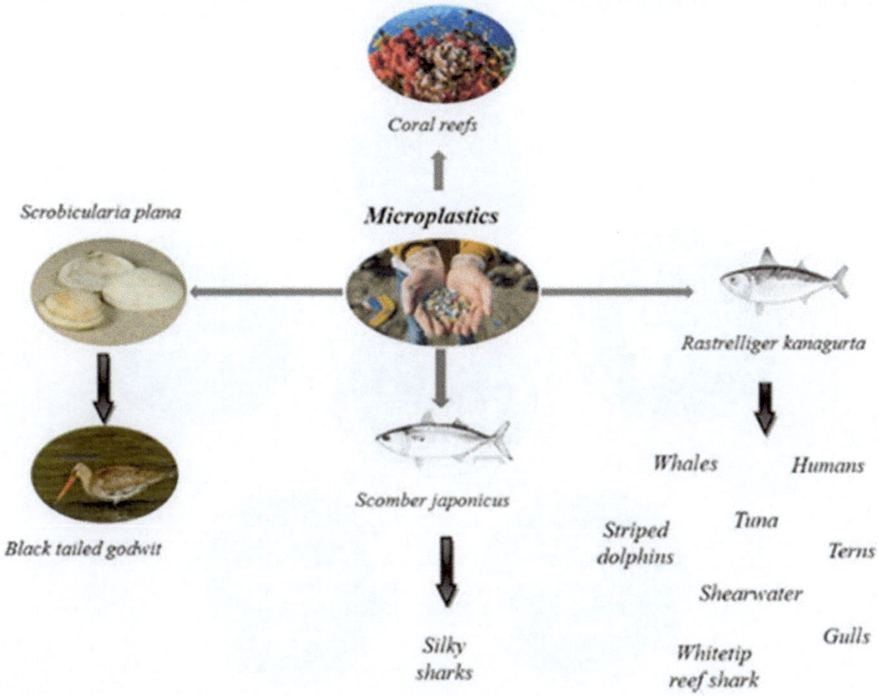

Fig. 4 Illustrates various species that are affected due to microplastics

sustainable practices to protect the environment and human health (Liang et al., 2021).

2.2 *Sources of Microplastics*

Microplastics are a global environmental issue that has received increasing attention in recent years. These small plastic particles, typically measuring less than 5 mm in diameter, can be found in various environments, including oceans, freshwater systems, and terrestrial ecosystems (Parekh et al., 2022; Xumiao et al., 2021). Microplastics are tiny plastic particles, measuring less than 5 mm in diameter, which have become a major environmental concern in recent years due to their widespread occurrence in water systems worldwide. These particles are often created from larger plastic items that have broken down over time, or from the manufacturing and disposal of smaller plastic products (Mason et al., 2022; Rozman et al., 2021). Studies have shown that microplastics are found in various bodies of water, including oceans, rivers, lakes, and even groundwater (Siegfried et al., 2017). They have been detected in both surface water and sediment, as well as in the tissues of aquatic organisms, such as fish and shellfish (Fig. 5).

Fig. 5 Source of microplastic

2.2.1 Plastic Litter

Plastic litter is a significant source of microplastics in the environment, with numerous studies highlighting the negative impacts on marine and terrestrial ecosystems (Yang et al., 2022; Gobel et al., 2021). These plastic pieces become smaller, they eventually reach a size where they are referred to as microplastics, for which the particles are less than 5 mm in size (Jambeck et al., 2015). Microplastics can persist in the environment for decades, and they can be ingested by wildlife, including fish and birds, and can enter the food chain. Additionally, plastic litter can also release microplastics directly into the environment through a process called leaching (Liu et al., 2022). This occurs when chemicals and additives in plastic, such as plasticizers and flame retardants, are released into the environment as the plastic breaks down. These chemicals can then bind to microplastics and enter the food chain as well. Overall, plastic litter is a significant source of microplastics in the environment, and reducing plastic waste and improving waste management practices can help to mitigate this issue (Schwabl et al., 2019; de Souza Machado et al., 2018). According to Roman et al. (2022) Three of India's major rivers, the Indus, Ganges, and Brahmaputra, are listed among the top 10 plastic waste carriers globally, as they carry water downstream with maximum disposal of plastic litter from various places as they flow.

One study conducted by Andrady (2011) estimates that as much as 5–10% of plastic litter in the ocean is converted to microplastics through natural weathering processes such as photodegradation and mechanical erosion. Whereas in another study conducted by Eriksen et al. (2014) estimated that between 15 and 51 trillion microplastic particles are present in the world's oceans. Plastic litter can also become a source of microplastics in terrestrial environments. In a study, researchers found that microplastic pollution was present in soil samples collected from different urban and rural locations in China, with plastic litter being a significant source of microplastics in soil. Moreover, microplastics have been found in freshwater systems as well, with plastic litter and wastewater treatment plants being the main sources of contamination (Guzzetti et al., 2018). Plastic litter can enter freshwater systems through stormwater runoff or direct disposal, which can then breakdown into smaller microplastic particles. According to Lu et al. (2016), plastic litter is a major source of microplastics in the environment, and reducing plastic waste is crucial to mitigate the harmful effects of microplastic pollution.

2.2.2 Personal Care Products

Personal care products such as exfoliants, body washes, toothpaste, and facial scrubs often contain tiny plastic beads known as microbeads, which can end up in the environment through various means such as wastewater treatment plants, and ultimately accumulate in water bodies, soil, and marine ecosystems (Naidu et al., 2022; Yurtsever & Yurtsever, 2019). A study conducted by the United Nations Environment Programme (UNEP) found that personal care products having microbeads were a major source of microplastics in marine ecosystems. The study estimated that up to

4,360 tons of microbeads were released into the environment each year in Europe alone (UNEP, 2015). When these personal care products are used and rinsed off, the microbeads can enter the wastewater system and eventually end up in rivers, lakes, and oceans. They are so small that they are not always effectively removed by wastewater treatment plants, and they can persist in the environment for hundreds of years. Once in the environment, microbeads can have a number of harmful effects on aquatic life. Fish and other marine organisms can mistake them for food, leading to ingestion and potential harm (Naidu et al., 2022). Additionally, microbeads can attract and absorb other pollutants, such as pesticides and heavy metals, which can then be transferred to the organisms that ingest them (Rochman et al., 2016). To reduce the impact of microbeads on the environment, many countries have started to ban the use of microbeads in personal care products, and some companies have voluntarily stopped using them in their products. Consumers can also choose to use products that do not have microbeads or switch to natural exfoliants, such as oatmeal or sugar, to reduce their environmental impact (Praveena et al., 2018).

Lake Erie had an estimated 1.1 million microplastic particles per square kilometre, with 85% of the particles identified as microbeads commonly found in personal care products (Eriksen et al., 2013). A study published in Environmental Science & Technology found that microbeads from personal care products were a significant contributor to microplastic pollution in the Great Lakes. The study estimated that approximately 8 trillion microbeads were released into waterways every day in the United States (Mason et al., 2016). A study conducted by the University of California, Santa Barbara found that microbeads from personal care products could be ingested by fish and other marine animals, potentially causing harm to the animals and their predators (Rochman et al., 2015a, 2015b). It was found that personal care products were a significant source of microplastics in wastewater treatment plants. The study estimated that up to 98% of microplastics found in wastewater were from personal care products (Hartmann et al., 2019). According to Alex et al. (2024) out of 50 different products which were used in India 45.00% of the products contained irregular or smooth microplastics, with Polyethylene being the dominant polymer, followed by Polypropylene, Polystyrene, Polyurethane, and Polycaprolactone (Paul-Pont et al., 2016). A significant number of products (23.33%) contained cellulose microbeads, with ambiguity regarding their biodegradability (Napper al., 2015).

Overall, these studies supply evidence that personal care products having microbeads are a significant source of microplastics in the environment and highlight the need for more sustainable alternatives to these products.

2.2.3 Synthetic Textiles

Polyester, nylon, and acrylic, are a significant source of microplastics in the environment. During the washing and drying of synthetic textiles, microfibers are shed from the fabric and are released into wastewater (Stapleton et al., 2023). These microfibers are too small to be filtered out by wastewater treatment plants and can end up in waterways and oceans. A study published in Environmental Science & Technology estimated that a single garment made from synthetic fibers can release more than 1,900 microfibers per wash. This means that a single load of laundry can release hundreds of thousands of microfibers into the environment (Yu et al., 2021). The problem is intensified by the fact that synthetic fibers are widely used in the textile industry (Bourgeois et al., 2020). According to a report by the Ellen MacArthur Foundation, synthetic fibers currently make up 60% of all textile production. As a result, the textile industry is a significant contributor to microplastic pollution in the environment. Synthetic fibers shed from clothing during washing, or the utilization can be released into the air and transported over long distances (Prata et al., 2020).

2.2.4 Paintings and Coatings

Painting and coating are potential sources of microplastic pollution in the environment, particularly using microplastic-containing coatings and the disposal of waste paint products (Wu et al., 2023). Many coatings, particularly those used for marine applications, have microplastics as a filler material. When these coatings wear down or are removed, the microplastics can be released into the environment (D'Hont et al., 2021). According to a study it was found that cleaning paint brushes and rollers with plastic-based solvents could release microplastics into the environment, and that using alternative cleaning methods, such as soap and water, could help to reduce this pollution (Liang et al., 2021; Praveena et al., 2018). Acrylic and vinyl resins act as binders, providing adhesion, flexibility, and durability to most paints in India. During the painting process, small particles of paint can become airborne and settle on surfaces or be released into the environment. These particles may have microplastics if the paint contains plastic-based polymers. A study found that paint chips from boats and other marine structures had microplastics, suggesting that these structures may be a source of microplastic pollution in marine environments (Horton et al., 2017). A study published in the Journal Science of the Total Environment found that microplastics were present in marine coatings used on ships and offshore structures, and that these coatings were a potential source of microplastic pollution in marine environments (Harrison et al., 2019).

The disposal of paint products, including cans, brushes, and rollers, can also contribute to microplastic pollution. If these products hold plastic-based materials, they may release microplastics as they degrade in the environment. Some paints and coatings contain microplastic particles as additives, which can then be released into the environment during use and disposal. These microplastic additives accounted for 10% of the total microplastics found in the ocean (Kershaw et al., 2016).

Cleaning paint brushes and rollers can also contribute to microplastic pollution if plastic-based materials are used in the cleaning process. For example, using plastic-based solvents to clean paint tools can release microplastics into the environment. Another study published in the Journal Environmental Pollution found that airborne microplastic particles were released during the sanding of painted surfaces, with plastic-based coatings being a significant contributor to the overall microplastic particle release (Choi et al., 2022).

2.2.5 Tire Wear Particles

Tire wear is a significant source of microplastic pollution in the environment. As vehicles travel on roads, the friction between the tires and the road surface causes small particles to be worn off the tires. These particles, which can have synthetic rubber and plastic-based materials, can then be released into the environment through a variety of mechanisms (Yu et al., 2021). India's economic boom has led to a dramatic increase in vehicles on the road. Millions of cars, two-wheelers, and trucks contribute to a substantial generation of TWPs (Kole et al., 2017).

As tire particles are worn off, they can be carried away by runoff water and enter water systems. This can result in the accumulation of tire-derived microplastics in rivers, lakes, and oceans. Tire wear can also result in the release of airborne particles, which can settle on nearby surfaces and enter the environment through inhalation (Naidu et al., 2022; Wu et al., 2022). These particles can be transported long distances and potentially affect ecosystems far from their source. A study found that microplastic particles from tire wear were present in roadside soils and that these particles were being transported into nearby rivers and streams (Luo et al., 2021). Tires that are disposed of in landfills can also contribute to microplastic pollution. As the tires degrade, microplastics can be released into the environment through leachate or air emissions. These are microplastics resulting from car tires' wear and tear, as tires are used, they release small particles of rubber into the environment, which can then become microplastics tire wear particles were the second largest contributor to microplastics in the ocean, accounting for 28% of the total (Kole et al.,.2017). While global research on TWPs is ongoing, data specific to India remains scarce. However, studies suggest TWPs can be a major source of microplastics in urban environments, particularly in runoff from heavily trafficked roads (Rasmussen et al., 2023).

In some cases, tire crumbs, which are produced by shredding tires, may be used as a soil amendment in agriculture. This can introduce tire-derived microplastics into the soil, which is produced by shredding tires, has elevated levels of microplastics, and can introduce these particles into the environment when used as a soil amendment (Olubusoye et al., 2024). Tire wear is a significant source of microplastic pollution in urban environments, with concentrations of microplastics in stormwater runoff from roads being up to 500 times higher than in other types of urban runoff (Nizzetto et al., 2016).

2.2.6 Agriculture

Agriculture can be a source of microplastic pollution in the environment through a variety of mechanisms. Plastic mulch is commonly used in agriculture to suppress weeds and keep moisture. When this plastic mulch is discarded or degraded, it can release microplastics into the environment (Wu et al., 2020). Mulch that is used in agriculture can be a significant source of microplastic pollution in soils, with up to 45 kg of microplastics per hectare being released from degraded mulch (Wang et al., 2018). Another study published in the same Journal found that microplastics from plastic mulch were present in agricultural soils, with concentrations of up to 5,600 particles per kilogram of soil (Li et al., 2020; Wu et al., 2020). A substantial portion of agricultural plastic waste, estimated to be over 60%, escapes proper collection and disposal in India (Pandey et al., 2009). Techniques like mulching with plastic films help to retain moisture, suppress weeds, and regulate soil temperature, leading to increased crop yields. A study published in the Journal of the International Society for Horticultural Science found that plastic mulch could increase Bt (Bacillus Thrungenisis) cotton yields by up to 42% compared to no mulch conditions (Singh et al., 2011).

Some irrigation systems, such as drip irrigation, may use plastic tubing that can degrade over time and release microplastics into the soil and water and this is being introduced into waterways through agricultural runoff, particularly from the use of plastic mulch films and microplastic-containing fertilizers (Sanz-Lázaro et al., 2020). Some synthetic fertilizers may have plastic-based materials, which can be released into the environment through runoff or leaching. A study published in the Journal of Science of the Total Environment found that synthetic fertilizers used in agriculture can have microplastics and that these particles can be released into the environment through runoff or leaching (Liu et al., 2020; Ng et al., 2018). The use of plastic components in agricultural machinery, such as seed drills or tillage equipment, can also contribute to microplastic pollution when these components degrade. In animal agriculture, microplastics may be present in animal feed or bedding materials, which can then be excreted in animal waste and enter the environment through runoff or land application. A study published in the Journal Chemosphere found that microplastics were present in animal feed and bedding materials in livestock farming and that these particles were being excreted in animal waste, which can then enter the environment through land application (Besseling et al., 2017).

2.2.7 Landfills and Waste Management

Landfills and waste management can be sources of microplastic pollution in the environment through a variety of mechanisms. Landfills are a major source of plastic waste and can release microplastics into the surrounding environment as plastic degrades over time (Dagwar & Dutta, 2024; Pivokonsky et al., 2018). According to a study by Horton et al. (2017), plastic waste in landfills can undergo biodegradation, photodegradation, and fragmentation, which can lead to the release of microplastics into the environment.

Sewage sludge, which is generated during the wastewater treatment process, can have microplastics that are flushed down the drain from personal care products, clothing, and other sources. A study by Rolsky et al. (2020) found that sewage sludge can have microplastics, and that the level of microplastic contamination in sewage sludge varies depending on the type of treatment process used. Tire shredding is a process used to recycle old tires, but this process can also release microplastics into the environment. Many Indian households and communities lack proper waste segregation practices (Sandu et al., 2020; Nandan et al., 2017). This blending of organic and inorganic waste hinders plastic separation and recycling efforts (Chatterjee et al., 2020). The infrastructure for efficient waste collection and transportation, particularly in rural regions, remains inadequate. Consequently, a substantial portion of plastic waste ends up in landfills (Mittal et al., 2019; He et al., 2019).

Landfills can also be a destination for tire waste, which can hold microplastics that are released during the tire shredding process. A study by Knight et al. (2020) found that tire wear particles can be a significant source of microplastics in the environment. Textile waste, such as clothing and other fabric products, can also hold microplastics that are released into the environment as they degrade. Plastic packaging waste, such as plastic bags and food packaging, can also contribute to microplastic pollution in the environment when improperly disposed off as improper disposal of plastic packaging waste, such as plastic bags and food packaging, can also contribute to microplastic pollution in the environment. A study by Derraik (2002) found that plastic bags can degrade into microplastics and that they are a usual form of litter in marine and terrestrial environments.

2.2.8 Road Marking

Road marking can be a source of microplastic pollution in the environment. Microplastics from road markings can be released into the environment through various mechanisms such as abrasion, wear and tear, and weathering. The most common types of road markings in India utilize thermoplastic or cold plastic paints. These paints incorporate plastic resins for better adhesion, weather resistance, and reflectivity compared to traditional paints (Indian Paint Association, "Road Marking Paints"). The white and yellow lines on roads are typically painted with a type of paint that holds microplastics, specifically thermoplastic polymers. Over time, the paint can wear away due to factors such as weathering, traffic, and road maintenance, which can result in the release of microplastics into the environment (Bergmann et al., 2015). Additionally, road marking materials may contain additives that can also contribute to microplastic pollution. A study by Enfrin et al. (2022) found that road marking paint contains up to 25% of microplastics by weight, and that abrasion from vehicles and weathering can result in the release of these microplastics into the environment. Another study by Jeong and Choi (2020) found that road dust samples collected from areas with heavy traffic contained microplastics, with the majority of the particles coming from road marking paint.

2.2.9 Packaging Materials

Packaging materials, particularly single-use plastics, can be a significant source of microplastic pollution in the environment (Singh & Ordonez, 2016). Microplastics can be released from packaging materials through various mechanisms such as abrasion, fragmentation, and weathering (Jahandari, 2023). A substantial portion of plastic packaging waste in India, estimated at over 60%, escapes proper collection and disposal, polluting landfills, waterways, and agricultural fields (Jhala & Reddy, 2021).

When plastics are used for packaging, they are often subjected to mechanical forces during transportation, storage, and disposal, which can result in the release of microplastics into the environment (Liu et al., 2022). Additionally, exposure to sunlight and other environmental factors can cause plastics to degrade into smaller and smaller particles, increasing the likelihood of microplastic pollution (Paul-Pont et al., 2016). A study by Rochman et al. (2015a, 2015b) found that plastic packaging was the most common source of microplastic pollution in the ocean, and that microplastics from packaging materials were also found in freshwater ecosystems. Another study by Napper and Thompson (2019) found that plastic packaging materials were a major contributor to microplastic pollution in the environment, with single-use items such as plastic bags and food containers being particularly problematic.

2.2.10 Atmospheric Degradation

Atmospheric degradation can also be a significant source of microplastic pollution in the environment. When plastic waste is exposed to the atmosphere, it can undergo various forms of degradation such as photo-oxidation, fragmentation, and abrasion, resulting in the release of microplastics into the air (Sheriff et al., 2023; Guo & Wang, 2019). These microplastics can then be transported over long distances and deposited in remote locations. As per Allen et al. (2019) found that microplastic particles were present in the air in urban and rural locations, with the highest concentrations found in urban areas. The study also found that microplastics were present in rainfall samples, showing that microplastics can be transported through the atmosphere and deposited in remote locations via rainfall. Microplastic particles can be transported over long distances and deposited in remote locations, such as mountainous regions and the Arctic (Bergmann et al., 2015; Zhang et al., 2020a, 2020b).

Figure 6 shows the main sources of microplastics, tiny plastic bits harming the environment. Synthetic clothes (35%) shed the most microplastics, followed by tires (28%) and road markings (14%) wearing down and other sources include plastic beads in cosmetics, paint flakes, and bits from packaging (Holmes et al., 2012).

3 Types of Microplastics in Environment

Microplastics are a global environmental issue that has received increasing attention in recent years. These small plastic particles, typically measuring less than 5 mm in diameter, can be found in various environments, including oceans, freshwater systems, and terrestrial ecosystems (Xumiao et al., 2021; He et al., 2018). Microplastics are tiny plastic particles, measuring less than 5 mm in diameter, which

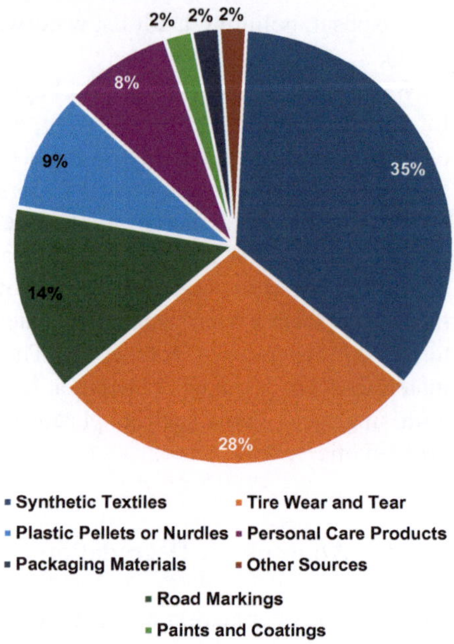

Fig. 6 Different sources of microplastics

have become a major environmental concern in recent years due to their widespread occurrence in water systems worldwide (Fig. 7).

Primary Microplastics: These are manufactured to be small, such as microbeads in personal care products, pellets used in plastic manufacturing, and fibers from synthetic clothing.

Secondary Microplastics: These are the result of larger plastic debris breaking down over time due to natural weathering processes, such as sunlight and wave action, or due to human activities, such as mechanical grinding or thermal treatments. Examples of secondary microplastics include fragments from plastic bags, bottles, and fishing nets (Table 2).

Microplastics can also be classified based on their shape and size, which can affect their behavior in the environment and their potential to cause harm to living

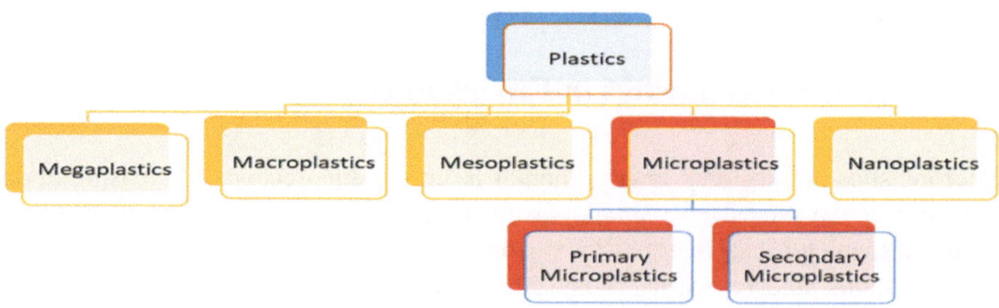

Fig. 7 Types of microplastics

Table 2 Types of plastics with size variations (Mashirin & Chitra, 2022; Tirkey & Upadhyay, 2021).

Type of plastics	Size Range	‘
Mega-kplastics	**≥100 mm**	Koelmans (2015)
Macro-plastics	**20–100 mm**	Turra et al. (2014)
Meso-plastics	**5–20 mm**	Sarkar et al. (2019)
Microplastics	**≤5 mm**	Oceanic et al. (2023); Liu et al. (2022)
Nano-plastics	**0.001–0.1 mm**	Paul et al. (2020); Koelmans (2015)

organisms. Some common shapes and sizes of microplastics include (Fig. 8 and Table 3).

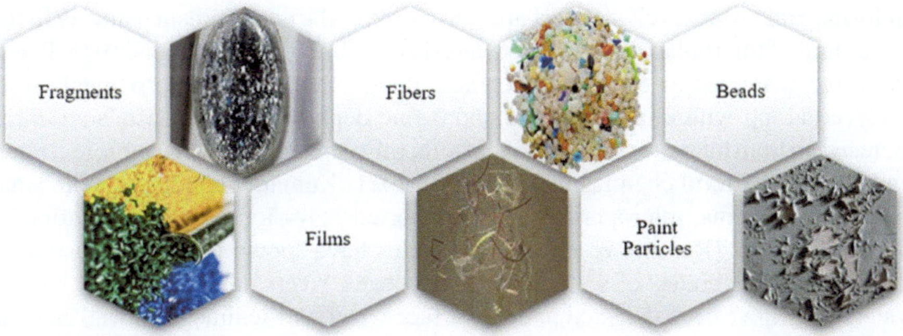

Fig. 8 Distinct types of microplastics found in the environment (Adopted from Lusher et al., 2020)

Table 3 Types of plastics with size variations

Type of microplastic	Description	References
Fibers:	These are thread-like pieces of plastic that can be released from synthetic textiles during laundering or from abrasion of clothing during wear	Rebelein et al. (2021)
Films:	These are thin sheets of plastic that can be used in packaging materials or as a coating on other materials	Ronca (2017)
Beads:	These are small, spherical pieces of plastic that are used in a variety of personal care products, such as exfoliating scrubs and toothpaste	Andrady (2017)
Fragments:	These are irregularly shaped pieces of plastic that can result from the breakdown of larger plastic items, such as bottles and bags	Derraik (2002)
Paint particles:	Paint particles are those released into the environment during sand blasting and other abrasive cleaning processes	Gaylarde et al. (2021)

4 Occurrences of Microplastics

Microplastics are the ubiquitous and slow-going process for degradation of contaminants with properties such as long residence time, great stability, high potential to be fragmented and an ability to absorb other contaminants (Padervand et al., 2020). The damage to the aquatic environment is a huge contribution done by human activities which has certainly affected the water. As per reports, annually 4.8 to 12.7 million metric tonnes of untreated plastics end up in the ocean and coastal countries (Jambeck et al., 2015). The Microplastics generated on the land are generally discarded straightaway into the water bodies in the form of effluent coming out of WWTP or industries. Depending upon the weight and size, the debris are lightweight which are mostly suspended on the surface of water bodies and the heavier or larger debris are deposited in the beds of water bodies (Huang et al., 2021). The predominant nature of microplastics marks their presence in both natural water cycle and anthropogenic water cycles. Rivers are considered as the predominant pathway for the discarding of microplastics into the oceans (Liu et al., 2021; Rochman, 2018). Plastic waste is considered as one of the potential dangers to aquatic environments when analyzed along with climate change and ozone depletion. Microplastics are threat inclusive to both human and environment affecting negatively by ingestion, and retention through harmful chemicals (Silva et al., 2021; Kumar et al., 2019). With water or aquatic systems, microplastics are transported to various ecosystems affecting the environment (Rezania et al., 2018). Studies have shown that microplastics are found in various bodies of water, including oceans, rivers, lakes, and even groundwater. They have been detected in both surface water and sediment, as well as in the tissues of aquatic organisms, such as fish and shellfish. Figure 9 shows Microplastics found in water bodies due to various sources: agricultural runoff (10%), industrial effluent (15%), atmospheric deposition (20%), urban runoff, and sewage (50%), with the remaining 5% from other sources (Van Sebille et al., 2015). Urban runoff and sewage represent the most significant contributor to microplastic pollution in water. This highlights the importance of addressing urban runoff and sewage management to mitigate microplastic contamination in aquatic ecosystems. Additionally, efforts should be made to control agricultural and industrial sources, as well as atmospheric deposition, to effectively reduce overall microplastic pollution in water bodies.

4.1 *Occurrence of Microplastics in Oceans*

The oceans are one of the primary environments where microplastics are found (Masura et al., 2015). Approximately 8 million metric tons of plastic waste enter the oceans each year (Liu et al., 2022; Sheriff et al., 2023). This plastic waste can breakdown into smaller particles, including microplastics, which can be ingested by marine organisms (Jambeck et al., 2015). Microplastics were present in 36.5% of

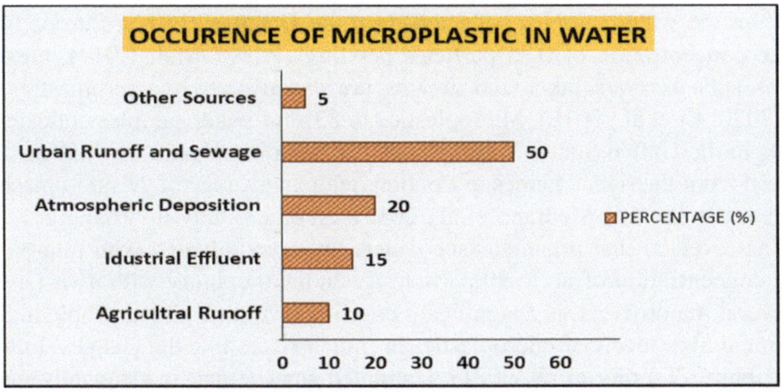

Fig. 9 Occurrence of microplastics in water from different sources, measured in percentages

fish samples collected from the English Channel, with an average of 5.5 microplastics per fish. Another study by (Rochman et al., 2015a, 2015b) found microplastics in 90% of seabirds sampled globally, with ingestion being the primary route of exposure (Lusher et al., 2013; Wilcox et al., 2015). One study found that microplastics were present in 100% of the water samples collected from the North Atlantic Ocean, with an average concentration of 1.06 particles per cubic meter (Cózar et al., 2014). India generates a substantial amount of plastic waste annually, estimated at around 3.5 million tons (Central Pollution Control Board, "Annual Report 2019–20"). Inefficient waste management practices lead to a significant portion of this plastic ending up in landfills, rivers, and ultimately, the ocean (Mittal et al., 2019). Studies have shown concerning levels of microplastics in the Indian Ocean, with an average concentration of 50 particles per cubic meter of water in near-surface waters (Eriksen et al., 2014). Another study found that microplastics were present in 83% of tap water samples collected from various countries, including the United States, India, and Lebanon, with an average concentration of 4.45 particles per liter (Kosuth et al., 2018). Microplastics can harm aquatic organisms by blocking their digestive tracts, reducing their feeding efficiency, and increasing their mortality rates (Marine Debris Program, 2015). They can also cause reproductive and developmental disorders and alter behavior and metabolism in marine animals (Cole et al., 2011).

4.2 Occurrence of Microplastics in Freshwater Systems

In freshwater systems, microplastics have been found in high concentrations in rivers, lakes, and wetlands (Mason et al., 2022; Xie et al., 2022). One study found that microplastics were present in 90% of the water samples collected from the River Thames in the United Kingdom, with an average concentration of 0.55 particles per litre (Horton et al., 2017). Another study found that microplastics were present in 96% of the water samples collected from the Danube River in Europe, with an average concentration of 0.75 particles per litre (Ivleva et al., 2017). Freshwater systems, such as rivers, lakes, and streams, are also affected by microplastics (Wang et al., 2020; Li et al., 2018). Microplastics in 83% of water samples collected from 29 sites in the United States. Wagner et al. (2014) microplastics in 96% of samples collected from the river Thames in London, with an average of 21 microplastics per litre of water (Eerkes-Medrano et al., 2015). Another study by Zhang et al., 2018 in China revealed that urban surface waters are contaminated with microplastics, with a concentration of up to 99 particles/L. India, a country with a vast coastline and several major rivers, is a significant contributor to this global problem. Studies have found high levels of microplastics in Indian rivers like the Ganges, Indus, and Brahmaputra (Tsering et al., 2021). Untreated sewage and inadequately managed waste disposal are identified as major sources of microplastic pollution in Indian water bodies (Mittal et al., 2019)."

4.3 Occurrence of Microplastics in Terrestrial Ecosystems

In terrestrial ecosystems, microplastics have been found in soils, sediments, and even the air. It is found that microplastics were present in 75% of the soil samples collected from urban parks in Beijing, China, with an average concentration of 110 particles per kilogram (Zhang et al., 2018). Another study found that microplastics were present in 89% of the sediment samples collected from the Tamar Estuary in the United Kingdom, with an average concentration of 1,037 particles per kilogram (Wang et al., 2019; Horton et al., 2017). In addition, microplastics have been found in the air in urban areas, with one study reporting an average concentration of 1.9 microplastic particles per cubic meter of air in Paris, France (Dris et al., 2015). Microplastics have also been detected in terrestrial ecosystems. microplastics in soil samples collected from urban and rural areas in Mexico, with an average of 440 microplastics per kilogram of soil Huerta (Lwanga et al., 2016). A study found microplastics in 20 out of 24 samples of household dust collected in Germany, with an average of 1.5 microplastics per gram of dust (Imhof et al., 2016). Sewage sludge, sometimes used as fertilizer, can be a source of microplastics in soil. Microplastics present in untreated sewage can persist in the sludge and contaminate agricultural land when applied (Rochman et al., 2015a, b). Studies have detected microplastics in various terrestrial environments across India, including agricultural soils, landfills, and even remote areas like the Himalayas (Debnath et al., 2020; Pandit et al., 2019). Microplastics can act as a vector for the transport of harmful chemicals, such as persistent organic pollutants (POPs), which can accumulate in the food chain (Wright et al., 2013). In terrestrial systems, microplastics have been shown to affect soil biota and plant growth (Ding et al., 2019).

5 Conclusion

Microplastics are found to be the most pressing issue of modern times due to its prevalence in the environment and constant seepage into soil and aquatic sources which leads to the harmful effects ultimately affecting the ecosystem. This chapter has covered the various aspects of microplastics presence in the environment by determining its types, sources, and possible occurrences in the environment. The rise in population and along with it the continuous rise in the standard has led to the increased production and unaccountable discarding of plastic waste ultimately leading to environmental pollution caused by degradation due to various environmental factors. Cooperation between the public, businesses, and government is crucial to solving this pressing problem. The solution must include stricter rules for plastic manufacture and disposal, creative recycling methods, and increased public awareness. To figure out the magnitude of microplastic contamination, its effect on ecological systems, and potential health consequences, further research and monitoring are required. To reduce the negative impacts of microplastics on the ecosystems of our world and the welfare of future generations, we must act collaboratively during this global crisis.

References

Abbas, Q., Yousaf, B., Amina, Ali, M. U., Munir, M. A. M., El-Naggar, A., Rinklebe, J., & Naushad, M. (2020). Transformation pathways and fate of engineered nanoparticles (ENPs) in distinct interactive environmental compartments: A review. *Environment International, 138*, 105646. https://doi.org/10.1016/J.ENVINT.2020.105646.

Akdogan, Z., & Guven, B. (2019). Microplastics in the environment: A critical review of current understanding and identification of future research needs. *Environmental Pollution, 254*, 113011. https://doi.org/10.1016/J.ENVPOL.2019.113011

Allen, S., Allen, D., Phoenix, V. R., Le Roux, G., Durántez Jiménez, P., Simonneau, A., ... & Galop, D. (2019). Atmospheric transport and deposition of microplastics in a remote mountain catchment. *Nature Geoscience, 12*(5), 339–344.

Alex, R. K., Maes, T., & Devipriya, S. P. (2024). Clean, but not green: Emission assessment, forecast modelling and policy solutions for plastic microbeads from personal care products in India. *Emerging Contaminants*, 100326.

Anandhan, K., Tharini, K., Thangal, S. H., Yogeshwaran, A., & Muralisankar, T. (2022). Occurrence of microplastics in the gastrointestinal tracts of edible fishes from South Indian Rivers. *Bulletin of Environmental Contamination and Toxicology, 109*(6), 1023–1028.

Andrady, A. L. (2011). Microplastics in the marine environment. *Marine Pollution Bulletin, 62*(8), 1596–1605.

Andrady, A. L. (2015). Persistence of plastic litter in the oceans. *Marine Anthropogenic Litter*, 57–72. https://doi.org/10.1007/978-3-319-16510-3_3

Andrady, A. L. (2017). The plastic in microplastics: A review. *Marine Pollution Bulletin, 119*(1), 12–22.

Atugoda, T., Vithanage, M., Wijesekara, H., Bolan, N., Sarmah, A. K., Bank, M. S., You, S., & Ok, Y. S. (2021). Interactions between microplastics, pharmaceuticals and personal care products: Implications for vector transport. *Environment International, 149*, 106367. https://doi.org/10.1016/j.envint.2020.106367

Bergmann, M., Gutow, L., & Klages, M. (2015). *Marine anthropogenic litter*. Springer International Publishing. https://doi.org/10.1007/978-3-319-16510-3

Bergmann, M., Tekman, M. B., & Gutow, L. (2017). Marine litter: Sea change for plastic pollution. *Nature, 544*(7650), 297. https://doi.org/10.1038/544297a

Besseling, E., et al. (2017). *Reducing potential sources of marine litter: A report to DG Environment, European Commission.* https://ec.europa.eu/environment/marine/good-environmental-status/descriptor-10/pdf/Sources_of_marine_litter.pdf.

Bourgeois, J. P., van der Voet, E., van der Zande, M., & Bouwmeester, H. (2020). Microplastics from consumer plastic food containers: Are we consuming it? *Current Opinion in Food Science, 32*, 22–29.

Boyle, K., & Örmeci, B. (2020). Microplastics and nanoplastics in the freshwater and terrestrial environment: A review. *Water, 12*(9), 2633. https://doi.org/10.3390/w12092633

Brooks, A. L., Wang, S., & Jambeck, J. R. (2018). The Chinese import ban and its impact on global plastic waste trade. *Science Advance, 4*(6), eaat0131. https://doi.org/10.1126/sciadv.aat0131.

Carbery, M., O'Connor, W., & Palanisami, T. (2018). Trophic transfer of microplastics and mixed contaminants in the marine food web and implications for human health. *Environment International, 115*, 400–409. https://doi.org/10.1016/j.envint.2018.03.007

Central Pollution Control Board: "Annual Report 2019–20" [Central Pollution Control Board, "Annual Report 2019–20"] (This is a governmental report by India's pollution control board).

Chang, X., Xue, Y., Li, J., Zou, L., & Tang, M. (2020). Potential health impact of environmental micro- and nanoplastics pollution. *Journal of Applied Toxicology, 40*(1), 4–15. https://doi.org/10.1002/jat.3915

Chatterjee, P., Dutta, S., & Singh, A. K. (2020). A review on municipal solid waste management in India. *Journal of Material Cycles and Waste Management, 22*(3), 879–904.

Chen, P., Xie, Q., Addy, M., Zhou, W., Liu, Y., Wang, Y., Cheng, Y., Li, K., & Ruan, R. (2016). Utilization of municipal solid and liquid wastes for bioenergy and bioproducts production. *Bioresource Technology, 215*, 163–172. https://doi.org/10.1016/j.biortech.2016.02.094

Choi, H., Im, D. H., Park, Y. H., Lee, J. W., Yoon, S. J., & Hwang, U. K. (2022). Ingestion and egestion of polystyrene microplastic fragments by the Pacific oyster, Crassostrea gigas. *Environmental Pollution, 307*, 119217.

Cole, M., Lindeque, P., Halsband, C., & Galloway, T. S. (2011). Microplastics as contaminants in the marine environment: A review. *Marine Pollution Bulletin, 62*(12), 2588–2597. https://doi.org/10.1016/j.marpolbul.2011.09.025

Cózar, A., Echevarría, F., González-Gordillo, J. I., Irigoien, X., Ubeda, B., Hernández-León, S., Palma, A. T., Navarro, S., García-de-Lomas, J., Ruiz, A., Fernández-de-Puelles, M. L., & Duarte, C. M. (2014). Plastic debris in the open ocean. *Proceedings of the National Academy of Sciences of the United States of America, 111*(28), 10239–10244. https://doi.org/10.1073/pnas.1314705111

D'Hont, A., Gittenberger, A., Leuven, R. S., & Hendriks, A. J. (2021). Dropping the microbead: Source and sink related microplastic distribution in the Black Sea and Caspian Sea basins. *Marine Pollution Bulletin, 173*, 112982.

Danso, D., Chow, J., & Streit, W. R. (2019). Plastics: environmental and biotechnological perspectives on microbial degradation (H. L. Drake (Ed.)). *Applied and Environmental Microbiology, 85*(19), e01095–19. https://doi.org/10.1128/AEM.01095-19.

Dagwar, P. P., & Dutta, D. (2024). Landfill leachate a potential challenge towards sustainable environmental management. *Science of The Total Environment*, 171668.

De Bhowmick, G., Sarmah, A. K., & Dubey, B. (2021). Microplastics in the NZ Environment: Current Status and Future Directions. *Case Studies in Chemical and Environmental Engineering, 3*, 100076. https://doi.org/10.1016/j.cscee.2020.100076

De Side, G. N., Widiyanti, A., Rancak, G. T., Aprianto, R., Widhiantari, I. A., & Sutawijaya, I. B. (2021). Correlation analysis of leachate in final disposal sites on groundwater and surface water quality. *IOP Conference Series: Earth and Environmental Science, 913*(1), 012048. https://doi.org/10.1088/1755-1315/913/1/012048.

De Souza Machado, A. A., Kloas, W., Zarfl, C., Hempel, S., & Rillig, M. C. (2018). Microplastics as an emerging threat to terrestrial ecosystems. *Global Change Biology, 24*(4), 1405–1416. https://doi.org/10.1111/gcb.14020

Debnath, S., Sarma, J. C., Sarma, H. K., & Sarma, P. (2020). Microplastic pollution in agricultural soils of Northeast India: A threat to soil health and food safety. *Science of the Total Environment, 724*, 138347.

DEPI-MarineLitter_FINAL_highres. (n.d.).

Derraik, J. G. B. (2002). The pollution of the marine environment by plastic debris: A review. *Marine Pollution Bulletin, 44*(9), 842–852. https://doi.org/10.1016/S0025-326X(02)00220-5

Di, M., & Wang, J. (2018). Microplastics in surface waters and sediments of the Three Gorges Reservoir, China. *Science of the Total Environment, 616–617*, 1620–1627. https://doi.org/10.1016/J.SCITOTENV.2017.10.150

Dijkstra, H., van Beukering, P., & Brouwer, R. (2020). Business models and sustainable plastic management: A systematic review of the literature. *Journal of Cleaner Production, 258*, 120967. https://doi.org/10.1016/j.jclepro.2020.120967

Ding, J., Jiang, F., Li, J., Wang, Z., Sun, C., Wang, Z., Fu, L., Ding, N. X., & He, C. (2019). Microplastics in the Coral reef systems from Xisha Islands of South China Sea. *Environmental Science and Technology, 53*(14), 8036–8046. https://doi.org/10.1021/acs.est.9b01452

Dowarah, K., & Devipriya, S. P. (2019). Microplastic prevalence in the beaches of Puducherry, India and its correlation with fishing and tourism/recreational activities. *Marine Pollution Bulletin, 148*, 123–133. https://doi.org/10.1016/j.marpolbul.2019.07.066

Dris, R., Gasperi, J., Mirande, C., Mandin, C., Guerrouache, M., Langlois, V., & Tassin, B. (2015). A first overview of textile fibers, including microplastics, in indoor and outdoor environments. *Environmental Pollution, 204*, 385–393. https://doi.org/10.1016/j.envpol.2015.05.044

Eerkes-Medrano, D., Thompson, R. C., & Aldridge, D. C. (2015). Microplastics in freshwater systems: A review of the emerging threats, identification of knowledge gaps and prioritisation of research needs. *Water Research, 75*, 63–82. https://doi.org/10.1016/j.watres.2015.02.012

Eerkes-Medrano, D., Leslie, H. A., & Quinn, B. (2019). Microplastics in drinking water: A review and assessment. *Current Opinion in Environmental Science & Health, 7*, 69–75. https://doi.org/10.1016/j.coesh.2018.12.001

Enfrin, M., Myszka, R., & Giustozzi, F. (2022). Paving roads with recycled plastics: Microplastic pollution or eco-friendly solution? *Journal of Hazardous Materials, 437*, 129334.

Eriksen, M., Lebreton, L. C., Carson, H. S., Thiel, M., Moore, C. J., Borerro, J. C., Galgani, F., Ryan, P. G., & Reisser, J. (2014). Plastic pollution in the world's oceans: More than 5 trillion plastic pieces weighing over 250,000 tons afloat at sea. *PLoS ONE, 9*(12), e111913. https://doi.org/10.1371/Journal.pone.0111913

Eriksen, M., Mason, S., Wilson, S., Box, C., Zellers, A., Edwards, W., Farley, H., & Amato, S. (2013). Microplastic pollution in the surface waters of the Laurentian Great Lakes. *Marine Pollution Bulletin, 77*(1–2), 177–182. https://doi.org/10.1016/J.MARPOLBUL.2013.10.007

Evode, N., Qamar, S. A., Bilal, M., Barceló, D., & Iqbal, H. M. N. (2021). Plastic waste and its management strategies for environmental sustainability. *Case Studies in Chemical and Environmental Engineering, 4*, 100142. https://doi.org/10.1016/j.cscee.2021.100142

Gaylarde, C. C., Neto, J. A. B., & da Fonseca, E. M. (2021). Paint fragments as polluting microplastics: A brief review. *Marine Pollution Bulletin, 162*, 111847.

Gallo, F., Fossi, C., Weber, R., Santillo, D., Sousa, J., Ingram, I., Nadal, A., & Romano, D. (2018). Marine litter plastics and microplastics and their toxic chemicals components: the need for urgent preventive measures. In *Environmental Sciences Europe* (Vol. 30, Issue 1). Springer. https://doi.org/10.1186/s12302-018-0139-z.

Geyer, R., Jambeck, J. R., & Law, K. L. (2017). Production, use, and fate of all plastics ever made. *Science Advances, 3*(7), e1700782. https://doi.org/10.1126/sciadv.1700782

Ghosh, D., & Kumari, M. (2021). Microplastic detection and analysis in River Yamuna, Delhi. In *Advances in Water Resources and Transportation Engineering: Select Proceedings of TRACE 2020* (pp. 233–240). Springer.

Göbel, A., Thiel, M., & Gerdts, G. (2021). Methodology for the detection of microplastics—A critical appraisal. *Marine Pollution Bulletin, 164*, 112011. https://doi.org/10.1016/j.marpolbul.2021.112011

Golwala, H., Zhang, X., Iskander, S. M., & Smith, A. L. (2021). Solid waste: An overlooked source of microplastics to the environment. *Science of the Total Environment, 769*, 144581.

Guo, X., & Wang, J. (2019). The chemical behaviors of microplastics in marine environment: A review. *Marine Pollution Bulletin, 142*, 1–14. https://doi.org/10.1016/j.marpolbul.2019.03.019

Gupta, P., Saha, M., Naik, A., Kumar, M. M., Rathore, C., Vashishth, S., ... & Thukral, H. (2024). A comprehensive assessment of macro and microplastics from Rivers Ganga and Yamuna: Unveiling the seasonal, spatial and risk factors. *Journal of Hazardous Materials, 469*, 133926.

Guzzetti, E., Bailly, D., Zhang, H., Frka, S., Jonsson, S., Kalogerakis, N., ... & Dachs, J. (2018). Fate of microplastics and other small anthropogenic litter (SAL) in wastewater treatment plants depends on unit processes employed. *Water Research, 141*, 307–316.

Harrison, J. P., Sapp, M., & Schratzberger, M. (2019). Impacts of marine debris on benthic fauna: A global review. *Marine Pollution Bulletin, 146*, 763–766.

Hartmann, N. B., Hüffer, T., Thompson, R. C., Hassellöv, M., Verschoor, A., & Daugaard, A. E. (2019). Are we speaking the same language? Recommendations for a definition and categorization framework for plastic debris. *Environmental Science & Technology, 53*(3), 1039–1047.

He, D., Luo, Y., Lu, S., Liu, M., Song, Y., & Lei, L. (2018). Microplastics in soils: Analytical methods, pollution characteristics and ecological risks. *TrAC Trends in Analytical Chemistry, 109*, 163–172. https://doi.org/10.1016/j.trac.2018.10.006

He, P., Chen, L., Shao, L., Zhang, H., & Lü, F. (2019). Municipal Solid Waste (MSW) landfill: A source of microplastics? Evidence of microplastics in landfill leachate. *Water Research, 159*, 38–45. https://doi.org/10.1016/j.watres.2019.04.060

Holmes, L. A., Turner, A., & Thompson, R. C. (2012). Adsorption of trace metals to plastic resin pellets in the marine environment. *Environmental Pollution, 160*(1), 42–48. https://doi.org/10.1016/J.ENVPOL.2011.08.052

Horton, A. A., Walton, A., Spurgeon, D. J., Lahive, E., & Svendsen, C. (2017). Microplastics in freshwater and terrestrial environments: Evaluating the current understanding to identify the knowledge gaps and future research priorities. *Science of the Total Environment, 586*, 127–141. https://doi.org/10.1016/J.SCITOTENV.2017.01.190

Huang, D., Tao, J., Cheng, M., Deng, R., Chen, S., Yin, L., & Li, R. (2021). Microplastics and nanoplastics in the environment: Macroscopic transport and effects on creatures. *Journal of Hazardous Materials, 407*, 124399. https://doi.org/10.1016/j.jhazmat.2020.124399

Imhof, H. K., Ivleva, N. P., Schmid, J., Niessner, R., & Laforsch, C. (2016). Contamination of beach sediments of a subalpine lake with microplastic particles. *Current Biology, 26*(19), R867–R868. https://doi.org/10.1016/j.cub.2016.08.041

Ivleva, N. P., Wiesheu, A. C., Niessner, R., & Baumann, T. (2017). Highly selective, rapid, and portable smartphone-based quantitation of mercury(ii) ions using hydrophobic polymerization and optofluidics. *Analytical Chemistry, 89*(9), 4758–4764. https://doi.org/10.1021/acs.analchem.7b00136

Jahandari, A. (2023). Microplastics in the urban atmosphere: Sources, occurrences, distribution, and potential health implications. *Journal of Hazardous Materials Advances, 12*, 100346. https://doi.org/10.1016/J.HAZADV.2023.100346

Jambeck, J. R., Geyer, R., Wilcox, C., Siegler, T. R., Perryman, M., Andrady, A., Narayan, R., & Law, K. L. (2015). Plastic waste inputs from land into the ocean. *Science, 347*(6223), 768–771. https://doi.org/10.1126/science.1260352

Jayasiri, H. B., Vennila, A., & Purushothaman, C. S. (2014). Spatial and temporal variability of metals in inter-tidal beach sediment of Mumbai, India. *Environmental Monitoring and Assessment, 186*(2), 1101–1111. https://doi.org/10.1007/s10661-013-3441-7

Jeong, J., & Choi, J. (2020). Development of AOP relevant to microplastics based on toxicity mechanisms of chemical additives using ToxCast™ and deep learning models combined approach. *Environment International, 137*, 105557.

Jha, A., Gomathi, R., Raghuvanshi, R., & Verma, S. K. (2019). Microplastic pollution in coastal and marine environments of India: Sources, distribution, and impacts—A review. *Marine Pollution Bulletin, 142*(2), 103–111.

Jhala, A. D., & Reddy, M. S. (2021). Plastic waste management in India: Challenges and solutions. *Environmental Science and Pollution Research, 28*(32), 49342–49352.

John, P. M., Murali, V., Chakraborty, K., Lotlikar, A., Shameem, K., Habeeb Rahman, K., & Gopinath, A. (2022). Spatial and seasonal trends of trace metals in the surficial sediments from off Kochi—Geochemistry and environmental implications. *Marine Pollution Bulletin, 182*, 114029. https://doi.org/10.1016/J.MARPOLBUL.2022.114029

Karuppasamy, P. K., Ravi, A., Vasudevan, L., Elangovan, M. P., Mary, P. D., Vincent, S. G., & Palanisami, T. (2020). Baseline survey of micro and mesoplastics in the gastro-intestinal tract of commercial fish from Southeast coast of the Bay of Bengal. *Marine Pollution Bulletin, 153*, 110974.

Kershaw, P. J., Turra, A., & Galgani, F. (2016). Marine anthropogenic litter on beaches: A 1977–2012 global review. *Environmental Science & Technology, 50*(20), 10725–10727.

Khalid, N., Aqeel, M., Noman, A., Hashem, M., Mostafa, Y. S., Alhaithloul, H. A. S., & Alghanem, S. M. (2021). Linking effects of microplastics to ecological impacts in marine environments. *Chemosphere, 264*, 128541.

Klingelhöfer, D., Braun, M., Quarcoo, D., Brüggmann, D., & Groneberg, D. A. (2020). Research landscape of a global environmental challenge: Microplastics. *Water Research, 170*, 115358. https://doi.org/10.1016/j.watres.2019.115358

Knight, L. J., Parker-Jurd, F. N., Al-Sid-Cheikh, M., & Thompson, R. C. (2020). Tyre wear particles: an abundant yet widely unreported microplastic?. *Environmental Science and Pollution Research, 27*, 18345–18354. https://doi.org/10.1007/s11356-020-08187-4

Koelmans, A. A. (2015). Modeling the role of microplastics in bioaccumulation of organic chemicals to marine aquatic organisms. A critical review. *Marine Anthropogenic Litter*, 309–324.

Kosuth, M., Mason, S. A., & Wattenberg, E. V. (2018a). Anthropogenic contamination of tap water, beer, and sea salt. *PLoS ONE, 13*(4), e0194970. https://doi.org/10.1371/Journal.pone.0194970

Kosuth, M., Mason, S. A., & Wattenberg, E. V. (2018b). Anthropogenic contamination of tap water, beer, and sea salt (Z. Zhou (Ed.)). *PLoS ONE, 13*(4), e0194970. https://doi.org/10.1371/Journal.pone.0194970

Kole, P. J., Löhr, A. J., Van Belleghem, F. G., & Ragas, A. M. (2017). Wear and tear of tyres: A stealthy source of microplastics in the environment. *International Journal of Environmental Research and Public Health, 14*(10), 1265.

Kumar, A., & Agrawal, A. (2020). Recent trends in solid waste management status, challenges, and potential for the future Indian cities—A review. *Current Research in Environmental Sustainability, 2*, 100011. https://doi.org/10.1016/j.crsust.2020.100011

Kumar, S., Ashwini & Varghese, G. (2019). Environmental forensic analysis of the microplastic pollution at "Nattika" Beach, Kerala Coast, India. *Environmental Forensics*, 1–16. https://doi.org/10.1080/15275922.2019.1693442.

Lebreton, L., Slat, B., Ferrari, F., Sainte-Rose, B., Aitken, J., Marthouse, R., & Reisser, J. (2018). Evidence that the Great Pacific Garbage Patch is rapidly accumulating plastic. *Scientific Reports, 8*(1), 1–15.

Lambert, S., Vercauteren, M., Catarino, A. I., Li, Y., Van Landuyt, J., Boon, N., … & Asselman, J. (2024). Aerosolization of micro-and nanoplastics via sea spray: Investigating the role of polymer type, size, and concentration, and potential implications for human exposure. *Environmental Pollution, 351*, 124105.

Li, J., et al. (2020). Microplastics in surface waters and sediments of the three gorges reservoir, China: Composition, source, and risk assessment. *Science of the Total Environment, 726*, 138532.

Li, J., Liu, H., & Paul Chen, J. (2018). Microplastics in freshwater systems: A review on occurrence, environmental effects, and methods for microplastics detection. *Water Research, 137*, 362–374. https://doi.org/10.1016/j.watres.2017.12.056

Li, W., Qamar, S. A., Qamar, M., Basharat, A., Bilal, M., & Iqbal, H. M. N. (2021). Carrageenan-based nano-hybrid materials for the mitigation of hazardous environmental pollutants. *International Journal of Biological Macromolecules, 190*, 700–712. https://doi.org/10.1016/j.ijbiomac.2021.09.039

Liang, Y., Tan, Q., Song, Q., & Li, J. (2021). An analysis of the plastic waste trade and management in Asia. *Waste Management, 119*, 242–253. https://doi.org/10.1016/j.wasman.2020.09.049

Liu, S., Chen, H., Wang, J., Su, L., Wang, X., Zhu, J., & Lan, W. (2021). The distribution of microplastics in water, sediment, and fish of the dafeng river, a remote river in China. *Ecotoxicology and Environmental Safety, 228*, 113009. https://doi.org/10.1016/j.ecoenv.2021.113009

Liu, S., Huang, J., Zhang, W., Shi, L., Yi, K., Yu, H., … & Li, J. (2022). Microplastics as a vehicle of heavy metals in aquatic environments: A review of adsorption factors, mechanisms, and biological effects. *Journal of Environmental Management, 302,* 113995.

Liu, K., Wang, X., Song, Z., Wei, N., & Li, D. (2020). Terrestrial plants as a potential temporary sink of atmospheric microplastics during transport. *Science of the Total Environment, 742,* 140523.

Lu, L., Wan, Z., Luo, T., Fu, Z., Jin, Y., & Liu, Y. (2016). The accumulation and fragmentation of plastic debris in global environments. *Environmental Science & Technology, 50*(21), 12186–12194.

Lusher, A. L., McHugh, M., & Thompson, R. C. (2013). Occurrence of microplastics in the gastrointestinal tract of pelagic and demersal fish from the English Channel. *Marine Pollution Bulletin, 67*(1–2), 94–99. https://doi.org/10.1016/J.MARPOLBUL.2012.11.028

Lusher, A. L., Munno, K., Hermabessiere, L., & Carr, S. (2020). Isolation and extraction of microplastics from environmental samples: An evaluation of practical approaches and recommendations for further harmonization. *Applied Spectroscopy, 74*(9), 1049–1065.

Luo, Z., Zhou, X., Su, Y., Wang, H., Yu, R., Zhou, S., … & Xing, B. (2021). Environmental occurrence, fate, impact, and potential solution of tire microplastics: Similarities and differences with tire wear particles. *Science of the Total Environment, 795*, 148902.

Lwanga, E. H., Gertsen, H., Gooren, H., Peters, P., Salánki, T., van der Ploeg, M., & Besseling, E. (2016). Microplastics in the sediments of a freshwater urban pond and its inflowing streams (The Netherlands). *Environmental Science & Technology, 50*(20), 10788–10796. https://doi.org/10.1021/acs.est.6b03150

Maharana, D., Saha, M., Dar, J. Y., Rathore, C., Sreepada, R. A., Xu, X. R., Koongolla, J. B., & Li, H. X. (2020). Assessment of micro and macroplastics along the west coast of India: Abundance, distribution, polymer type and toxicity. *Chemosphere, 246*, 125708. https://doi.org/10.1016/J.CHEMOSPHERE.2019.125708

Marine Debris Program, N. (2015). Laboratory Methods for the Analysis of Microplastics in the Marine Environment: Recommendations for quantifying synthetic particles in waters and sediments.

Mashirin, K. R., & Chitra, K. C. (2022). Microplastics–an emerging threat in the Indian waterbodies. In *Marine Biology Research* (Vol. 18, Issues 1–2, pp. 1–12). Taylor and Francis Ltd. https://doi.org/10.1080/17451000.2022.2096905.

Mason, V. G., Skov, M. W., Hiddink, J. G., & Walton, M. (2022). Microplastics alter multiple biological processes of marine benthic fauna. *Science of The Total Environment, 845*, 157362.

Mason, S. A., Garneau, D., Sutton, R., Chu, Y., Ehmann, K., Barnes, J., Fink, P., Papazissimos, D., & Rogers, D. L. (2016). Microplastic pollution is widely detected in US municipal wastewater treatment plant effluent. Environmental pollution (Barking, Essex : 1987), 218, 1045–1054. https://doi.org/10.1016/j.envpol.2016.08.056

Masura, J., Baker, J., Foster, G., & Arthur, C. (2015). Laboratory methods for the analysis of microplastics in the marine environment: Recommendations for quantifying synthetic particles in waters and sediments. NOAA Technical Memorandum NOS-OR&R-48.

Mittal, A., Mani, S., & Bawa, K. (2019). Challenges and opportunities in municipal solid waste management in India. *Journal of Air Pollution and Control, 1*(2), 142–150.

Mohan, P. M., Tiwari, S., Karuvelan, M., Malairajan, S., Mageswaran, T., & Sachithanandam, V. (2022). A baseline study of meso and microplastic predominance in pristine beach sediment of the Indian tropical island ecosystem. *Marine Pollution Bulletin, 181,* 113825.

Naidu, B. C., Xavier, K. M., Shukla, S. P., Jaiswar, A. K., & Nayak, B. B. (2022). Comparative study on the microplastics abundance, characteristics, and possible sources in yellow clams of different demographic regions of the northwest coast of India. *Journal of Hazardous Materials Letters, 3,* 100051.

Nandan, A., Yadav, B.P., Baksi, S., & Bose, D. (2017). Recent scenario of solid waste management in India.

Napper, I. E., Bakir, A., Rowland, S. J., & Thompson, R. C. (2015). Characterisation, quantity and sorptive properties of microplastics extracted from cosmetics. *Marine Pollution Bulletin, 99*(1–2), 178–185. https://doi.org/10.1016/J.MARPOLBUL.2015.07.029

Napper, I. E., & Thompson, R. C. (2019). Marine plastic pollution: Other than microplastic. In *Waste* (pp. 425–442). Academic Press.

Ng, E.-L., Lwanga, E. H., Eldridge, S. M., Johnston, P., Hu, H.-W., Geissen, V., & Chen, D. (2018). An overview of microplastic and nanoplastic pollution in agroecosystems. *Science of the Total Environment, 627*, 1377–1388. https://doi.org/10.1016/j.scitotenv.2018.01.341

Nizzetto, L., Futter, M., & Langaas, S. (2016). Are agricultural soils dumps for microplastics of urban origin?

Olubusoye, B. S., Cizdziel, J. V., Wontor, K., Heinen, E., Grandberry, T., Bennett, E. R., & Moore, M. T. (2024). Removal of microplastics from agricultural runoff using biochar: A column feasibility study. *Frontiers in Environmental Science, 12,* 1388606.

Padervand, M., Lichtfouse, E., Robert, D., & Wang, C. (2020). Removal of microplastics from the environment. A review. *Environmental Chemistry Letters, 18*(3), 807–828. https://doi.org/10.1007/s10311-020-00983-1

Pandey, S., Pandey, V., & Kumar, L. C. S. (2009). Effects of irrigation methods and nitrogen levels on growth, yield and water use efficiency of winter wheat (Triticum aestivum L.) in mediterranean climate. *Agricultural Water Management, 96*(11), 1553–1564. https://doi.org/10.1016/j.agwat.2009.06.002.

Pandit, R., Raghuwanshi, R., & Verma, S. K. (2019). Microplastic contamination in remote areas of the Indian Himalayas. *Environmental Pollution, 252*(Pt 1), 1214–1223. https://doi.org/10.1016/j.envpol.2019.06.061

Patchaiyappan, A., ZakiAhmed, S., Dowarah, K., Khadanga, S. S., Singh, T., Jayakumar, S., … & Devipriya, S. P. (2021). Prevalence of microplastics in the sediments of Odisha beaches, southeastern coast of India. *Marine Pollution Bulletin, 167*, 112265.

Chaudhari, P. S., & Surjushe, A. M. (n.d.). *International Journal for Multidisciplinary Research International Conference on Multidisciplinary Research & Studies 2023*. E-ISSN: 2582–2160.

Parekh, D., Vaidh, S., Patel, D., & Vishwakarma, G. S. (2022). Temporal study of physicochemical parameters of leachates and assessment of solid waste and ground water quality of pirana dumping site, Ahmedabad, India. *Environmental Monitoring and Assessment, 194*(10), 708. https://doi.org/10.1007/s10661-022-10402-3

Rasmussen, L. A., Lykkemark, J., Andersen, T. R., & Vollertsen, J. (2023). Permeable pavements: A possible sink for tyre wear particles and other microplastics? *Science of the Total Environment, 869,* 161770.

Rolsky, C., Kelkar, V., Driver, E., & Halden, R. U. (2020). Municipal sewage sludge as a source of microplastics in the environment. *Current Opinion in Environmental Science & Health, 14,* 16–22.

Silva, P., Ana, L., Prata, J. C., Walker, T. R., Campos, D., Duarte, A. C., Soares, A. M. V. M., Barcelò, D., & Rocha-Santos, T. (2020). Rethinking and optimising plastic waste management under COVID-19 pandemic: Policy solutions based on redesign and reduction of single-use plastics and personal protective equipment. *Science of the Total Environment, 742*, 140565. https://doi.org/10.1016/j.scitotenv.2020.140565

Paul, M. B., Stock, V., Cara-Carmona, J., Lisicki, E., Shopova, S., Fessard, V., Braeuning, A., Sieg, H., & Böhmert, L. (2020). Micro- and nanoplastics—Current state of knowledge with the focus on oral uptake and toxicity. *Nanoscale Advances, 2*(10), 4350–4367. https://doi.org/10.1039/D0NA00539H

Paul-Pont, I., Lacroix, C., González Fernández, C., Hégaret, H., Lambert, C., Le Goïc, N., Frère, L., Cassone, A. L., Sussarellu, R., Fabioux, C., Guyomarch, J., Albentosa, M., Huvet, A., & Soudant, P. (2016). Exposure of marine mussels Mytilus spp. to polystyrene microplastics: Toxicity and influence on fluoranthene bioaccumulation. *Environmental Pollution, 216*, 724–737. https://doi.org/10.1016/J.ENVPOL.2016.06.039

Perumal, K., & Muthuramalingam, S. (n.d.). Microplastics pollution studies in India: A recent review of sources, abundances and research perspectives. https://doi.org/10.21203/rs.3.rs-535083/v1.

Pivokonsky, M., Cermakova, L., Novotna, K., Peer, P., Cajthaml, T., & Janda, V. (2018). Occurrence of microplastics in raw and treated drinking water. *The Science of the Total Environment, 643*, 1644–1651. https://doi.org/10.1016/j.scitotenv.2018.08.102

Prata, J. C., da Costa, J. P., Lopes, I., Duarte, A. C., & Rocha-Santos, T. (2020). Environmental exposure to microplastics: An overview on possible human health effects. *The Science of the Total Environment, 702*, 134455. https://doi.org/10.1016/j.scitotenv.2019.134455

Prateek, R., Ameen, K., Kaushik. (2018, June). Challenges and opportunities: Plastic waste management in India. http://www.teriin.org/sites/default/files/2018-06/plastic-waste-management_0.pdf.

Praveena, S. M., Shaifuddin, S. N. M., & Akizuki, S. (2018). Exploration of microplastics from personal care and cosmetic products and its estimated emissions to marine environment: An evidence from Malaysia. *Marine Pollution Bulletin, 136*, 135–140. https://doi.org/10.1016/j.marpolbul.2018.09.012

Raha, U. K., Ramesh Kumar, B., & Sarkar, S. K. (2021). Policy framework for mitigating land-based marine plastic pollution in the gangetic delta region of Bay of Bengal-a review. *Journal of Cleaner Production, 278*, 123409. https://doi.org/10.1016/j.jclepro.2020.123409

Rabari, V., Patel, K., Patel, H., & Trivedi, J. (2022). Quantitative assessment of microplastic in sandy beaches of Gujarat state, India. *Marine Pollution Bulletin, 181*, 113925.

Rebelein, A., Int-Veen, I., Kammann, U., & Scharsack, J. P. (2021). Microplastic fibers—Underestimated threat to aquatic organisms? *Science of the Total Environment, 777*, 146045.

Rezania, S., Park, J., Din, M. F. M., Taib, S. M., Talaiekhozani, A., Yadav, K. K., & Kamyab, H. (2018). Microplastics pollution in different aquatic environments and biota: A review of recent studies. *Marine Pollution Bulletin, 133*, 191–208. https://doi.org/10.1016/j.marpolbul.2018.05.022

Robin, R. S., Karthik, R., Purvaja, R., Ganguly, D., Anandavelu, I., Mugilarasan, M., & Ramesh, R. (2020). Holistic assessment of microplastics in various coastal environmental matrices, southwest coast of India. *Science of the Total Environment, 703*, 134947. https://doi.org/10.1016/J.SCITOTENV.2019.134947

Rochman, C. M., Browne, M. A., Halpern, B. S., Hentschel, B. T., Hoh, E., Karapanagioti, H. K., Rios-Mendoza, L. M., Takada, H., & Teh, S. (2015a). Policy: Classify plastic waste as hazardous. *Nature, 494*(7436), 169–171. https://doi.org/10.1038/494169a

Rochman, C. M., Cook, A. M., & Koelmans, A. A. (2016). Plastic debris and policy: Using current scientific understanding to invoke positive change. *Environmental Toxicology and Chemistry, 35*(7), 1617–1626. https://doi.org/10.1002/etc.3408

Rochman, C., Tahir, A., Williams, S., et al. (2015b). Anthropogenic debris in seafood: Plastic debris and fibers from textiles in fish and bivalves sold for human consumption. *Science and Reports, 5*, 14340. https://doi.org/10.1038/srep14340

Rochman, C. M. (2018). Microplastics research—From sink to source. *Science, 360*(6384), 28–29. https://doi.org/10.1126/science.aar7734

Roman, L., Hardesty, B. D., & Schuyler, Q. (2022). A systematic review and risk matrix of plastic litter impacts on aquatic wildlife: A case study of the Mekong and Ganges River Basins. *Science of the Total Environment, 843*, 156858.

Ronca, S. (2017). Polyethylene. In *Brydson's plastics materials* (pp. 247–278). Elsevier. https://doi.org/10.1016/B978-0-323-35824-8.00010-4.

Rozman, U., Turk, T., Skalar, T., Zupančič, M., Korošin, N. Č., Marinšek, M., ... & Kalčíková, G. (2021). An extensive characterization of various environmentally relevant microplastics–material properties, leaching and ecotoxicity testing. *Science of the Total Environment, 773*, 145576.

Krishnakumar, S., Anbalagan, S., Kasilingam, K., Smrithi, P., Anbazhagi, S., & Srinivasalu, S. (2020). Assessment of plastic debris in remote islands of the Andaman and Nicobar Archipelago, India. *Marine Pollution Bulletin, 151*, 110841. https://doi.org/10.1016/j.marpolbul.2019.110841

Sandu, C., Takacs, E., Suaria, G., Borgogno, F., Laforsch, C., Löder, M.M.G.J., Tweehuysen, G. & Florea, L. (2020). Society role in the reduction of plastic pollution. In F. Stock, G. Reifferscheid, N. Brennholt & E. Kostianaia (Eds.), Plastics in the Aquatic Environment-Part II (Vol. 112, pp. 39–65). Springer International Publishing. https://doi.org/10.1007/698_2020_483.

Sanz-Lázaro, C., et al. (2020). Microplastics in the North Atlantic atmospheric boundary layer. *Environmental Science & Technology Letters, 7*(6), 388–394.

Sarkar, D. J., Das Sarkar, S., Das, B. K., Manna, R. K., Behera, B. K., & Samanta, S. (2019). Spatial distribution of meso and microplastics in the sediments of river Ganga at eastern India. *The Science of the Total Environment, 694*, 133712. https://doi.org/10.1016/j.scitotenv.2019.133712

Sathish, M. N., Jeyasanta, I., & Patterson, J. (2020). Microplastics in salt of Tuticorin, Southeast Coast of India. *Archives of Environmental Contamination and Toxicology, 79*(1), 111–121. https://doi.org/10.1007/s00244-020-00731-0

Schwabl, P., Köppel, S., Königshofer, P., Bucsics, T., Trauner, M., Reiberger, T., & Liebmann, B. (2019). Detection of various microplastics in human stool: A prospective case series. *Annals of Internal Medicine, 171*(7), 453–457. https://doi.org/10.7326/M19-0618

Shams, M., Alam, I., & Mahbub, M. S. (2021). Plastic pollution during COVID-19: Plastic waste directives and its long-term impact on the environment. *Environmental Advances, 5*, 100119. https://doi.org/10.1016/j.envadv.2021.100119

Sharma, K. D., & Jain, S. (2019). Overview of municipal solid waste generation, composition, and management in India. *Journal of Environmental Engineering, 145*(3), 04018143. https://doi.org/10.1061/(ASCE)EE.1943-7870.0001490

Sheriff, I., Yusoff, M. S., Manan, T. S. B. A., & Koroma, M. (2023). Microplastics in manure: Sources, analytical methods, toxicodynamic, and toxicokinetic endpoints in livestock and poultry. *Environmental Advances, 12*, 100372. https://doi.org/10.1016/J.ENVADV.2023.100372

Siegfried, M., Koelmans, A. A., Besseling, E., & Kroeze, C. (2017). Export of microplastics from land to sea. A modelling approach. *Water Research, 127*, 249–257. https://doi.org/10.1016/J.WATRES.2017.10.011

Silori, R., Shrivastava, V., Mazumder, P., Mootapally, C., Pandey, A., & Kumar, M. (2023). Understanding the underestimated: Occurrence, distribution, and interactions of microplastics in the sediment and soil of China, India, and Japan. *Environmental Pollution, 320*, 120978. https://doi.org/10.1016/j.envpol.2022.120978

Silva, C. J. M., Beleza, S., Campos, D., Soares, A. M. V. M., Silva, A. L. P., Pestana, J. L., & Gravato, C. (2021). Immune response triggered by the ingestion of polyethylene microplastics in the dipteran larvae chironomus riparius. *Journal of Hazardous Materials, 414*, 125401. https://doi.org/10.1016/j.jhazmat.2021.125401

Singh, J., & Ordoñez, I. (2016). Resource recovery from post-consumer waste: Important lessons for the upcoming circular economy. *Journal of Cleaner Production, 134*, 342–353. https://doi.org/10.1016/j.jclepro.2015.12.020

Singh, R. K., Singh, D. K., & Singh, K. P. (2011). Effect of mulch and weed management practices on yield, water use efficiency and water productivity of Bt cotton (Gossypium hirsutum L.) under semi-arid condition. *Journal of the International Society for Horticultural Science, 78*(1–2), 127–134.

Sri Sasi Jyothsna, T., & Chakradhar, B. (2020). Current scenario of plastic waste management in India: Way forward in turning vision to reality. In S. K. Ghosh (Ed.), *Urban Mining and Sustainable Waste Management* (pp. 203–218). Springer. https://doi.org/10.1007/978-981-15-0532-4_21.

Stapleton, M. J., Ansari, A. J., Ahmed, A., & Hai, F. I. (2023). Evaluating the generation of microplastics from an unlikely source: The unintentional consequence of the current plastic recycling process. *Science of the Total Environment, 902*, 166090. https://doi.org/10.1016/J.SCITOTENV.2023.166090

Sundar, S., Chokkalingam, L., Roy, P. D., & Usha, T. (2021). Estimation of microplastics in sediments at the southernmost coast of India (Kanyakumari). *Environmental Science and Pollution Research International, 28*(15), 18495–18500. https://doi.org/10.1007/s11356-020-10333-x

Syamala Devi, K., & Swamy Shaheda Nilofer, A. V. V. S. (n.d.). Municipal solid waste management in India—An overview. https://web.archive.org/web/20180425152949id_/http://www.apjor.com/downloads/3105201615.pdf.

The National Oceanic and Atmospheric Administration (NOAA) and European Chemical Agency (ECA) 2023.

Thompson, R. C., Olsen, Y., Mitchell, R. P., Davis, A., Rowland, S. J., John, A. W. G., McGonigle, D., & Russell, A. E. (2004). Lost at sea: Where is all the plastic? *Science, 304*(5672), 838–938. https://doi.org/10.1126/science.1094559

Tirkey, A., & Upadhyay, L. S. B. (2021). Microplastics: An overview on separation, identification and characterization of microplastics. *Marine Pollution Bulletin, 170*, 112604. https://doi.org/10.1016/J.MARPOLBUL.2021.112604

Tiwari, M., Rathod, T. D., Ajmal, P. Y., Bhangare, R. C., & Sahu, S. K. (2019). Distribution and characterization of microplastics in beach sand from three different Indian coastal environments. *Marine Pollution Bulletin, 140*, 262–273.

Tsering, K., Ramesh, M., & Sivakumar, K. (2021). Microplastic pollution in freshwater ecosystems of India: Current trends and future perspectives. *Environmental Science and Pollution Research*, 1–14.

Turra, A., Manzano, A. B., Dias, R. J. S., Mahiques, M. M., Barbosa, L., Balthazar-Silva, D., & Moreira, F. T. (2014). Three-dimensional distribution of plastic pellets in sandy beaches: Shifting paradigms. *Scientific Reports, 4*, 4435. https://doi.org/10.1038/srep04435

UNEP. (2015). *Marine plastic debris and microplastics–global lessons and research to inspire action and guide policy change* (Vol. 428). United Nations Environment Programme.

Vaid, M., Mehra, K., Sarma, K., & Gupta, A. (2022). Investigations on the co-occurrence of microplastics and other pollutants in the River Yamuna, Delhi. *Water Supply, 22*(12), 8767–8777. https://doi.org/10.2166/ws.2022.408

van Sebille, E., Wilcox, C., Lebreton, L., Maximenko, N., Hardesty, B. D., van Franeker, J. A., Eriksen, M., Siegel, D., Galgani, F., & Law, K. L. (2015). A global inventory of small floating plastic debris. *Environmental Research Letters, 10*(12), 124006. https://doi.org/10.1088/1748-9326/10/12/124006.

Vikas Madhav, N., Gopinath, K. P., Krishnan, A., Rajendran, N., & Krishnan, A. (2020). A critical review on various trophic transfer routes of microplastics in the context of the Indian coastal ecosystem. *Watershed Ecology and the Environment, 2*, 25–41. https://doi.org/10.1016/J.WSEE.2020.08.001

Wang, F., Wong, C. S., Chen, D., Lu, X., Wang, F., & Zeng, E. Y. (2018). Interaction of toxic chemicals with microplastics: A critical review. *Water Research, 139*, 208–219. https://doi.org/10.1016/j.watres.2018.04.003

Wang, W., Gao, H., Jin, S., Li, R., & Na, G. (2019). The ecotoxicological effects of microplastics on aquatic food web, from primary producer to human: A review. *Ecotoxicology and Environmental Safety, 173*, 110–117. https://doi.org/10.1016/j.ecoenv.2019.01.113

Wang, Z., Lin, T., & Chen, W. (2020). Occurrence and removal of microplastics in an Advanced Drinking Water Treatment Plant (ADWTP). *Science of the Total Environment, 700*, 134520. https://doi.org/10.1016/j.scitotenv.2019.134520

Wagner, M., Scherer, C., Alvarez-Muñoz, D., Brennholt, N., Bourrain, X., Buchinger, S., … & Reifferscheid, G. (2014). Microplastics in freshwater ecosystems: What we know and what we need to know. *Environmental Sciences Europe, 26*(1), 1–9.

Waring, R. H., Harris, R. M., & Mitchell, S. C. (2018). Plastic contamination of the food chain: A threat to human health? *Maturitas, 115*, 64–68. https://doi.org/10.1016/j.maturitas.2018.06.010

Weinstein, J. E., Crocker, B. K., & Gray, A. D. (2016). From macroplastic to microplastic: Degradation of high-density polyethylene, polypropylene, and polystyrene in a salt marsh habitat. *Environmental Toxicology and Chemistry, 35*(7), 1632–1640.

Wilcox, C., Van Sebille, E., & Hardesty, B. D. (2015). Threat of plastic pollution to seabirds is global, pervasive, and increasing. *Proceedings of the National Academy of Sciences of the United States of America, 112*(38), 11899–11904. https://doi.org/10.1073/pnas.1502108112

Wright, S. L., & Kelly, F. J. (2017a). Plastic and human health: A micro issue? *Environmental Science & Technology, 51*(12), 6634–6647. https://doi.org/10.1021/acs.est.7b00423

Wright, S. L., & Kelly, F. J. (2017b). Plastic and human health: A micro issue? *Environmental Science and Technology, 51*(12), 6634–6647. https://doi.org/10.1021/acs.est.7b00423

Wright, S. L., Rowe, D., Thompson, R. C., & Galloway, T. S. (2013). Microplastic ingestion decreases energy reserves in marine worms. *Current Biology: CB, 23*(23), R1031–R1033. https://doi.org/10.1016/j.cub.2013.10.068

Wu, C., et al. (2020). Macroplastics in riverine environments: Occurrence and sources in a developing region of China. *Environmental Pollution, 258*, 113736.

Wu, X., Zhao, X., Chen, R., Liu, P., Liang, W., Wang, J., … & Gao, S. (2022). Wastewater treatment plants act as essential sources of microplastic formation in aquatic environments: A critical review. *Water Research, 221*, 118825.

Wu, H., Hou, J., & Wang, X. (2023). A review of microplastic pollution in aquaculture: Sources, effects, removal strategies and prospects. *Ecotoxicology and Environmental Safety, 252*, 114567. https://doi.org/10.1016/J.ECOENV.2023.114567

Xie, Q., Li, H., Li, Z., Zhang, H., Yuan, M., Wu, M., … & Xu, X. (2022). Accumulation, chemical speciation and ecological risks of heavy metals on expanded polystyrene microplastics in seawater. *Gondwana Research, 108*, 181–192.

Xumiao, L., Prata, J. C., Alves, J. R., Duarte, A. C., Rocha-Santos, T., & Cerqueira, M. (2021). Airborne microplastics and fibers in indoor residential environments in Aveiro, Portugal. *Environmental Advances, 6*, 100134.

Yang, L., Zhang, Y., Kang, S., Wang, Z., & Wu, C. (2021). Microplastics in freshwater sediment: A review on methods, occurrence, and sources. *Science of the Total Environment, 754*, 141948. https://doi.org/10.1016/j.scitotenv.2020.141948

Yang, X., An, C., Feng, Q., Boufadel, M., & Ji, W. (2022). Aggregation of microplastics and clay particles in the nearshore environment: Characteristics, influencing factors, and implications. *Water Research, 224*, 119077.

Yaranal, N. A., Subbiah, S., & Mohanty, K. (2021). Distribution and characterization of microplastics in beach sediments from Karnataka (India) coastal environments. *Marine Pollution Bulletin, 169*, 112550.

Yogeshwaran, A., Gayathiri, K., Muralisankar, T., Gayathri, V., Monica, J. I., Rajaram, R., Marimuthu, K., & Bhavan, P. S. (2020). Bioaccumulation of heavy metals, antioxidants, and metabolic enzymes in the crab Scylla serrata from different regions of Tuticorin, Southeast Coast of India. *Marine Pollution Bulletin, 158*, 111443. https://doi.org/10.1016/j.marpolbul.2020.111443

Yukioka, S., Tanaka, S., Nabetani, Y., Suzuki, Y., Ushijima, T., Fujii, S., Takada, H., Van Tran, Q., & Singh, S. (2020). Occurrence and characteristics of microplastics in surface road dust in Kusatsu (Japan), Da Nang (Vietnam), and Kathmandu (Nepal). *Environmental Pollution, 256*, 113447. https://doi.org/10.1016/j.envpol.2019.113447

Yurtsever, M., & Yurtsever, U. (2019). Use of a convolutional neural network for the classification of microbeads in urban wastewater. *Chemosphere, 216*, 271–280. https://doi.org/10.1016/j.chemosphere.2018.10.084

Yu, Z. F., Song, S., Xu, X. L., Ma, Q., & Lu, Y. (2021). Sources, migration, accumulation and influence of microplastics in terrestrial plant communities. *Environmental and Experimental Botany, 192*, 104635.

Zhang, K., Su, J., Xiong, X., Wu, H., Wu, C., & Liu, J. (2020a). Microplastic pollution in the surface waters of the Bohai Sea, China. *Science of the Total Environment, 709*, 136050.

Zhang, X., Li, J., Li, Q., Li, X., & Wang, X. (2018). Microplastic pollution of inland water in China: A case study in urban surface water of Nanjing. *Ecotoxicology and Environmental Safety, 152*, 312–318. https://doi.org/10.1016/j.ecoenv.2018.01.040

Zhang, Y., Kang, S., Allen, S., Allen, D., Gao, T., & Sillanpää, M. (2020b). Atmospheric microplastics: A review on current status and perspectives. *Earth-Science Reviews, 203*, 103118. https://doi.org/10.1016/J.EARSCIREV.2020.103118

Zhao, Y., Qamar, S. A., Qamar, M., Bilal, M., & Iqbal, H. M. N. (2021). Sustainable remediation of hazardous environmental pollutants using biochar-based nanohybrid materials. *Journal of Environmental Management, 300*, 113762. https://doi.org/10.1016/j.jenvman.2021.113762

Chapter 3
Global Risks of Microplastics to Ecosystem and Human Health: An Emerging Environmental Disaster

Avinash Pratap Gupta, Joystu Dutta, Aseem Kerketta, Piyush Pandey, and Tarun Kumar Thakur

Abstract Microplastic contamination is an emerging environmental issue of current times. It has been shown to have negative impacts on numerous levels of biological organization. Research on the distribution and effects of microplastics in the environment has been gaining momentum since few years, especially post COVID-19 pandemic. Scientists have used a wide range of studies to demonstrate that tiny, insoluble polymeric particles may have the ability to have widespread effects on ecosystems. In ecosystems, microplastics offer a distinctive and substantial surface for microbial colonization. The rate of organic matter decomposition and horizontal gene transfer is increased to the greatest extent by the creation of microorganism-microplastic complexes, such as biofilms. The present scenario entails the influence of microplastics on the makeup and functioning of microbial communities, which in turn affects microplastics physical and chemical fate. For the fields of microbiology, ecology, and ecotoxicology, this new paradigm presents difficulties. There is a requirement for the establishment of a process that is capable of assessing the efficacy of experimental research. This mechanism is necessary to accurately assess the possible risks that are connected with plastic particles within the context of risk assessment. To improve comprehension of the methodologies utilized and their potential ramifications for the capacity of environmental risk evaluations to assess

A. P. Gupta · J. Dutta (✉) · P. Pandey
Department of Environmental Science, Sant Gahira Guru University, Sarguja 497001, Chhattisgarh, India
e-mail: joystu.dutta@gmail.com

A. Kerketta
Department of Biotechnology, Sant Gahira Guru University, Sarguja 497001, Chhattisgarh, India

T. K. Thakur
Department of Environmental Science, Indira Gandhi National Tribal University (IGNTU), Amarkantak, Anuppur 484887, Madhya Pradesh, India

J. Dutta · T. K. Thakur
IUCN Commission On Ecosystem Management (CEM), South Asia Regional Chapter, New Delhi, India

A. K. Mishra et al. (eds.), *Microplastics*, https://doi.org/10.1007/978-981-97-6461-7_3

the risk of microplastics, an in-depth assessment was conducted to investigate the existing standards and extent of research concerning environmental microplastics.

Keywords Microplastics · Ecology · Ecotoxicology · Environmental health · Microorganisms · Risk evaluation

1 Introduction

The low manufacturing input cost, light weight, adaptability, universality and durability of plastics have opened ways for their widespread application in a variety of sectors and in many facets of human existence (Wang et al., 2016). Polymers, both natural and synthetic, make up plastic. Plastics such as the polyethylene (PE), poly styrene (PS), polyethylene terephthalate (PET), polypropylene (PP), and polyvinyl chloride (PVC) serve as fundamental constituents of synthetic materials. The annual use of plastic has exhibited a consistent and gradual increase since the 1950s, culminating in the production of around 359 million tonnes of plastics in the year 2018 (Plasticseurope, 2019). Our blue planet Earth is 4.5 billion years old. In just over 100 years, anthropogenic plastic pollution has reached every continent and ocean (Ritchie & Roser, 2018). The COVID-19 pandemic since late 2019 through 2022 had a notable impact on global plastic waste generation and management. While the full extent of these effects may vary by region and time period, several general trends and observations emerged. Materials such as disposable masks, gloves, face shields, and packaging for takeaway food and online shopping. This surge in demand for disposable items was driven by hygiene concerns and changes in consumer behavior (de Sousa, 2021; Filho et al., 2021).

Lockdowns, restrictions on movement, and disruptions in supply chains hampered recycling efforts in many places. Recycling facilities face challenges in collecting, processing, and transporting plastic waste due to workforce shortages and other logistical issues (Shams et al., 2021). The pandemic generated a significant amount of medical waste, including personal protective equipment (PPE) and medical supplies. Improper disposal and mismanagement of medical waste raised environmental and health concerns in some areas (Igalavithana et al., 2022).

Informal waste pickers and recyclers faced economic hardships and health risks due to reduced waste collection and recycling activities. Many rely on collecting and selling recyclable materials, and the decrease in waste generation affected their livelihoods (Kulkarni and Anantharama, 2020).

Several regions postponed or suspended plastic bans and regulations to prioritize public health and safety during the pandemic. This delay slowed down efforts to reduce plastic consumption and pollution (Ebner and Lacovidou, 2021).

Shifts in consumer behavior, such as increased online shopping and demand for packaged goods, led to changes in the types and amounts of plastic packaging being used (Oliviera et al., 2021).

The increased visibility of plastic waste, especially single-use plastics like masks and gloves littering public spaces, raised awareness about the broader plastic pollution issue. This prompted renewed calls for sustainable and responsible plastic use (Vanapalli et al., 2021).

It's important to note that while the pandemic had short-term effects on plastic waste generation and management, the long-term impacts on plastic pollution and waste management strategies may vary depending on how societies and governments respond to the challenges posed by the crisis. Many organizations and governments are now focusing on sustainable recovery and strategies to manage plastic waste more effectively, taking into account the lessons learned during the pandemic.

Globally, the number of plastics manufactured for their different industrial applications has increased exponentially; as a result, the trash created by these polymers has also increased exponentially. Plastic garbage has subsequently been identified as widespread, largely non-biodegradable, and ubiquitously scattered over our planet's ecosystems (Amobonye et al., 2020). Plastic garbage, on the other hand, has become one of the most important environmental challenges due to inappropriate management and limited biodegradability (Krüger et al., 2020). According to the research conducted by Geyer et al. (2017), it is evident that approximately 79% of plastic trash is deposited in the natural environment, while 12% is sent to burning, and a meager 9% enters the process of recycling. Plastics within the natural environment have the capacity to undergo gradual fragmentation into minuscule particles as a result of many factors, including exposure to sunlight, microbial activity, and mechanical abrasion. Plastic exhibits a high degree of resilience, with the bulk of its many kinds characterized by remarkably slow rates of disintegration. This characteristic enables plastic to persist for extended periods and accumulate in terrestrial, freshwater, and marine environments (Barnes et al., 2009; Geyer et al., 2017). According to Thompson (2004), microplastics (MPs), which are tiny particulates with a size of 5000 μm, can undergo degradation in aquatic settings by mechanical, chemical, and biological processes.

According to the definition provided by Frias and Nash (2019), microplastics can be described as synthesized robust polymer particles or polymer matrices that are insoluble in water, showing a range of dimensions spanning from 1 μm to 5 mm. Microplastics are also purposefully made (primary microplastics) for a variety of applications, such as microbeads used in household cleaning as well as industrial and medicinal items (Cole et al., 2011). There are various forms of microplastics, including particles, fragments, and fibers. In recent times, there has been an increased level of concern regarding the issue of microscopic particles in the environment, gaining the interest of both the general populace and the scholarly community. This attention has primarily been directed toward aquatic environments, particularly the seas, as highlighted by Auta et al., (2017a, 2017b) and Rezania et al. (2018). According to Jambeck et al. (2015), it is widely thought that the majority of microplastics originate from terrestrial sources, a phenomenon attributed to the continuous release and scattering of large quantities of polymer waste (Auta et al., 2017a, 2017b). Anthropogenic pollution has become ubiquitous across the globe, with its presence documented in several environmental compartments such as soils

(Garcés-Ordóñez et al., 2019), surface-level water (Wang et al., 2019a, b), surroundings (Vianello et al., 2019), deposits (Prata et al., 2020a, b), the seas and oceans (Aliabad et al., 2019), and biota (Digka et al., 2018). According to Eriksen et al. (2014), it is estimated that there are more than 5 trillion fragments of plastic that are presently suspended in the Earth's oceans. This estimation is projected to increase due to the transfer of uncontrolled plastic products from land to sea through rivers, particularly from areas with high populations or highly industrialized regions (Lebreton et al., 2017). According to Lebreton et al. (2017), the presence of microscopic particles in the oceans can be attributed to various factors, including sewage, effluents, the release of beach sediments, the proximity of fossil fuel and natural gas production sites, the decomposition of fisheries nets, the generation of waste during marine operations, and the impact of tourism.

Soil and other terrestrial domains are more vulnerable to plastic pollution than aquatic ones. The incorporation of sewage solids or biological products into agricultural land has emerged as a notable factor in the proliferation of microplastics within the environment. Multiple studies have provided evidence suggesting that the yearly consumption of microscopic plastic particles resulting from this activity has the potential to equal the combined quantity of microplastics already existing in the oceans of the planet (Nizzetto et al., 2016a, 2016b).

2 Ecotoxicology of Microplastics and Environmental Safety:

The widespread distribution of microplastics has become a subject of interest among ecotoxicologists in recent years, primarily due to its potential risks to safety and toxicological consequences, according to the study conducted by Wright et al. (2013). Additionally, there is a growing interest in exploring the potential connection between plastic products and the transmission of invasive species, which has emerged as a significant ecological concern. Microplastics have a negative influence on the environment for two reasons which include the microscopic dimensions of these entities that facilitate internalization by micro-organisms and their subsequent accumulation within the food chain and the organisms have the ability to absorb pollutants present on their external layers (Rillig, 2012).

Koelmans et al. (2016) proposed various processes that impact the function of microplastics as a transporter of hydrophobic organic pollutants. This pertains to the consumption of micro plastics and their ensuing discharge of chemicals inside the human body via the mechanism of absorption. In specific instances, the consumption of plastic may result in an increased excretion of organic contaminants. In the interim, it is probable that organic pollutants may infiltrate the human body through both direct and indirect means of nutritional consumption. Furthermore, a multitude of research has presented empirical evidence indicating that microplastics provide a potential risk to human health when they traverse the food web through the consumption of multiple bivalves, crustaceans, and fish species (Browne et al., 2008;). Previous studies have demonstrated the detrimental effects of microplastics on living creatures. Following oral exposure, microplastics (MPs) have been observed to have inhibitory effects on growth, feeding, survival, and fertility in several aquatic organisms, including zooplankton, crustaceans, micro algae, and fish (Cole et al., 2013; Gardon et al., 2018; Garrido et al., 2019; Goldstein et al., 2012; Lee et al., 2013). The global prevalence of microplastic pollution has emerged as a significant concern within the realm of environmental studies, prompting numerous scholars to delve into the examination of the detrimental impacts associated with microplastics. The introduction of microplastics into organisms poses a significant threat to their biological well-being, as evidenced by several detrimental processes including oxidative stress, neurological impairment, endocrine disruption, immunological impairment, and other adverse effects. In addition, the existence of micro plastics in the environment leads to unknown consequences on organisms, predominantly adverse, as a result of the correlation among persistent organic pollutants (POPs) and minuscule plastic particles. The introduction of plastic products into the ecosystem by people, followed by their disposal, can have detrimental effects on organisms. These effects can then impact humans through their food (Hidalgo-Ruz et al., 2012). As previously stated, several systems can generate microplastics, which can then be transferred across different environmental compartments. Ultimately, these microplastics can infiltrate the food web and subsequently reach the human body. Figure 1 depicts the principal pathways that lead to the contamination of food. The issue of human intake of infected food and beverages is a matter of worry in academic literature (Guo et al., 2020a, b; Wright & Kelly, 2017). However, it is observed that this problem is often overlooked in comparison to the broader environmental ramifications.

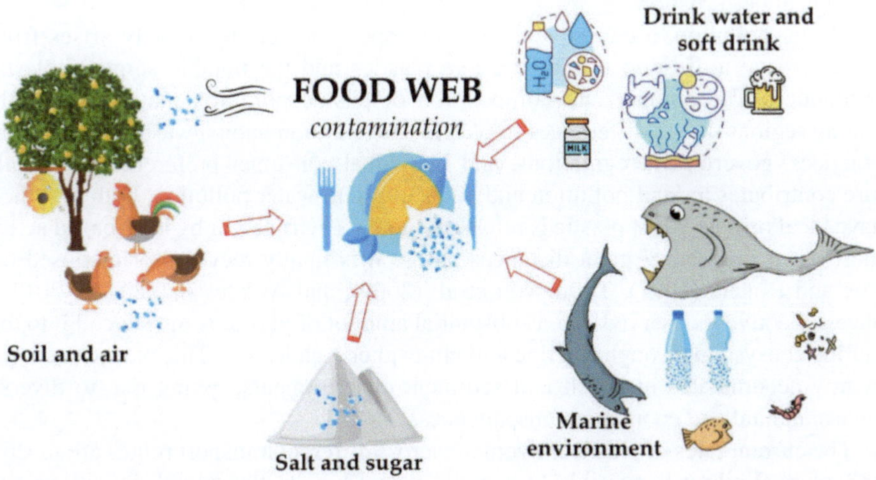

Fig. 1 Microplastics contamination at different trophic levels of the ecosystem

3 Microplastics Sources and Pathways in the Environment

Primary microplastics enter the environment through discharge from households, agriculture runoff, and industrial waste, and they account for 15–30% of all plastics released into the environment (Oliveira et al., 2022; Ter Halle & Ghiglione, 2021; Toussaint et al., 2019). Microplastics possess the capacity to enter the terrestrial ecosystem through various means, such as landfills (Geyer et al., 2017), atmospheric contamination (Evangeliou et al., 2020) with a worldwide yearly approximation of 0.343 million tonnes, and agricultural practices like the use of plastic planting mulch (Büks & Kaupenjohann, 2020; Crossman et al., 2020). Agro-ecosystems serve as significant pathways for the introduction of microscopic particles into the terrestrial realm, particularly in cases when wastewater treatment sludge, also known as biosolids, is spread to the surface of the earth (Corradini et al., 2019). The sediment produced during wastewater treatment has been found to effectively retain approximately 99% of microplastics. According to Nizzetto et al., (2016a, 2016b), the annual consumption of sewage sludge for soil amendment in agricultural environments throughout Europe and North America is projected to contribute from 63–430 to 44–300 thousand tonnes of tiny particles (MPs). According to Van Sebille et al. (2015), the use of wastewater sludge in terrestrial farming operations surpasses the estimated annual quantity (93–236 thousand tonnes) of tiny particles that are discharged into the oceans. Microplastics exhibit unique characteristics that distinguish them from other chemical contaminants. Unlike other contaminants, microplastics are made up of a wide variety of polymers and can vary in size, ranging from the nanoscale (0.1 m). Additionally, microplastics come in various shapes and are constructed of separate chemical components (Rochman et al., 2019). These identified traits enhance the probability of microplastics exerting adverse effects on ecosystems and triggering ecological perturbations.

Plastic pollution, a consequence of anthropogenic activity, mostly arises from the extensive utilization of single-usage plastics and the notable surge in plastic production. The amount and composition of plastic pollution may vary greatly among regions due to differences in plastic consumption habits, waste management practices, government regulations, and individual consumer preferences. Agriculture contributes to land pollution and aquaculture to water pollution, both of which have local origins. Most plastic is released into the environment by land-based activities, but it may travel great distances before it is finally recycled or disposed of. Coe and Rogers (1997), Thompson et al. (2009), and Auta et al., (2017a, 2017b) have collectively observed that a substantial amount of plastic is introduced into the marine ecosystem through riverine and atmospheric pathways. This plastic predominantly accumulates in beach and sediment compartments, giving rise to diverse environmental and economic consequences.

The current rates of plastic movement across different transport routes are mostly unknown. While it is possible to visually detect bigger marine debris without the aid of magnification, the task of collecting, sorting, identifying, and quantifying minuscule particles within complex environmental matrices poses a greater level of difficulty. Every year, it is believed that a significant quantity of plastic, amounting to several hundred thousand of tonnes, is deposited into the Earth's oceans. Most of it is found in the environment as macroplastic products, which might eventually break down into microplastic. On the contrary, microplastics are unintentionally released into the environment, and it is widely agreed upon that the majority of microplastics found in aquatic ecosystems come from terrestrial places, rather than being produced through the breakdown of larger plastic items. According to Ivleva et al. (2017), there is a notable disparity in the abundance of minuscule particles of plastic and the nanoscale plastic in environmental samples when compared to larger plastic items. This discrepancy suggests a huge augmentation in particle amount as the size diminishes.

Discharges from WWTPs and sewage treatment facilities are a major contributor to microplastic pollution in aquatic ecosystems. Despite the fact that 70% of households in the European Union have access to wastewater treatment plants (WWTPs), the majority of these facilities are not specifically designed to effectively capture microplastics. Furthermore, the efficiency of microplastic removal varies depending on the specific characteristics and features of each WWTP. Therefore, the capacity of waste water treatment plants (WWTPs) to eradicate microplastics from the entering wastewater can be ascribed to the structural and compositional characteristics of these facilities. In accordance with the findings of Sun et al. (2019), it has been shown that wastewater treatment facilities (WWTPs) have the capacity to eradicate microplastics from the influents and efficiently keep them inside the sludge, assuming ideal conditions. The practice of utilizing sewage sludge as a fertilizer on agricultural lands is commonly seen. It is widely believed that this sludge contains substantial amounts of microplastics, making it a notable source of microplastic pollution in terrestrial ecosystems (Horton et al., 2017). Although only a small proportion of particles that enter these facilities are ultimately found in the water from the industries, the daily discharge of microplastic particles in wastewater treatment plant (WWTP) effluents can reach levels ranging from 100,000 to 10,000,000 particles, as reported by Sun et al. (2019). The fate of microplastics in the terrestrial ecosystem and the different modes of interaction at different trophic levels of ecosystem and primarily with the soil components have been detailed in a study conducted by (Rai et al., 2023).

The prevalence of microplastic fibers in aquatic environments, namely in the water's column and sediments, has garnered significant attention due to their high abundance, accounting for up to 80% (Browne et al., 2011; Kanhai et al., 2018; Mathalon & Hill, 2014). This discovery has raised concerns among researchers and the public alike. Microplastic fibers have the ability to permeate the natural environment through multiple pathways, such as the presence of fishing gear that is either utilized or left behind in natural habitats, the inclusion of these fibers in sewage sludge, the dispersion of these fibers through air pollution originating from urban areas, and the release of these fibers into sewage systems and wastewater treatment plants through domestic and industrial textile washing. According to a study conducted by Browne et al. (2011), the concentration of microplastic fibers in wastewater treatment facilities has been shown to exceed 100 fibers per litre. This high concentration is mostly attributed to the shedding of over 1900 fibers from a single piece of synthetic clothing during each washing cycle. The absence or limited presence of wastewater treatment plants (WWTP) in certain areas may result in substantial discharge of microplastic fibers into nearby ecosystems. Preliminary studies indicate that microplastic fibers tend to accumulate in close proximity to their point of origin. Because Svalbard, an Arctic archipelago, does not treat its wastewater, microplastic fibers from the laundering of synthetic garments end up in the marine environment. (MPs) may enter the region following ocean and maritime transports, with implications on Arctic biota, human health, and socioeconomic issues related to the exploitation of marine resources (Caruso et al., 2023).

4 The Consequences of Microplastic Contamination

4.1 Potential Effects of Microplastics on the Environment

In France, the concentration of plastic fibers in outdoor air is estimated to be approximately 0.9 per cubic metre (Dris et al., 2016, 2017). The recorded values for this concentration range from 2 to 355 fibers per square metre. Airborne microplastics have the potential to enter the marine environment. Indeed, it has been demonstrated in controlled laboratory conditions that microplastics can yield diverse detrimental effects. Nevertheless, it is crucial to acknowledge that these effects manifest at concentrations that are far higher than those commonly found in the natural environment. Consequently, the potential ecological implications of microplastics, even in minimal concentrations, remain uncertain. Numerous research have yielded empirical evidence indicating that microplastics possess the capacity to become accumulated within the food system chain via marine organisms and table salt. Since soils aren't exposed to the sun and experience little mechanical wear, plastic degrades slowly (Rillig et al., 2019). Microplastics possess the capacity to accumulate deeper inside the soil, so establishing a microplastic reservoir. According to Guo et al., (2020a, b), the presence of microplastics has the potential to modify various soil characteristics, including bulk density, permeability, capacity for holding water, and resilience to water aggregation, when soil particles surround or combine with them. Anderson et al. (de Souza Machado et al., 2018) conducted a study to examine the impact of four distinct microplastic variants, namely PP and PET, on the bulk density, capacity for water retention, and the stability of water agglomeration of sandy soil. It was found that only PET exhibited a negative association among bulk density of soil and the resilience of water aggregation. Contrary to previous findings, Zhang et al. (2019) reported contrasting results, indicating that PET had a negative impact on the soil's water retention capacity while having no effect on its bulk density. The observed discrepancy could potentially be attributed to alterations in the physicochemical properties of the soil employed in the investigation. Microplastics have the potential to modify soil porosity by affecting permeability, hence influencing soil permeability and perhaps leading to subsequent impacts on water evaporation.

Further research is necessary to validate the aforementioned ideas, as there is a lack of existing studies in this particular field.

4.2 Possible Peril to Aquatic Organisms and Vegetation

The presence of floating MPs might result in the direct or indirect exposure of aquatic organisms to these particles. The consumption of microplastics (MPs) through this pathway is prevalent among aquatic organisms (Cole et al., 2013; Watts et al., 2014; Wright et al., 2013). The aforementioned organisms encompass organisms such as mussels, crabs, marine parasites, and many species of fish. When there is a high concentration of these MPs in the water, they accumulate in the organs of aquatic animals. It is generally agreed that ingestion is the principal route of exposure for MPs in aquatic species. Members of Parliament (MPs) have the capacity to exert chemical, physical, and biological impacts on the environment. The obstruction of food flow in the alimentary tract by MP has led to several severe consequences, such as heightened levels of starvation, diminished capacity for feeding and digestion, and elevated rates of sickness and mortality (Van Franeker, 1985). The ingestion of larger plastic debris by aquatic species might result in the obstruction of their digestive tract, hence causing malnutrition and illness. There exist two potential consequences resulting from the physical influence on animals, namely entanglement and ingestion. The encirclement and entanglement of marine animals in plastic rings, fibers, or tyres has been identified as a prominent factor contributing to their mortality (Udyawer et al., 2013). Aquatic Mammals, ocean turtles, pinniped (seals and sea lions), and marine snakes, among other marine organisms, have been documented to experience entanglement in or ingestion of plastic waste, resulting in heightened rates of morbidity and mortality (Alimba & Faggio, 2019). The confinement within plastic rings has been associated with various detrimental effects on individuals, including scoliosis of the spinal skeletons, wounds to the skin, injury to tissue, undernourishment, inadequate ventilation, reduced foraging opportunities, and eventually, hunger and mortality (Alimba & Faggio, 2019). Furthermore, plastics can be ingested either directly or indirectly. Plankton ingest microplastics (MPs), which subsequently undergo bioaccumulation and biomagnification as they progress through the marine trophic levels, ultimately reaching humans. The act of directly consuming plastics leads to a range of adverse effects, such as impairments in intestinal function, diminished eating ability, depletion of energy stores, hindered reproductive capabilities, and other related consequences. Unfortunately, the existing body of knowledge regarding the effects of ingesting microplastics on aquatic creatures is still inadequate.

MPs are also dangerous to fish because they disrupt the organ structure. Fish accumulate MPs in a variety of tissues, including their gills, muscles, digestive tract, and liver (Abbasi et al., 2018; Akhbarizadeh et al., 2018; Collard et al., 2017; Jin et al., 2018). Furthermore, research has demonstrated that MPs have the capacity to disrupt the balance of microorganisms in the microbiota, leading to a condition known as dysbiosis, as evidenced by studies conducted by Grigorakis et al. (2017), Jin et al. (2018), Lei et al. (2018), and Lu et al., (2018). The function of particle size in the formation and scattering sequences of microplastics, or MPs, inside the body tissues of aquatic creatures is of considerable importance, as evidenced by reported variances among various species. As an example, it has been observed that zebrafish (Danio rerio) typically exhibit hepatic accumulation of 5-μm microplastics (MPs) (Lu et al., 2016). Conversely, *Carcinus maenas* (Crabs) tends to accumulate MPs mostly in its gastrointestinal tract and gills, with 8-μm MPs being the most prevalent (Browne et al., 2008; Watts et al., 2014). Additionally, *Mytilus edulis* has been found to accumulate 10-μm MPs (microplastics) (Browne et al., 2008). MPs can cause inflammation, oxidative stress, growth suppression, toxicity to reproduction, and lipid metabolic impairment in a variety of aquatic organisms, including zebrafish, daphnia, pulex, ceriodaphnia, and paracentrotus. Numerous investigations pertaining to the toxicological impacts of microplastics (MPs) have consistently demonstrated their propensity to induce oxidative stress via alterations in gene expression and the generation of free radicals. Additionally, MPs have been observed to elicit immune responses, exhibit genotoxic properties, disrupt the endocrine system, exert neurotoxic effects, induce reproductive abnormalities, and potentially transmit adverse effects across generations in marine organisms. Cellular macromolecules including DNA, carbohydrates, lipids, and proteins are vulnerable to damage when oxidative stress causes an excess generation of reactive oxygen species. When experimented with Micro-plastics (50 mg/L, diameter 1.0 m) for 6, 12, and 24 h, scleractinian coral (*Pocillopora damicornis*) showed signs of immune and detoxification system suppression via the Jun N-terminal kinases signaling pathways (Tang et al., 2018). MPs play a vital role in the context of marine toxicity, since they accumulate in various tissues and subsequently undergo biomagnification along the food chain. Based on these results, it is recommended that future toxicological research take into account the possible dangers associated with MP absorption by humans through the food web.

The investigation of the impacts of microplastics (MPs) has predominantly been around the proliferation of toxins in aquatic flora. A limited number of aquatic plant species have been investigated, such as *Chlorella pyrenoidosa*, *Microcystis sp.*, *Myriophyllum spicatum* (commonly known as earthworms), and *Lemna minor*. According to Kalcikova et al., (2017), microplastics (MPs) have been found to decrease photosynthetic activity. (Mao et al., 2018) have demonstrated that MPs hinder the process of photosynthesis. Furthermore, (Wu et al., 2019) have observed that MPs also modify plant development. However, the degree of exposure greatly modifies the toxicity of MPs. Davarpanah and Guilhermino (2019) undertook a study on the toxicity of MPs, whereby they observed that doses of MP ranging from 3 to 4 mg/L did not exhibit any discernible impact on the mean specific rate of growth of *Tetraselmis chuii*. The present lack of risk assessment investigations pertaining to the toxicity of plants and the challenges associated with detecting microplastics (MPs) in aquatic plant material contribute to the restricted availability of relevant data.

The manner in which microplastics engage with aquatic species is largely unexplored and lacks comprehensive understanding. There exists a dearth of research pertaining to the ramifications of microplastics on invertebrate organisms, specifically amphipods (Carrasco et al., 2019). The predominant focus of research has mostly centered on investigating the impacts of microplastics on mortality rates, growth inhibition, trophic actions, bleaching and necrosis, and reproductive capacity. The findings of empirical investigations have revealed that the harmful effects of microplastics against invertebrates are contingent upon both the duration of the interaction and the degree of concentration of MPs. Mateos-Cardenas et al. (2019) demonstrated that the death rate or commuting of *Gammarus duebeni* did not exhibit any discernible impact following exposure to microplastics (MPs) for either 24 or 48 h. In a study conducted by Bruck and Ford (2018), it was shown that both short- and long-term exposure to microplastics (MPs) did not exhibit any observable detrimental impact on the developmental process of *Echinogammarus marinus* or *Gammarus pulex*. The study conducted by Au et al. (2015) found that both the development and breeding of *Hyalella Azteca* decreased substantially when exposed to low and intermediate dosages of polyethylene microplastics over an extended period of time. The findings presented in this study align with the research conducted by Aljaibachi and Callaghan (2018), which shows that MPs exacerbate the death rate of malnourished daphnids, while having little impact on daphnids that are fed with sufficient food. The negative health consequences of MP exposure have been shown in in vivo investigations, although only in aquatic species. Current scientific investigations have mostly focused on the modifications occurring in mammals, animal, and plant organisms, as significant indicators of the in vivo toxicities associated with microplastics. However, there exists an acute lack of understanding pertaining to the destiny of microplastics throughout the upper echelons of the water-based trophic pyramid.

4.3 Possibility of Harm to Terrestrial Organisms and Vegetation

The deleterious ramifications of accumulated plastic in terrestrial environments have been underscored, and multiple studies have revealed that the magnitude of pollution from plastic on land is considerably greater than that in the oceans (Horton et al., 2017). Major sources of plastic pollution in agricultural areas include sewage from homes, fertilizers, litter, and mulch made from recycled plastic. Nizzetto et al., (2016a, 2016b) found that the usage of vinyl mulch contributed more to the contamination of agricultural soil than the contamination of water. One of the major contributors to plastic toxicity in terrestrial animals is the animals' habit of ingesting plastic particles as food. Multiple recent studies have offered the theory that soil-dwelling creatures consume, carry, and pull MPs, causing them to become embedded in the soil food web. It is possible for MPs in soil surface litter treatments to be carried by invertebrates. It has been demonstrated that Collembola and subsoil dwellers (anecic earthworms) may transport MPs across distances of several centimetres in the soil in a couple of days (Lwanga et al., 2017; Maass et al., 2017). Furthermore, the transportation of microplastics (MPs) by soil microarthropods has had an impact on the exposure of various soil organisms to MPs and has also resulted in modifications to soil's physical characteristics (Zhu et al., 2018). Most plastic polymers have hydrophobic properties that negatively impact the structural aspects of soil, including bulk density and particle aggregations, as well as the metabolic activities of soil-dwelling organisms. According to Zhu et al. (2018), empirical research has demonstrated that microplastic particles (MPs) induce avoidance behaviors, diminish reproductive potential, and disrupt the gut ecology of bacteria in springtails and Collembolan species, earthworms (Rodriguez-Seijo et al., 2017), and other animals. Soil organisms' susceptibility to MPs' harmful effects may be evaluated by looking at how they respond to oxidative stress, intestinal damage, and aberrant immune systems (Song et al., 2019).

High-trophic-level mammals are especially vulnerable to MP exposure because they are more likely to consume them in food or water. The digestive tract, kidneys, and liver are prominent storage sites for MPs in animals. The impact of microplastics (MPs) on animals is also contingent upon the size of the particles. In a study conducted by Deng et al. (2017), it was found that the dimensions of microplastic fragments had a significant impact upon how much accumulation of these lethal fragments in several organs of mice, such as the liver, kidney, and gut. Additionally, the researchers found that the size of MPs affected the movement and distribution of these particles throughout the tissues. The results are consistent with the study conducted by Yang et al. (2019), which indicated that the toxicokinetic and toxicodynamic characteristics in mice may be affected by the size of microplastic particles. Jin et al. (2019) have previously conducted studies indicating a decrease in intestine mucus secretion, compromised function of the intestinal barrier, and the presence of dysbiosis within the gut microbiota.

4.4 *Effects of Microplastics on Soil Function and Structure*

4.4.1 Alterations in Soil Physicochemical Properties

Soil is a crucial component of the environmental matrix, according to Bardgett and van der Putten (2014) because it provides ecological services and functions. The presence of microplastics can have a substantial influence on various physicochemical characteristics of soil, such as the dispersion of water-stable aggregates, bulk density, water-holding capacity, and pH levels. The alterations in physicochemical characteristics are influenced by the morphology, polymer composition, and concentration of microplastics. According to De Souza Machado et al. (2019), the effects of fibers on bulk density of soil, distribution of soil agglomerates and development of macropores (>30 m) have been found to be more significant compared to other evaluated soil additions. Furthermore, it has been observed that the presence of soil-dwelling organisms, including plants and fungi, can have a moderating effect on alterations in the physicochemical properties of soil (De Souza Machado 2018; Liang et al., 2019; Wang et al., 2020).

4.4.2 Interactions with Soil Organisms

The possibility exists for alterations in the physical and chemical characteristics of soil as a result of the interactions between soil biota and tiny particles, specifically plastics. Figure 2 highlights the different modes of interaction of microplastics (MPs) with soil organisms ultimately leading to modification of soil structure and function. Liang et al. (2019) conducted a study that demonstrated the influence of diverse saprobic fungus strains on water-stable assemblages in the soil when exposed to microplastics. The research conducted in this study documented a significant reduction in the quantity of water-stable aggregates due to the associations involving microplastics fragments and two distinct fungus strains. Conversely, the opposite effect was observed when a third strain was included. Soil pH was shown to be drastically lowered when microplastics (polyethylene polymer) were added, according to research by Boots et al. (2019). Microplastic addition alone resulted in a drop in pH, however microplastic addition in conjunction with a plant (perennial ryegrass) resulted in no change in pH. The study conducted by De Souza Machado et al. (2019) examined the impact of various microplastics materials (polyamide, polyester, polypropylene, polyethylene, and polystyrene) on the characteristics of soil, namely the bulk density and capacity for holding water, in conjunction with the cultivation of spring onions. The study conducted by de Souza Machado et al. (2019) showed that the interactions between microplastics and plants yielded outcomes that differed from those observed when microplastics were examined in isolation. According to de Souza Machado et al. (2019), the identification of plastic particles (mostly MPs) has been associated with a reduction in soil density as a whole, except in the case of polypropylene. Additionally, the water-holding capacity of the soil tends to rise due to the interactions between microplastics and plants. The findings of this study illustrate the possibility of unforeseen ecological consequences arising from the introduction of microplastics into soil ecosystems. In order to comprehend the potential effects of alterations in soil physicochemical characteristics on the integrity of agroecosystems, it is imperative to examine the interplay between microplastics, plants, fungus, and other soil organisms. This task is challenging due to the intricate nature of these interactions, even within controlled experimental environments.

4.4.3 Relating the Presence of Microplastics to Modifications in Soil Function and Structure

The primary agents responsible for the cycling of nitrogen (Rong et al., 2021) and carbon-based material extraction (Rillig et al., 2021), among other essential biogeochemical processes for human sustenance (Bender et al., 2016), are soil microbial communities. The occurrence of microplastics can have a direct impact

Fig. 2 Interaction of Microplastics (MPs) with soil microorganisms leading to alterations in soil structure and function

on these processes by causing changes in soil structures, including bulk density, water-stabilized agglomerates, retention of water, and pore volume. The existence of microplastics has been observed to induce changes in soil microbial activity, potentially leading to significant implications for the cycling of nutrients (De Souza Machado et al., 2018, 2019; Liu et al., 2017; Rong et al., 2021). The research carried out by Liu et al. (2017) showed a favorable association between enhanced microbial activity, as assessed using a fluorescein diacetate hydrolase enzyme, and the existence of microplastics. This increases microbial activity which was found to be associated with greater quantities of soluble nutrients in the soil water. Based on the findings of Liu et al. (2017), it has been deduced that the existence of microplastics possesses the capacity to influence the functioning of microbe communities and the formation of nutrients in the water column. Several researchers have reported contrasting findings about the impact of microplastics on respiration and nitrification. For instance, Judy et al. (2019) and Yan et al. (2020) have presented divergent results in their respective investigations. The aforementioned investigations also observed alterations in the variety of microbial communities, however no discernible impact on soil nutrient levels was detected. Moreover, it is noteworthy to consider the reaction of microbial populations to reduced levels of microplastics, ranging from 0.05% to 0.25% weight/weight. It has been observed that these lower levels of exposure have more pronounced effects compared to greater concentrations, as evidenced by studies conducted by De Souza Machado et al. (2018) and Judy et al. (2019). The mechanism underlying this reaction remains poorly comprehended; nonetheless, it is plausible that it is associated with the ability of microbes to inhabit microplastic particles and the alterations in soil structure induced by microplastics (De Souza Machado et al., 2018).

The available literature presents varying perspectives on the role of microplastics in the terrestrial carbon cycle (Gao et al., 2021; Ren et al., 2020; Rillig et al., 2021). Microplastics are predominantly composed of carbon, so setting them apart from the other sources of carbon that arise from natural mechanisms like photosynthesis or biomass generation is difficult. Unlike these more transient carbon sources, microplastics possess a refractory carbon pool that exhibits greater resistance to degradation. Consequently, the presence of microplastics can have implications for the storage of carbon in soil (Rillig & Lehmann, 2020). Because microplastics may stay in the soil for a while, microorganisms have time to adapt to and make use of the extra carbon they provide (Rillig, 2018). The ability of microbial communities to breakdown recalcitrant organic compounds, such as aromatic polycyclic hydrocarbons (PAHs), via multiple pathways of metabolism (Posada-Baquero & Ortega-Calvo, 2011), implies that microplastics may also undergo degradation by similar communities. The aforementioned findings underscore the significance of examining the impacts of microscopic particles on the biogeochemical cycle mechanisms of microbial communities in the soil. Insufficient attention has been given to the quantitative ramifications of these discoveries with regard to the carbon cycle process.

4.4.4 Microplastics' Detrimental Effects on Human Health

The multifarious impacts of micro and nano plastics on human health and ecosystems have been documented in scientific literature but still a lot of research scope is pending (Campanale et al., 2020). The utilization of plastics has had a notable influence on the enhancement of food safety through facilitating more effective packaging methods, hence prolonging the shelf life of perishable items and reducing instances of food wastage. Disposable medical devices, including facial masks, syringes, and surgical tubes, contribute to the transmission of infectious illnesses, but this material helps keep communities healthy by preventing this spread (Prata et al., 2019). However, the exposure of people to microplastics is inevitable as a result of the extensive use of plastics in everyday activities and the subsequent pollution of the environment caused by the disposal of plastic waste. Cox et al. (2019) estimate that people eat 39,000 to 52,000 particles per year via this route; According to Catarino et al. (2018), the aforementioned figure may potentially increase when accounting for the deposition of dust onto plates during meal consumption. The widespread presence of microplastics in several food sources, such as salts (Seth & Shriwastav, 2018), water used for consumption, packaged water, a beverage (Wiesheu et al., 2016), mussels as well as shellfish, indicates that their significant intake is not surprising. The extent to which additional factors, such as packing containers, contribute remains uncertain. The regulation of microplastics in food inside the European Union is overseen by the European Food Safety Authority (EFSA) and the European Commission's Rapid Alert System for Food and Feed (RASFF) (Barboza et al., 2018).

Particles, once ingested, make their way to the colon via the digestive tract. Intestinal cells can alone be accessed by particles of sufficiently small size or those that have acquired a biocompatible coating, which enables them to penetrate the mucus layer lining the intestines. Microplastics have the ability to enter the bloodstream through two primary mechanisms. Firstly, they can be ingested by specialized M-cells that are present in Peyer's patches, a type of lymphatic tissue (Ensign et al., 2012). Secondly, microplastics can also enter the bloodstream by persorption, which refers to the para-cellular transportation of particulates (Volkheimer, 1977). Once in the bloodstream, these microplastics can circulate throughout the body. Forte et al. (2016) conducted a study that revealed that in vitro investigations have demonstrated the capacity of microplastics to alter gene expression, impact cell viability, and elicit pro-inflammatory reactions. In contrast, a study conducted by Senathirajah and Palanisami (2019) demonstrated that in vivo testing on mice revealed a significant reduction in spermatogenic cells along with a drop in sperm count over a 42-day period, following the intake of polystyrene tiny fragments at a dosage as low as 0.01 mg of 5 m per day. The observed quantity is lower than the expected human intake of 5 g of microplastics on a weekly basis. Inhaling microplastics floating in the air is another method of exposure. Vianello et al. (2019) in his research inferred that individuals are typically exposed to an estimated 272 microplastics on a daily basis by inhalation. This exposure is attributed to the widespread presence of microplastics and synthetic fibers in the atmosphere. The emissions of these particulates are subject to significant alterations in their concentrations and characteristics as a result of the materials and activities they encounter (Prata, 2018).

The depth to which inhaled particles deposit is proportional to their size and density. Increased vascular permeability and blocked macrophage migration result in chronic inflammation if air particle concentrations are high enough to trigger a dust overload response (Donaldson et al., 2000). Based on the findings of a study conducted by Brown et al. (2001), it has been determined that the presence of microplastics inside the respiratory system leads to inflammation and significant oxidative stress. The inhalation of significant amounts of airborne microplastics due to occupational exposure might result in the development of malignant lesions in the respiratory tract and interstitial lung tissue (Prata, 2018). Last but not least, because microplastics are so pervasive, cutaneous exposure can't be ruled out. Plastics less than 100 nm in size may be able to penetrate the protective epithelium layer that separates cells. In addition, this is an important area to research because many skin care products still include microplastics (Godoy et al., 2019). Previous research has demonstrated that the internalization of human epithelial cells might result in cytotoxicity when exposed to oxidative stress (Schirinzi et al., 2017). According to Jeong and Choi (2019), a comprehensive analysis of the existing toxicological information using Adverse Outcome Pathways has determined that the primary toxicological risks to human and animal well-being arise from the generation of reactive oxygen species in reaction to microplastics, leading to oxidative stress and inflammation.

4.4.5 Microorganisms that Degrade Microplastics and Counter Plastic Pollution

Microorganisms that degrade microplastics are of great interest in addressing the global problem of plastic pollution (Jain et al., 2023). These microorganisms have the ability to break down or degrade plastic materials into smaller fragments or even completely mineralize them, reducing their impact on the environment. Here are a few types of microorganisms that have been identified for their potential to degrade microplastics:

1. **Bacteria:** Certain species of bacteria, such as *Ideonella sakaiensis*, have been discovered to possess enzymes capable of breaking down certain types of plastics, like polyethylene terephthalate (PET) (Yoshida et al., 2016). These bacteria produce enzymes called PETases and MHETases that work together to degrade PET into its constituent components.
2. **Fungi:** Some fungi, like *Aspergillus* and *Penicillium* species, have shown the ability to degrade plastics. These fungi can produce enzymes like esterases and lipases that breakdown plastic polymers (Okal et al., 2023).
3. **Algae:** Certain types of microalgae have been found to attach to microplastic surfaces and potentially contribute to their degradation. Algae can produce enzymes that help weaken the plastic surface, making it more susceptible to other microbial degradation processes (Moog et al., 2019)

Several examples of microalgae that have been found to attach to microplastic surfaces and potentially contribute to their degradation include:

- *Chlorella vulgaris*: This is a common freshwater microalga that has been observed to attach to microplastic particles. Researchers have found that *Chlorella vulgaris* can colonize plastic surfaces and produce enzymes that may participate in breaking down the plastic. (Govindan et al., 2021)
- *Scenedesmus spp.*: Various species of the microalga *Scenedesmus* have been shown to adhere to microplastic surfaces. These microalgae can form biofilms on the plastic, which can lead to physical and chemical changes on the plastic surface, potentially aiding in degradation.
- *Chlamydomonas reinhardtii*: Another freshwater microalga, *Chlamydomonas reinhardtii*, has been found to interact with microplastic particles. While its exact role in degradation is still being studied, its ability to attach to plastic surfaces suggests a potential contribution.
- *Navicula sp.*: Diatoms like *Navicula* have been observed to colonize microplastic particles. Diatoms are known for their silica-rich cell walls and can potentially produce enzymes that interact with plastics (Behera and Das, 2023).
- *Phaeodactylum tricornutum*: This diatom species has shown the ability to colonize and interact with microplastics. Diatoms like *Phaeodactylum tricornutum* are known for their potential to produce extracellular enzymes that might aid in plastic degradation.

- *Isochrysis sp.*: Some studies have reported the attachment of microalgae from the genus *Isochrysis* to microplastic surfaces. These microalgae can contribute to the formation of biofilms and potentially impact plastic degradation.

It's important to note that while these microalgae have been observed to attach to microplastics and potentially play a role in their degradation, the exact mechanisms and contributions are still areas of active research. Microalgae can influence plastic degradation through various processes, including physical attachment, production of enzymes, and alteration of the plastic surface chemistry. However, the overall effectiveness of microalgae in plastic degradation and their ecological implications require further investigation.

4. **Actinomycetes**: These are a group of bacteria known for their ability to degrade a wide range of organic materials, including plastics. They produce various enzymes that can break down different types of plastics.
5. **Biofilms:** Microorganisms can form biofilms on the surface of plastics, which can accelerate degradation processes. Biofilms consist of diverse microbial communities working together to break down the plastic (Rather et al., 2021).
6. **Marine Microbes:** Marine microbes play a significant role in the degradation of microplastics in ocean environments (Boucher & Friot, 2017). They can colonize, interact with, and contribute to the breakdown of plastic particles through various mechanisms (Zhai et al., 2023). Here are some examples of marine microbes that have been identified for their potential role in microplastic degradation:

Marine Bacteria: Many types of marine bacteria have been found to colonize and attach to microplastic surfaces. These bacteria can produce enzymes, such as lipase and esterase, that help break down the plastic polymers. Some examples of marine bacteria associated with microplastic degradation include *Vibrio*, *Pseudomonas*, and *Marinobacter* species.

Marine Fungi: Studies on Marine fungi are generally rare. Fungi on marine plastic debris (MPD) have been observed (Jacquin et al., 2019). Oberbeckmann et al. (2016) conducted a seasonal comparison of microbial communities including both bacteria and fungi on submerged PET bottles, glass slides, and seawater in the North Sea. Similar to bacteria, marine fungi have also been implicated in the degradation of microplastics. They can produce enzymes like lignin-degrading enzymes, which may have the potential to break down plastic polymers. *Zalerion maritimum* is one example of a marine fungus that has been found on plastic debris.

Marine Algae: Certain marine algae, including *diatoms* and *dinoflagellates*, can attach to microplastic surfaces and potentially contribute to their degradation. These algae may release extracellular enzymes that interact with plastics and facilitate degradation processes.

Protozoa: Some marine protozoa, such as *ciliates* and *flagellates*, have been observed interacting with microplastic particles. These microorganisms can graze on bacteria colonizing the plastic surface, leading to changes in the plastic's physical and chemical properties over time (Bulannga and Schmidt, 2024).

Biofilm Formation: Microbes in marine environments often form biofilms on microplastic surfaces. Biofilms consist of complex communities of microorganisms that work together to break down and utilize organic matter, including plastic debris.

Marine Viruses: Viruses that infect marine bacteria, known as bacteriophages, can influence microbial communities on microplastic surfaces. They may impact the metabolism and activity of bacterial populations, which can indirectly affect plastic degradation (Cai et al., 2023).

Synergistic Microbial Communities: Microbes in the marine environment often interact in complex ways. Synergistic interactions between different microbial species can enhance the degradation of microplastics by facilitating the exchange of nutrients, enzymes, and metabolites.

Environmental Factors: Environmental conditions, such as temperature, salinity, and nutrient availability, can influence the activity and composition of marine microbial communities involved in plastic degradation (Urbanek et al., 2018). These factors can impact the efficiency of plastic degradation processes.

Research into the roles of marine microbes in microplastic degradation is ongoing, and scientists are working to better understand the specific mechanisms by which these microorganisms interact with plastics and contribute to their breakdown. Harnessing the potential of marine microbes for plastic pollution mitigation requires a comprehensive understanding of their ecology and interactions within marine ecosystems (Yang et al., 2023).

It's important to note that while these microorganisms show potential for plastic degradation, there is still ongoing research to fully understand their mechanisms and optimize their effectiveness. Additionally, the degradation process can be influenced by factors such as plastic type, environmental conditions, and the microbial community present.

Efforts are being made to harness the potential of these microorganisms for practical applications, such as bioremediation strategies to clean up plastic waste or the development of bio-based plastic materials that are more easily biodegradable.

5 Conclusion

Microplastic Pollution is the rapidly emerging and evolving environmental crisis of recent times. Ecosystems across the globe are now threatened by MPs. Understanding the dynamics of MPs and designing a mitigation plan necessitates research on exposure akin toxicity in combination with hazardous chemicals. In this chapter, we present a comprehensive survey of the existing body of research pertaining to the ecotoxicological implications of microplastics (MPs). This review encompasses investigations into the impacts of MPs on various organisms, such as insects, plants, and other members of the animal kingdom. In addition, we examine the collective toxicity of microplastics (MPs) and the objective of the research, alongside their inherent presence and environmental transport. The role of micro-organisms in biodegradation of microplastics has also been explored in this chapter in detail.

Various types of microplastics (MPs) have been inadvertently transported from one habitat to another due to anthropogenic activities involving manufacture and processing. These MPs encompass fragments of fishing gear, packing materials, water bottles, synthetic textiles, automobile tyres, paints, cosmetics, facial cleansers, and bath gels, among others. However, microplastics may still be harmful due to their potential to cause indirect impacts and a chronic and cumulative load on organisms. Some examples of such oblique effects are shifts in population size and composition, modifications to habitat features, and effects on vital ecological processes. Food security, the spread of infections, and the possibility of floods are just a few examples of how microplastics' effects on ecosystem services can pose risks to human health. The potential ramifications of both direct and indirect effects should be a matter of worry, as microplastics have a persistent and accumulative nature within the natural system, lacking effective removal mechanisms.

References

Abbasi, S., Soltani, N., Keshavarzi, B., Moore, F., Turne, R. A., & Hassanaghaei, M. (2018). Microplastics in different tissues of fish and prawn from the Musa Estuary, Persian Gulf. *Chemosphere, 205*, 80–87. https://doi.org/10.1016/j.chemosphere.2018.04.076

Akhbarizadeh, R., Moore, F., & Keshavarzi, B. (2018). Investigating a probable relationship between microplastics and potentially toxic elements in fish muscles from northeast of Persian Gulf. *Environmental Pollution, 232*, 154–163. https://doi.org/10.1016/j.envpol.2017.09.028

Alimba, C. G., & Faggio, C. (2019). Microplastics in the marine environment: Current trends in environmental pollution and mechanisms of toxicological profile. *Environmental Toxicology and Pharmacology, 68*, 61–74. https://doi.org/10.1016/j.etap.2019.03.001

Aljaibachi, R., & Callaghan, A. (2018). Impact of polystyrene microplastics on Daphnia magna mortality and reproduction in relation to food availability. *PeerJ, 6*, e4601. https://doi.org/10.7717/peerj.4601

Anderson, A. G., Grose, J., Pahl, S., Thompson, R. C., Wyles, K. J. (2016a). Microplastics in personal care products: Exploring perceptions of environmentalists, beauticians and students. *Marine Pollution Bulletin, 113*(1–2), 454–460, ISSN 0025-326X, https://doi.org/10.1016/j.marpolbul.2016.10.048

Anderson, J. C., Park, B. J., & Palace, V. P. (2016b). Microplastics in aquatic environments: Implications for Canadian ecosystems. *Environmental Pollution, 218*, 269–280.

Anderson, A. S. M., Wai, L. C., Jennifer, T., Werner, K., Anika, L., & Roland, B. R. (2018a). Impacts of Microplastics on the Soil Biophysical Environment. *Environmental Science & Technology., 52*, 9656–9665.

Anderson, Z. T., Cundy, A. B., Croudace, I. W. et al. (2018b). A rapid method for assessing the accumulation of microplastics in the sea surface microlayer (SML) of estuarine systems. *Scientific Reports, 8*, 9428. https://doi.org/10.1038/s41598-018-27612-w

Au, S. Y., Bruce, T. F., Bridges, W. C., & Klaine, S. J. (2015). Responses of Hyalella azteca to acute and chronic microplastic exposures. *Environmental Toxicology and Chemistry, 34*, 2564–2572. https://doi.org/10.1002/etc.3093

Auta, H. S., Emenike, C. U., & Fauziah, S. H. (2017a). Distribution and importance of microplastics in the marine environment: A review of the sources, fate, effects, and potential solutions. *Environment International, 102*, 165–176. https://doi.org/10.1016/j.envint.2017.02.013

Auta, H. S., Emenike, C. U., & Fauziah, S. H. (2017b). Distribution and importance of microplastics in the marine environment: A review of the sources, fate, effects, and potential solutions. *Environment International, 102*, 165–176.

Barboza, L. G. A., Vethaak, A. D., Lavorante, B. R. B. O., Lundebye, A.-K., & Guilhermino, L. (2018). Marine microplastic debris: An emerging issue for food security, food safety and human health. *Marine Pollution Bulletin, 133*, 336–348. https://doi.org/10.1016/j.marpolbul.2018.05.047

Bardgett, R. D., & van der Putten, W. H. (2014). Belowground biodiversity and ecosystem functioning. *Nature, 515*(7528), 505–511. https://doi.org/10.1038/nature13855

Barnes, D. K., Galgani, F., Thompson, R. C., & Barlaz, M. (2009). Accumulation and fragmentation of plastic debris in global environments. *Philosophical Transactions of the Royal Society of London. Series B, Biological Sciences, 364*, 1985–1998. https://doi.org/10.1098/rstb.2008.0205

Behera, S., & Das, S. (2023). Environmental impacts of microplastic and role of plastisphere microbes in the biodegradation and upcycling of microplastic. *Chemosphere, 334*, 138928.

Bender, S. F., Wagg, C., & van der Heijden, M. G. A. (2016). An underground revolution: Biodiversity and soil ecological engineering for agricultural sustainability. *Trends in Ecology & Evolution, 31*(6), 440–452. https://doi.org/10.1016/j.tree.2016.02.016

Boots, B., Russell, C. W., & Green, D. S. (2019). Effects of microplastics in soil ecosystems: Above and below ground. *Environmental Science & Technology, 53*(19), 11496–11506. https://doi.org/10.1021/acs.est.9b03304

Boucher, J., & Friot, D. (2017). *Primary Microplastics in the Oceans: A Global Evaluation of Sources* (p. 43). IUCN.

Brown, D. M., Wilson, M. R., MacNee, W., & Donaldson, K. (2001). Size-dependent proinflammatory effects of ultrafine polystyrene particles: a role for surface area and oxidative stress in the enhanced activity of ultrafines. *Toxicology and Applied Pharmacology, 175*, 191–199. https://doi.org/10.1006/taap.2001.9240

Browne, M. A., Crump, P., Niven, S. J., Teuten, E., Tonkin, A., Galloway, T., & Thompson, R. (2011). Accumulation of microplastic on shorelines worldwide: Sources and sinks. *Env Sci Technol, 45*(21), 9175–9179. https://doi.org/10.1021/es201811s

Browne, M. A., Dissanayake, A., Galloway, T. S., Lowe, D. M., & Thompson, R. C. (2008). Ingested microscopic plastic translocates to the circulatory system of the mussel, Mytilus edulis (L). *Environmental Science & Technology, 42*, 5026–5031. https://doi.org/10.1021/es800249a

Bruck, S., & Ford, A. T. (2018). Chronic ingestion of polystyrene microparticles in low doses has no effect on food consumption and growth to the intertidal amphipod *Echinogammarus marinus*? *Environmental Pollution, 233*, 1125–1130. https://doi.org/10.1016/j.envpol.2017.10.015

Bulannga, R. B., & Schmidt, S. (2024). Micro- and nanoplastics in freshwater ecosystems—interaction with and impact upon bacterivorous ciliates. *Frontiers in Earth Science, Section Hydrosphere*. https://doi.org/10.3389/feart.2024.1349865.

Büks, F., & Kaupenjohann, M. (2020). Global concentrations of microplastic in soils—A review. *Soil, 2020*, 1–26. https://doi.org/10.5194/soil-2020-50

Cai, Z., Li, M., Zhu, Z., Wang, X., Huang, Y., Li, T., Gong, H., & Yan, M. (2023). Biological degradation of plastics and microplastics: A recent perspective on associated mechanisms and influencing factors. *Microorganisms, 11*, 1661. https://doi.org/10.3390/microorganisms11071661

Campanale, C., Massarelli, C., Savino, I., Locaputo, V., & Uricchio, V. F. (2020). A detailed review study on potential effects of microplastics and additives of concern on human health. *International Journal of Environmental Research and Public Health, 17*(4), 1212. https://doi.org/10.3390/ijerph17041212.PMID:32069998;PMCID:PMC7068600

Carrasco, A., Pulgar, J., Quintanilla-Ahumada, D., Perez-Venegas, D., Quijón, P. A., & Duarte, C. (2019). The influence of microplastics pollution on the feeding behavior of a prominent sandy beach amphipod, *Orchestoidea tuberculata* (Nicolet, 1849). *Marine Pollution Bulletin, 145*, 23–27.

Caruso, G., Iannilli, V., Vitale, G., Vardè, M., Oliverio, M., Bogialli, S., Litti, L., Setini, A., Rosso, B., & Corami, F. (2023). Small microplastics: A yet unknown threat in the Svalbard (Norway) Region. *Journal of Marine Science and Engineering, 11*(12), 2330. https://doi.org/10.3390/jmse11122330

Catarino, A. I., Macchia, V., Sanderson, W. G., Thompson, R. C., & Henry, T. B. (2018). Low levels of microplastics (MP) in wild mussels indicate that MP ingestion by humans is minimal compared to exposure via household fibres fallout during a meal. *Environmental Pollution, 237*, 675–684. https://doi.org/10.1016/j.envpol.2018.02.069

Coe, J. M., & Rogers, D. (1997). *Marine Debris: Sources, Impacts, and Solutions*. Springer.

Cole, M., Lindeque, P., Halsband, C., & Galloway, T. S. (2011). Microplastics as contaminants in the marine environment: A review. *Marine Pollution Bulletin, 62*(12), 2588–2597. https://doi.org/10.1016/j.marpolbul.2011.09.025

Cole, M., Lindeque, P., Fileman, E., Halsband, C., Goodhead, R., Moger, J., & Galloway, T. S. (2013). Microplastic ingestion by zooplankton. *Environmental Science & Technology, 47*, 6646–6655. https://doi.org/10.1021/es400663f

Collard, F., Gilbert, B., Compere, P., Eppe, G., Das, K., Jauniaux, T., & Parmentier, E. (2017). Microplastics in livers of European anchovies (Engraulis encrasicolus, L.). *Environmental Pollution, 229*, 1000–1005. https://doi.org/10.1016/j.envpol.2017.07.089

Cox, K. D., Covernton, G. A., Davies, H. L., Dower, J. F., Juanes, F., & Dudas, S. E. (2019). Human consumption of microplastics. *Environmental Science and Technology, 53*, 7068–7074. https://doi.org/10.1021/acs.est.9b01517

Corradini, F., Bartholomeus, H., Huerta Lwanga, E., Gertsen, H., & Geissen, V. (2019). Predicting soil microplastic concentration using vis-NIR spectroscopy. *Science of the Total Environment, 650*, 922–932.

Crossman, J., Hurley, R. R., Futter, M., & Nizzetto, L. (2020). Transfer and transport of microplastics from biosolids to agricultural soils and the wider environment. *Science of the Total Environment, 724*, 138334. https://doi.org/10.1016/j.scitotenv.2020.138334

Davarpanah, E., & Guilhermino, L. (2019). Are gold nanoparticles and microplastics mixtures more toxic to the marine microalgae Tetraselmis chuii than the substances individually? *Ecotoxicology and Environmental Safety, 181*, 60–68. https://doi.org/10.1016/j.ecoenv.2019.05.078

De Sousa F. D. B. (2021). Plastic and its consequences during the COVID-19 pandemic. *Environmental Science and Pollution Research International, 28*(33):46067–46078. https://doi.org/10.1007/s11356-021-15425-w. Epub 2021 Jul 19. PMID: 34279783; PMCID: PMC8287553.

De Souza Machado, A. A., Lau, C. W., Kloas, W., Bergmann, J., Bachelier, J. B., Faltin, E., Becker, R., Görlich, A. S., & Rillig, M. C. (2019). Microplastics can change soil properties and affect plant performance. *Environmental Science & Technology, 53*(10), 6044–6052. https://doi.org/10.1021/acs.est.9b01339

De Souza Machado, A. A., Lau, C. W., Till, J., Kloas, W., Lehmann, A., Becker, R., & Rillig, M. C. (2018). Impacts of microplastics on the soil biophysical environment. *Environmental Science & Technology, 52*(17), 9656–9665. https://doi.org/10.1021/acs.est.8b02212

Deng, Y. F., Zhang, Y., Lemos, B., & Ren, H. Q. (2017). Tissue accumulation of microplastics in mice and biomarker responses suggest widespread health risks of exposure. *Scientific Reports, 7*, 46687. https://doi.org/10.1038/Srep46687

Digka, N., Tsangaris, C., Torre, M., Anastasopoulou, A., & Zeri, C. (2018). Microplastics in mussels and fish from the Northern Ionian Sea. *Marine Pollution Bulletin, 135*, 30–40. https://doi.org/10.1016/j.marpolbul.2018.06.063

Donaldson, K., Stone, K., Gilmour, P. S., Brown, D. M., & MacNess, W. (2000). Ultrafine particles: mechanisms of lung injury. *Philosophical Transactions of the Royal Society A, 358*, 2741–2749. https://doi.org/10.1098/rsta.2000.0681

Dris, R., Gasperi, J., Mirande, C., Mandin, C., Guerrouache, M., Langlois, V., et al. (2017). A first overview of textile fibers, including microplastics, in indoor and outdoor environments. *Environmental Pollution, 221*, 453–458.

Dris, R., Gasperj, J., Saad, M., Mirande, C., & Tassin, B. (2016). Synthetie fibers in atmospheric fallout: a source of microplastics in the environment. *Marine Pollution Bulletin, 104*, 290–293.

Ebner, N., & Iacovidou, E. (2021). The challenges of Covid-19 pandemic on improving plastic waste recycling rates. *Sustainable Production and Consumption, 28*, 726–735. https://doi.org/10.1016/j.spc.2021.07.001. Epub 2021 Jul 6. PMID: 34722849; PMCID: PMC8536949.

Ensign, L. M., Cone, R., & Hanes, J. (2012). Oral drug delivery with polymeric nanoparticles: The gastrointestinal mucus barriers. *Advanced Drug Delivery Reviews, 64*, 557–570. https://doi.org/10.1016/j.addr.2011.12.009

Evangeliou, N., Grythe, H., Klimont, Z., Heyes, C., Eckhardt, S., Lopez-Aparicio, S., & Stohl, A. (2020). Atmospheric transport is a major pathway of microplastics to remote regions. *Nature Communications, 11*(1), 3381. https://doi.org/10.1038/s41467-020-17201-9

Filho, W.L., Salvia, A.L., Minhas, A., Paço, A., & Dias-Ferreira, C. (2021). The COVID-19 pandemic and single-use plastic waste in households: A preliminary study. *Science of The Total Environment, 793*, 148571. https://doi.org/10.1016/j.scitotenv.2021.148571, ISSN 0048-9697.

Forte, M., Iachetta, G., Tussellino, M., Carotenuto, R., Prisco, M., De Falco, M., Laforgia, V., & Valiante, S. (2016). Polystyrene nanoparticles internalization in human gastric adenocarci noma cells. *Toxicology in Vitro, 31*, 126–136. https://doi.org/10.1016/j.tiv.2015.11.006

Frias, J. P. G. L., & Nash, R. (2019). Microplastics: Finding a consensus on the definition. *Marine Pollution Bulletin, 138* , 145–147.

Galloway, T.S. (2015). Micro- and nano-plastics and human health. In M. Bergmann, L. Gutow, M. Klages (Eds.), *Marine Anthropogenic Litter* (pp. 343–366). Spr inger. https://doi.org/10.1007/978-3-319-16510-3_13.

Gao, B., Yao, H., Li, Y., & Zhu, Y. (2021). Microplastic addition alters the microbial community structure and stimulates soil carbon dioxide emissions in vegetable-growing soil. *Environmental Toxicology and Chemistry, 40*(2), 352–365. https://doi.org/10.1002/etc.4916

Gardon, T., Reisser, C., Soyez, C., Quillien, V., & Le Moullac, G. (2018). Microplastics affect energy balance and gametogenesis in the pearl oyster Pinctada margaritifera. *Environmental Science & Technology, 52*, 5277–5286. https://doi.org/10.1021/acs.est.8b00168

Garrido, S., Linares, M., Campillo, J. A., & Albentosa, M. (2019). Effect of microplastics on the toxicity of chlorpyrifos to the microalgae Isochrysis galbana, clone t-ISO. *Ecotoxicology and Environmental Safety, 173*, 103–109. https://doi.org/10.1016/j.ecoenv.2019.02.020

Geyer, R., Jambeck, J. R., & Law, K. L. (2017). Production, use, and fate of all plastics ever made. *Science Advances, 3*(7), e1700782.

Godoy, V., Martín-Lara, M. A., Calero, M., & Blázquez, G. (2019). Physical-chemical characterization of microplastics present in some exfoliating products from Spain. *Marin e Pollution Bulletin, 139*, 91–99. https://doi.org/10.1016/j.marpolbul.2018.12.026

Goldstein, M. C., Rosenberg, M., & Cheng, L. (2012). Increased oceanic microplastic debris enhances oviposition in an endemic pelagic insect. *Biology Letters, 8*, 817–820. https://doi.org/10.1098/rsbl.2012.0298

Govindan, N., Bhuyar, P., Feng, H. X., Rahim, M. H. A., Maniam, G. P., Sundararaju, S., & Muniyasamy, S. (2021). Evaluation of Microalgae's Plastic Biodeterioration Property by a Consortium of Chlorella sp. and Cyanobacteria sp. *Journal of Environmental Research, Engineering and Management., 77*(3), 86–98. https://doi.org/10.5755/j01.erem.77.3.25317

Grigorakis, S., Mason, S. A., & Drouillard, K. G. (2017). Determination of the gut retention of plastic microbeads and microfibers in goldfish (Carassius auratus). *Chemosphere, 169*, 233–238. https://doi.org/10.1016/j.chemosphere.2016.11.055

Guo, X. T., Hu, G. L., Fan, X., & Jia, H. (2020). Sorption properties of cadmium on microplastics: The common practice experiment and A two dimensional correlation spectroscopic study. *Ecotoxicology and Environmental Safety, 190*, 110118.

Guo, J.-J., Huang, X.-P., Xiang, L., Wang, Y.-Z., Li, Y.-W., Li, H., Cai, Q.-Y., Mo, C.-H., & Wong, M.-H. (2020). Source, migration and toxicology of microplastics in soil. *Environment International, 137*, 105263. [Google Scholar] [CrossRef].

Hidalgo-Ruz, V., Gutow, L., Thompson, R. C., & Thiel, M. (2012). Microplastics in the marine environment: A review of the methods used for identification and quantification. *Environmental Science & Technology, 46*, 3060–3075.

Horton, A. A., Walton, A., Spurgeon, D. J., Lahive, E., & Svendsen, C. (2017). Microplastics in freshwater and terrestrial environments: Evaluating the current understanding to identify the knowledge gaps and future research priorities. *Science of the Total Environment, 586*, 127–141. https://doi.org/10.1016/j.scitotenv.2017.01.190

Igalavithana, A. D., Yuan, X., Attanayake, C. P., Wang, S., You, S., Tsang, D. C. W., Nzihou, A., & Ok, Y. S. (2022). Sustainable management of plastic wastes in COVID-19 pandemic: The biochar solution. *Environmental Research, 212*(Pt E):113495. https://doi.org/10.1016/j.envres.2022.113495. Epub 2022 Jun 1. PMID: 35660402; PMCID: PMC9155208.

Ivleva, N. P., Wiesheu, A. C., & Niessner, R. (2017). Microplastic in aquatic ecosystems. *Angewandte Chemie International Edition, 56*(7), 1720–1739. https://doi.org/10.1002/anie.201606957

Jacquin, J., Cheng, J., Odobel, C., Pandin, C., Conan, P., Pujo-Pay, M., et al. (2019). Microbial ecotoxicology of marine plastic debris: A review on colonization and biodegradation by the "plastisphere." *Frontiers in Microbiology, 10*, 865. https://doi.org/10.3389/fmicb.2019.00865

Jaikumar, G., Baas, J., Brun, N. R., Vijver, M. G., & Bosker, T. (2018). Acute sensitivity of three Cladoceran species to different types of microplastics in combination with thermal stress. *Environmental Pollution, 239*, 733–740. https://doi.org/10.1016/j.envpol.2018.04.069. Epub 2018 May 1. PMID: 29723823

Jaikumar, G., Brun, N. R., Vijver, M. G., & Bosker, T. (2019). Reproductive toxicity of primary and secondary microplastics to three cladocerans during chronic exposure. *Environmental Pollution, 249*, 638–646. https://doi.org/10.1016/j.envpol.2019.03.085

Jain, R., Gaur, A., Suravajhala, R., Chauhan, U., Pant, M., Tripathi, V., & Pant, G. (2023). Microplastic pollution: Understanding microbial degradation and strategies for pollutant reduction. *Science of the Total Environment, 905*, 167098.

Jambeck, J. R., Geyer, R., Wilcox, C., Siegler, T. R., Perryman, M., Andrady, A., Narayan, R., & Law, K. L. (2015). Plastic waste inputs from land into the ocean. *Science, 347*, 768–771.

Jeong, J., & Choi, J. (2019). Adverse outcome pathways potentially related to hazard identification of microplastics based on toxicity mechanisms. *Chemosphere, 231*, 249–255. https://doi.org/10.1016/j.chemosphere.2019.05.003

Jin, Y. X., Lu, L., Tu, W. Q., Luo, T., & Fu, Z. W. (2019). Impacts of polystyrene microplastic on the gut barrier, microbiota and metabolism of mice. *Science of the Total Environment, 649*, 308–317. https://doi.org/10.1016/j.scitotenv.2018.08.353

Jin, Y. X., Xia, J. Z., Pan, Z. H., Yang, J. J., Wang, W. C., & Fu, Z. W. (2018). Polystyrene microplastics induce microbiota dysbiosis and inflammation in the gut of adult zebrafish. *Environmental Pollution, 235*, 322–329. https://doi.org/10.1016/j.envpol.2017.12.088

Judy, J. D., Williams, M., Gregg, A., Oliver, D., Kumar, A., Kookana, R., & Kirby, J. K. (2019). Microplastics in municipal mixed-waste organic outputs induce minimal short to long-term toxicity in key terrestrial biota. *Environmental Pollution, 252*, 522–531. https://doi.org/10.1016/j.envpol.2019.05.027

Kalcikova, G., Gotvajn, A. Z., Kladnik, A., & Jemec, A. (2017). Impact of polyethylene microbeads on the floating freshwater plant duckweed Lemna minor. *Environmental Pollution, 230*, 1108–1115. https://doi.org/10.1016/j.envpol.2017.07.050

Kanhai, L. D. K., Gårdfeldt, K., Lyashevska, O., Hassellöv, M., Thompson, R. C., & O'Connor, I. (2018). Microplastics in sub-surface waters of the Arctic Central Basin. *Marine Pollution Bulletin, 130*, 8–18. https://doi.org/10.1016/j.marpolbul.2018.03.011

Koelmans, A. A., Bakir, A., Burton, G. A., & Janssen, C. R. (2016). Microplastic as a vector for chemicals in the aquatic environment: Critical review and model-supported reinterpretation of empirical studies. *Environmental Science and Technology, 50*(7), 3315–3326. https://doi.org/10.1021/acs.est.5b06069

Krüger, L., Casado-Coy, N., Valle, C., Ramos, M., Sánchez-Jerez, P., Gago, J., Carretero, O., Beltran-Sanahuja, A., Sanz-Lazaro, C. (2020). Plastic debris accumulation in the seabed derived from coastal fish farming. *Environmental Pollution, 257,* 113336. ISSN 0269-7491. https://doi.org/10.1016/j.envpol.2019.113336

Kulkarni, B. N., Anantharama, V. (2020). Repercussions of COVID-19 pandemic on municipal solid waste management: Challenges and opportunities. *Science of the Total Environment, 743,* 140693. https://doi.org/10.1016/j.scitotenv.2020.140693. Epub 2020 Jul 2. PMID: 32663690; PMCID: PMC7331525.

Lee, K. W., Shim, W. J., Kwon, O. Y., & Kang, J. H. (2013). Size-dependent effects of micro polystyrene particles in the marine copepod Tigriopus japonicus. *Environmental Science & Technology, 47*, 11278–11283. https://doi.org/10.1021/es401932b

Lei, L. L., Wu, S. Y., Lu, S. B., Liu, M. T., Song, Y., Fu, Z. H., ... (2018). He, D. F. Microplastic particles cause intestinal damage and other adverse effects in zebrafish Danio rerio and nematode Caenorhabditis elegans. *Science of the Total Environment, 619*, 1–8. https://doi.org/10.1016/j.scitotenv.2017.11.103.

Lebreton, L. C. M., Van der Zwet, J., Damsteeg, J. W., Slat, B., Andrady, A., & Reisser, J. (2017). River plastic emissions to the world's oceans. *Nature Communications, 8,* 15611. https://doi.org/10.1038/Ncomms15611

Liang, Y., Lehmann, A., Ballhausen, M.-B., Muller, L., & Rillig, M. C. (2019). Increasing temperature and microplastic fibers jointly influence soil aggregation by saprobic fungi. *Frontiers in Microbiology, 10*, 2018. https://doi.org/10.3389/fmicb.2019.02018

Liu, H., Yang, X., Liu, G., Liang, C., Xue, S., Chen, H., Ritsema, C. J., & Geissen, V. (2017). Response of soil dissolved organic matter to microplastic addition in Chinese loess soil. *Chemosphere, 185*, 907–917.

Lu, L., Wan, Z. Q., Luo, T., Fu, Z. W., & Jin, Y. X. (2018). Polystyrene microplastics induce gut microbiota dysbiosis and hepatic lipid metabolism disorder in mice. *Science of the Total Environment, 631–632*, 449–458. https://doi.org/10.1016/j.scitotenv.2018.03.051

Lu, Y., Zhang, Y., Deng, Y., Jiang, W., Zhao, Y., Geng, J., ... Ren, H. (2016). Uptake and Accumulation of polystyrene microplastics in zebrafish (Danio rerio) and toxic effects in liver. *Environmental Science & Technology, 50*, 4054–4060. https://doi.org/10.1021/acs.est.6b00183.

Lwanga, E. H., Gertsen, H., Gooren, H., Peters, P., Salanki, T., van der Ploeg, M., Besseling, E., Koelmans, A. A., & Geissen, V. (2017). Incorporation of microplastics from litter into burrows of Lumbricus terrestris. *Environmental Pollution, 220*, 523–531. https://doi.org/10.1016/j.envpol.2016.09.096

Maass, S., Daphi, D., Lehmann, A., & Rillig, M. C. (2017). Transport of microplastics by two collembolan species. *Environmental Pollution, 225*, 456–459. https://doi.org/10.1016/j.envpol.2017.03.009

Mao, Y. F., Ai, H. N., Chen, Y., Zhang, Z. Y., Zeng, P., Kang, L., ... Li, H. (2018). Phytoplankton response to polystyrene microplastics: Perspective from an entire growth period. *Chemosphere, 208*, 59–68. https://doi.org/10.1016/j.chemosphere.2018.05.170.

Mateos-Cardenas, A., Scott, D. T., Seitmaganbetova, G., van Pelt, F. N. A. M., O'Halloran, J., & Jansen, M. A. K. (2019). Polyethylene microplastics adhere to Lemna minor (L), yet have no effects on plant growth or feeding by Gammarus duebeni (Lillj.). *Science of the Total Environment, 689*, 413–421. https://doi.org/10.1016/j.scitotenv.2019.06.359

Mathalon, A., & Hill, P. (2014). Microplastic fibers in the intertidal ecosystem surrounding Halifax Harbor, Nova Scotia. *Marine Pollution Bulletin, 81*(1), 69–79. https://doi.org/10.1016/j.marpolbul.2014.02.018

Moog, D., Schmitt, J., Senger, J., Zarzycki, J., Rexer, K. H., Linne, U., Erb, T., & Maier, U. G. (2019). Using a marine microalga as a chassis for polyethylene terephthalate (PET) degradation. *Microbial Cell Factories, 18*(1), 171. https://doi.org/10.1186/s12934-019-1220-z.Erratum.In:MicrobCellFact.2020Jan2;19(1):1.PMID:31601227;PMCID:PMC6786278

Nizzetto, L., Futter, M., & Langaas, S. (2016a). Are agricultural soils dumps for microplastics of urban origin? *Environmental Science & Technology, 50*, 10777–10779. https://doi.org/10.1021/acs.est.6b04140

Nizzetto, L., Futter, M., & Langaas, S. (2016b). Are agricultural soils dumps for microplastics of urban origin? *Environmental Science and Technology, 50*, 10777–10779.

Oberbeckmann, S., Osborn, A. M., & Duhaime, M. B. (2016). Microbes on a bottle: Substrate, season and geography influence community composition of microbes colonizing marine plastic debris. *PLoS ONE, 11*, e0159289. https://doi.org/10.1371/journal.pone.0159289

Okal, E. J., Heng, G., Magige, E. A., Khan, S., Wu, S., Ge, Z., Zhang, T., Mortimer, P. E., & Xu, J. (2023). Insights into the mechanisms involved in the fungal degradation of plastics. *Ecotoxicology and Environmental Safety, 262*, 115202.

Oliviero, M., Tato, T., Schiavo, S., Fernandez, V., Manzo, S., & Beiras, R. (2019). Leachates of micronized plastic toys provoke embryotoxic effects upon sea urchin Paracentrotus lividus. *Environmental Pollution, 247*, 706–715. https://doi.org/10.1016/j.envpol.2019.01.098

Oliveira, W. Q., Azeredo, H. M. C., Neri-Numa, I. A., & Pastore, G. M. (2021) Food packaging wastes amid the COVID-19 pandemic: Trends and challenges. *Trends in Food Science and Technology, 116*, 1195–1199. https://doi.org/10.1016/j.tifs.2021.05.027. Epub 2021 Jun 1. PMID: 34092920; PMCID: PMC8166460.

Oliveira, Y. M., Vernin, N. S., Maia Bila, D., Marques, M., & Tavares, F. W. (2022). Pollution caused by nanoplastics: Adverse effects and mechanisms of interaction via molecular simulation. *Peer J., 10*, e13618. https://doi.org/10.7717/peerj.13618

Paço, A., Duarte, K., da Costa, J. P., Santos, P. S. M., Pereira, R., Pereira, M. E., Freitas, A. C., Duarte, A. C., & Rocha-Santos, T. A. P. (2017). Biodegradation of polyethylene microplastics by the marine fungus Zalerion maritimum. *Science of the Total Environment, 586*, 10–15.

Plasticseurope. (2019). Plastics – The facts 2019 an analysis of European plastics production, Demand and Waste Data

Posada-Baquero, R., & Ortega-Calvo, J.-J. (2011). Recalcitrance of polycyclic aromatic hydrocarbons in soil contributes to background pollution. *Environmental Pollution, 159*(12), 3692–3699. https://doi.org/10.1016/j.envpol.2011.07.012

Prata, J. C. (2018). Airborne microplastics: Consequences to human health? *Environmental Pollution, 234*, 115–126.

Prata, J. C., Manana, M. J., da Costa, J. P., Duarte, A. C., & Rocha-Santos, T. (2020a). What is the minimum volume of sample to find small microplastics: Laboratory experiments and sampling of Aveiro Lagoon and Vouga River, Portugal. *Water, 12*, 1219. https://doi.org/10.3390/w12041219

Prata, J. C., Reis, V., Paço, A., Martins, P., Cruz, A., da Costa, J. P., Duarte, A. C., & Rocha-Santos, T. (2020b). Effects of spatial and seasonal factors on the characteristics and carbonyl index of (micro)plastics in a sandy beach in Aveiro, Portugal. *Science of the Total Environment, 709*, 135892. https://doi.org/10.1016/j.scitotenv.2019.135892

Prata, J. C., Silva, A. L. P., da Costa, J. P., Mouneyrac, C., Walker, T. R., Duarte, A. C., & Rocha-Santos, T. (2019). Solutions and integrated strategies for the control and mitigation of plastic and microplastic pollution. *International Journal of Environmental Research and Public Health, 16*, 2411. https://doi.org/10.3390/ijerph16132411

Qiao, R. X., Sheng, C., Lu, Y. F., Zhang, Y., Ren, H. Q., & Lemos, B. (2019a). Microplastics induce intestinal inflammation, oxidative stress, and disorders of metabolome and microbiome in zebrafish. *Science of the Total Environment, 662*, 246–253. https://doi.org/10.1016/j.scitotenv.2019.01.245

Qiao, R., Deng, Y., Zhang, S., Wolosker, M. B., Zhu, Q., Ren, H., & Zhang, Y. (2019b) Accumulation of different shapes of microplastics initiates intestinal injury and gut microbiota dysbiosis in the gut of zebrafish. *Chemosphere, 236*, 124334. https://doi.org/10.1016/j.chemosphere.2019.07.065. Epub 2019 Jul 10. PMID: 31310986

Rai, M., Pant, G., Pant, K., Aloo, B. N., Kumar, G., Singh, H. B., & Tripathi, V. (2023). Microplastic pollution in terrestrial ecosystems and its interaction with other soil pollutants: A potential threat to soil ecosystem sustainability. *Resources, 12*, 67. https://doi.org/10.3390/resources12060067

Rather, M. A., Gupta, K., & Mandal, M. (2021). Microbial biofilm: formation, architecture, antibiotic resistance, and control strategies. *Brazilian Journal of Microbiology, 52*(4), 1701–1718. https://doi.org/10.1007/s42770-021-00624-x. Epub 2021 Sep 23. PMID: 34558029; PMCID: PMC8578483.

Ren, X., Tang, J., Liu, X., & Liu, Q. (2020). Effects of microplastics on greenhouse gas emissions and the microbial community in fertilized soil. *Environmental Pollution, 256*, 113347. https://doi.org/10.1016/j.envpol.2019.113347

Rezania, S., Park, J., Md Din, M. F., Mat Taib, S., Talaiekhozani, A., Kumar Yadav, K., & Kamyab, H. (2018). Microplastics pollution in different aquatic environments and biota: A review of recent studies. *Marine Pollution Bulletin, 133*, 191–208.

Rillig, M. C., Lehmann, A., Machado, A., Yang, G. (2019). Micro-plastic effects on plants. *New Phytologist, 223*, 1066–1070.

Rillig, M. C. (2012). Microplastic in terrestrial ecosystems and the soil? *Environmental Science & Technology, 46*, 6453–6454. https://doi.org/10.1021/es302011r

Rillig, M. C. (2018). Microplastic disguising as soil carbon storage. *Environmental Science & Technology, 52*(11), 6079–6080. https://doi.org/10.1021/acs.est.8b02338

Rillig, M. C., & Lehmann, A. (2020). Microplastic in terrestrial ecosystems. *Science, 368*(6498), 1430–1431. https://doi.org/10.1126/science.abb5979

Rillig, M. C., Leifheit, E., & Lehmann, J. (2021). Microplastic effects on carbon cycling processes in soils. *PLOS Biology, 19*(3), e3001130. https://doi.org/10.1371/journal.pbio.3001130

Ritchie, H., & Roser, M. (2018). Plastic pollution. Our World in Data. https://ourworldindata.org/plastic-pollution.

Rochman, C. M., Brookson, C., Bikker, J., Djuric, N., Earn, A., Bucci, K., Athey, S., Huntington, A., McIlwraith, H., & Munno, K. (2019). Rethinking microplastics as a diverse contaminant suite. *Environmental Toxicology and Chemistry, 38*, 703–711.

Rodriguez-Seijo, A., Lourenco, J., Rocha-Santos, T. A. P., da Costa, J., Duarte, A. C., Vala, H., & Pereira, R. (2017). Histopathological and molecular effects of microplastics in Eisenia andrei Bouche. *Environmental Pollution, 220*, 495–503. https://doi.org/10.1016/j.envpol.2016.09.092

Rong, L., Zhao, L., Zhao, L., Cheng, Z., Yao, Y., Yuan, C., Wang, L., & Sun, H. (2021). LDPE microplastics affect soil microbial communities and nitrogen cycling. *Science of the Total Environment, 773*, 145640. https://doi.org/10.1016/j.scitotenv.2021.145640.

Schirinzi, G. F., Pérez-Pomeda, I., Sanchís, J., Rossini, C., Farré, M., & Barceló, D. (2017). Cytotoxic effects of commonly used nanomaterials and microplastics on cerebral and epithelial human cells. *Environmental Research, 159*, 579–587. https://doi.org/10.1016/j.envres.2017.08.043

Senathirajah, K., & Palanisami, T. (2019). How much microplastics are we ingesting? Estimation of the mass of microplastics ingested. [WWW document]. University of Newcastle. https://www.newcastle.edu.au/newsroom/featured/plastic-ingestion-by-people-could-be-equating-to-a-credit-card-a-week/how-much-microplastics-are-weingesting-estimation-of-the-mass-ofm icroplastics-ingested.

Seth, C. K., & Shriwastav, A. (2018). Contamination of Indian sea salts with microplastics and a potential prevention strategy. *Environmental Science and Pollution Research, 25*, 30122–30131. https://doi.org/10.1007/s11356-018-3028-5

Shams, M., Alam, I., & Mahbub, M. S. (2021). Plastic pollution during COVID-19: Plastic waste directives and its long-term impact on the environment. *Environmental Advances, 5*, 100119.

Song, Y., Cao, C. J., Qiu, R., Hu, J. N., Liu, M. T., Lu, S. B., … He, D. F. (2019). Uptake and adverse effects of polyethylene terephthalate microplastics fibers on terrestrial snails (Achatina fulica) after soil exposure. *Environmental Pollution, 250*, 447–455. https://doi.org/10.1016/j.envpol.2019.04.066.

Sun, J., Dai, X., Wang, Q., van Loosdrecht, M. C. M., & Ni, B.-J. (2019). Microplastics in wastewater treatment plants: Detection, occurrence and removal. *Water Research, 152*, 21–37. https://doi.org/10.1016/j.watres.2018.12.050

van Sebille, E., Wilcox, C., Lebreton, L., Maximenko, N., Hardesty, B. D., van Franeker, J. A., Eriksen, M., Siegel, D., Galgani, F., & Law, K. L. (2015). A global inventory of small floating plastic debris. *Environmental Research Letters, 10*, 124006. https://doi.org/10.1088/1748-9326/10/12/124006

Tang, J., Ni, X. Z., Zhou, Z., Wang, L. G., & Lin, S. J. (2018). Acute microplastic exposure raises stress response and suppresses detoxification and immune capacities in the scleractinian coral *Pocillopora damicornis*. *Environmental Pollution, 243*, 66–74. https://doi.org/10.1016/j.envpol.2018.08.045

Ter Halle, A., & Ghiglione, J. F. (2021). Nanoplastics: A complex, polluting *terra incognita*. *Environmental Science and Technology, 55*(21), 14466–14469. https://doi.org/10.1021/acs.est.1c04142

Thompson, R. C., Swan, S. H., Moore, C. J., & vom Saal, F. S. (2009). Our plastic age. *Philosophical Transactions of the Royal Society B, 364*(1526), 1973–1976. https://doi.org/10.1098/rstb.2009.0054

Toussaint, B., Raffael, B., Angers-Loustau, A., Gilliland, D., Kestens, V., Petrillo, M., et al. (2019). Review of micro- and nanoplastic contamination in the food chain. *Food Additives and Contaminants: Part A, 36*(5), 639–673. https://doi.org/10.1080/19440049.2019.1583381

Udyawer, V., Read, M. A., Hamann, M., Simpfendorfer, C. A., & Heupel, M. R. (2013). First record of sea snake (Hydrophis elegans, Hydrophiinae) entrapped in marine debris. *Marine Pollution Bulletin, 73*, 336–338. https://doi.org/10.1016/j.marpolbul.2013.06.023

Urbanek, A. K., Rymowicz, W., & Mirończuk, A. M. (2018). Degradation of plastics and plastic-degrading bacteria in cold marine habitats. *Applied Microbiology and Biotechnology, 102*(18), 7669–7678. https://doi.org/10.1007/s00253-018-9195-y. Epub 2018 Jul 11. PMID: 29992436; PMCID: PMC6132502

Vanapalli, K. R., Sharma, H. B., Ranjan, V. P., Samal, B., Bhattacharya, J., Dubey, B. K., & Goel, S. (2021). Challenges and strategies for effective plastic waste management during and post COVID-19 pandemic. *Science of the Total Environment, 750*, 141514. https://doi.org/10.1016/j.scitotenv.2020.141514. Epub 2020 Aug 4. PMID: 32835961; PMCID: PMC7399665.

Van Franeker, J. A. (1985). Plastic ingestion in the North Atlantic Fulmar. *Marine Pollution Bulletin, 16*, 367–369. https://doi.org/10.1016/0025-326X(85)90090-6

Vianello, A., Jensen, R. L., Liu, L., & Vollertsen, J. (2019). Simulating human exposure to indoor airborne microplastics using a breathing thermal manikin. *Science and Reports, 9*, 8670. https://doi.org/10.1038/s41598-019-45054-w

Volkheimer, G. (1977). Persorption of particles: Physiology and pharmacology. *Advances in Pharmacology, 14*, 163–187. https://doi.org/10.1016/S1054-3589(08)60188-X

Wang, F., Zhang, X., Zhang, S., Zhang, S., & Sun, Y. (2020). Interactions of microplastics and cadmium on plant growth and arbuscular mycorrhizal fungal communities in an agricultural soil. *Chemosphere, 254*, 126791. https://doi.org/10.1016/j.chemosphere.2020.126791

Wang, J., Liu, X., Li, Y., Powell, T., Wang, X., Wang, G., & Zhang, P. (2019a). Microplastics as contaminants in the soil environment: A mini-review. *Science of the Total Environment, 691*, 848–857.

Wang, J. D., Tan, Z., Peng, J. P., Qiu, Q. X., & Li, M. M. (2016). The behaviors of microplastics in the marine environment. *Marine Environmental Research, 113*, 7–17.

Wang, Z., Qin, Y., Li, W., Yang, W., Meng, Q., & Yang, J. (2019b). Microplastic contamination in freshwater: First observation in Lake Ulansuhai, Yellow River Basin, China. *Environmental Chemistry Letters, 17*, 1821–1830.

Watts, A. J., Lewis, C., Goodhead, R. M., Beckett, S. J., Moger, J., Tyler, C. R., & Galloway, T. S. (2014). Uptake and retention of microplastics by the shore crab Carcinus maenas. *Environmental Science & Technology, 48*(15), 8823–8830. https://doi.org/10.1021/es501090e

Wiesheu, A. C., Anger, P. M., Baumann, T., Niessner, R., & Ivleva, N. P. (2016). Raman microspectroscopic analysis of fibers in beverages. *Analytical Methods, 8*, 5722. https://doi.org/10.1039/c6ay01184e

Wright, S. L., Thompson, R. C., & Galloway, T. S. (2013). The physical impacts of microplastics on marine organisms: A review. *Environmental Pollution, 178*, 483–492. https://doi.org/10.1016/j.envpol.2013.02.031

Wright, S. L., & Kelly, F. J. (2017). Plastic and human health: A micro issue? *Environmental Science and Technology, 51*, 6634–6647. [Google Scholar] [CrossRef].

Wu, Y., Guo, P., Zhang, X., Zhang, Y., Xie, S., & Deng, J. (2019). Effect of microplastics exposure on the photosynthesis system of freshwater algae. *Journal of Hazardous Materials, 374*, 219–227. https://doi.org/10.1016/j.jhazmat.2019.04.039

Yan, Y., Chen, Z., Zhu, F., Zhu, C., Wang, C., & Gu, C. (2020). Effect of polyvinyl chloride microplastics on bacterial community and nutrient status in two agricultural soils. *Bulletin of Environmental Contamination and Toxicology*. https://doi.org/10.1007/s00128-020-02900-2

Yang, Y. F., Chen, C. Y., Lu, T. H., & Liao, C. M. (2019). Toxicity-based toxicokinetic/toxicodynamic assessment for bioaccumulation of polystyrene microplastics in mice. *Journal of Hazardous Materials, 366*, 703–713. https://doi.org/10.1016/j.jhazmat.2018.12.048

Yang, X. G., Wen, P. P., Yang, Y. F., Jia, P. P., Li, W. G., & Pei, D. S. (2023). Plastic biodegradation by in vitro environmental microorganisms and in vivo gut microorganisms of insects. *Frontiers in Microbiology, 6*(13), 1001750. https://doi.org/10.3389/fmicb.2022.1001750.PMID:36687617; PMCID:PMC9852869

Yoshida, S., Hiraga, K., Takehana, T., Taniguchi, I., Yamaji, H., Maeda, Y., Toyohara, K., Miyamoto, K., Kimura, Y., & Oda, K. (2016). A bacterium that degrades and assimilates poly(ethylene terephthalate). *Science, 351*(6278), 1196–1199. https://doi.org/10.1126/science.aad6359

Zhai, X., Zhang, X. H., & Yu, M. (2023). Microbial colonization and degradation of marine microplastics in the plastisphere: A review. *Frontiers in Microbiology, 17*(14), 1127308. https://doi.org/10.3389/fmicb.2023.1127308.PMID:36876073;PMCID:PMC9981674

Zhang, C., Zhou, H., Cui, Y., Wang, C., Li, Y., Zhang, D. (2019a). Microplastics in offshore sediment in the Yellow Sea and East China Sea, China, *Environmental Pollution, 244*, 2019, 827–833,ISSN 0269-7491. https://doi.org/10.1016/j.envpol.2018.10.102

Zhang, P., Yan, Z., Lu, G., & Ji, Y. (2019b). Single and combined effects of microplastics and roxithromycin on Daphnia magna. *Environmental Science and Pollution Research International, 26*, 17010–17020. https://doi.org/10.1007/s11356-019-05031-2

Zhang, X., Zheng, M., Yin, X., Wang, L., Lou, Y. H., Qu, L. Y., et al. (2019c). Sorption of 3,6-dibromocarbazole and 1,3,6,8-tetrabromocarbazole by microplastics. *Marine Pollution Bulletin, 138*, 458–463.

Zhu, D., Bi, Q. F., Xiang, Q., Chen, Q. L., Christie, P., Ke, X., … Zhu, Y. G. (2018). Trophic predator-prey relationships promote transport of microplastics compared with the single Hypoaspis aculeifer and Folsomia candida. *Environmental Pollution, 235*, 150–154. https://doi.org/10.1016/j.envpol.2017.12.058.

Chapter 4
The Environmental Fate of Microplastics

Naveen Bunekar, Tsai Tsung-yen, Basharat Hussain, and S. S. Hwang

Abstract The input of pollutants into the environment is increasing every year. Micropollutants are both inorganic and organic compounds that may negatively impact the environment even in minimal quantities. These are key contaminants in the soil, lakes, oceans, and rivers. Pollutants enter the environment from many anthropogenic sources and are spread out in many different areas of the environment. Micropollutants depend on their natural qualities and whether they are organic to inorganic by their sources like as pharmaceuticals, and micro and nano plastics, personal care products, pesticides, stimulants, persistent organic pollutants. Trace and heavy metals, etc., are inorganic. Micropollutants get into the air, water, and earth from a variety of sources. These chemicals pollute the soil and the living things that live in it. Plants also eat them and they get into the groundwater. Microorganisms, earthworms, and other animals are most strongly affected by pharmaceuticals. This book chapter focuses on the fate, behavior, the occurrence of the contaminants to environmental and health risks effect with emerging pharmaceuticals, endocrine disruptors, hormones, toxins, and biological micropollutants in soils and sediments, groundwater, industrial and municipal wastewaters, aquaculture effluents, and freshwater and marine ecosystems and geochemical environment. These micro-pollutants pose significant risks at every stage of their life cycle to the environment and human health. Additionally, some ideas for lowering the potential hazards of micropollutants have been put forward based on recent developments in this field of research.

Keywords Micropollutants · Environment · Contaminates · Pollutants and human life

N. Bunekar (✉) · T. Tsung-yen (✉) · B. Hussain
Department of Chemistry, Center for Nanotechnology, Chung Yuan Christian University, Chung-Li, Taoyuan City 32023, Taiwan
e-mail: naveen.bunekar@gmail.com

T. Tsung-yen
e-mail: yen@cycu.edu.tw

S. S. Hwang
Chien-Hsin University of Science and Technology, Chung-Li, Taoyuan City 32097, Republic of China

A. K. Mishra et al. (eds.), *Microplastics*, https://doi.org/10.1007/978-981-97-6461-7_4

1 Introduction

These days, the world endures hardship from the bad effects of uncontrolled urbanization and numerous human endeavors, which cause pollution and climate change. Micropollutants are chemical compounds that come from tiny quantities in the environment (Gavrilescu, 2010; Thomaidis et al., 2012). Higher living standards and more consumer desire have made pollution worse. In the last few decades, a lot more pharmaceutical and hygiene kits, pesticides, plastics, endocrine-disrupting chemicals, and other chemical products have been made. These products are light, cheap, flexible, long-lasting (they don't break down easily in heat or electricity), and have an enormous number of practical uses. There are more than 350000 chemicals and mixtures of chemicals that are listed to be made right now. The world's chemical business was worth more than $5 trillion in 2017. The Global Chemicals Outlook 2020 Report from the UN Environment Program says it will double in size by 2030. Over 70% of these are chemicals that have substantial environmental impacts on the world (http://epp.eurostat.ec.europa.eu/tgm/table.do?tab=table&init=1&plugin=1&language=en&pcode=ten00011; http://wwf.panda.org/about_our_earth/teacher_resources/webfieldtrips/toxics/).

It is becoming widely accepted that 99% of the water quantity around the world is not suitable for humans, and the fraction of the water source is shown in Fig. 1. Freshwater resources are very limited. In addition, the global population growth has resulted in an increasing shortage of water in several sectors, including home, industrial, and agricultural usage (Zimmerman et al., 2008). Moreover, on a worldwide scale, the predominant allocation of freshwater resources is directed toward agriculture irrigation (70% water), industrial use (20% water), and domestic utilization (10% water). At present, human activities and urbanization have resulted in an increasing number of contaminants in aquatic and soil systems, which have been widely confirmed. The extensiveness of biological micropollutants, such as viruses and bacteria, causes water pollution, a major worry worldwide. These small-scale contaminant substances have sparked renewed attention because of their potential to cause disease, and they are commonly known as emerging or reemerging pathogens. Bacteria containing genes resistant to antibiotics are also regarded as emerging pollutants (Yingfei et al., 2021). In recent years, there has been a significant rise in the production of anthropogenic chemicals. The increased presence of pollutants in aquatic systems is widely acknowledged to be primarily caused by the ongoing introduction and breakdown of large particles originating from land discharge. Consequently, the terrestrial environment serves as a crucial sink for pollutants. These micropollutants possess inherent toxicity, largely attributed to the inclusion of plastic, colorants or dyes, non-polymerized monomers, and other toxic substances. As plastic breaks down, this substance directly pollutes the environment, posing a threat to both ecosystems and human health (Amy et al., 2006).

Furthermore, the soil pollution is similar to the water pollution. Soils comprise complex systems with changing physical, chemical, and biological properties concerning time. Grassland soil encounters about 10% organic matter. Pollutants

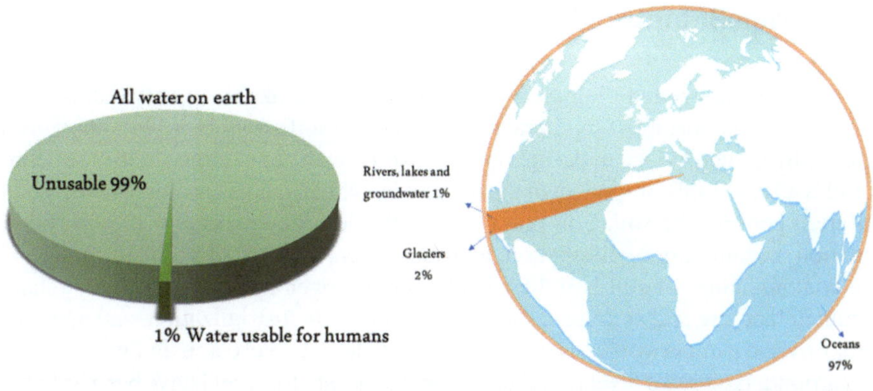

Fig. 1 Usable water and earth water (USGS,1998)

in the soil influence the structure, health, and function of the soil ecosystem. They also impact organisms in the soil, like plants, bacteria, fungi, and pathogen pollution passageways that let toxins get to foodstuff and water. All of them are hazardous in excess amounts and cause human illness, including carcinoma, heart disease, and hypertension. From the literature review, these soil pollutions affect the nutrient cycle and the physiochemical parameters such as water holding capacity, bulk density, pH value, electrical conductivity, porosity, and aggregation of the soil. Moreover, the emission of pollutants causes long-term enrichment and aging reactions that change the surface characteristics, size, and shape of soil. During these processes, pollutants can also take in and release other pollutants, like toxic substances and organic pollutants, which in turn affect soil organisms (Bowman & McGettigan, 1994; Byrne, 1979). The majority derived from decaying vegetation with significant amounts of animal and microbial biomass.

The current chapter aims to deliver a wide-ranging overview and challenges to the pollution grade of micropollutants. The existing review reports show that micropollutants are in soils, groundwater, sediments, municipal wastewater, industrial, freshwater, aquaculture effluents marine ecosystems, and so on. sources and ecological risks of micropollutants in soils. We identified and summarized the dominant partitioning processes for organic pollutants such as agrichemicals and pharmaceutical products to highlight their significant health and environmental effects. In addition, it lists upcoming problems that will need to be solved to lower the adverse effects that micropollutants have on the natural environment. We will address the existing gap in knowledge regarding micropollutants in soil and water, with a particular focus on leveraging innovative and advanced tools and technologies for monitoring, preventing, and mitigating environmental and health risks and pressures (Naveen & Tsai, 2020; Naveen et al., 2023).

2 Micropollutants in Soils

The physiochemical and biological properties of soils make them very complicated systems. Some of the most significant components in soil are silicon, aluminum, iron, along with oxygen, which make up a lot of the soil matrix. The unique matrix structure of soil is an essential component of the process by which metals and persistent organic pollution get into the soil. Soil plays a critical role in the processes of adsorption, retention, accumulation, dissipation, and subsequent transfer of pollutants to rivers. The contamination of soil is influenced by anthropogenic factors primarily related to human land use and activities, such as agriculture, industrialization, and activities within river basin networks, as well as natural factors like erosion and geochemical background. Excessive levels of all these substances are toxic and have been linked to various human illnesses, including cancers, hypertension, and heart disease. Figure 2 shows the soil pollution occurrences such as air-ground-water systems, illegal dumps, and sewage treatment stations (Table 1).

According to research findings, typical grassland topsoil contains approximately 10% organic matter, primarily originating from decomposing plant matter alongside substantial quantities of animal and microbial biomass. Additionally, soil receives a significant influx of synthetic organic compounds such as perfluoroalkylated substances (PFASs), and phthalates (PAEs), with lesser occurrences of polybrominated diphenyl ethers (PBDEs) and alkylphenols (APs). There is cause for concern when these compounds exhibit resistance to degradation and demonstrate biological activity. It is important to evaluate the effects of organochlorine insecticides

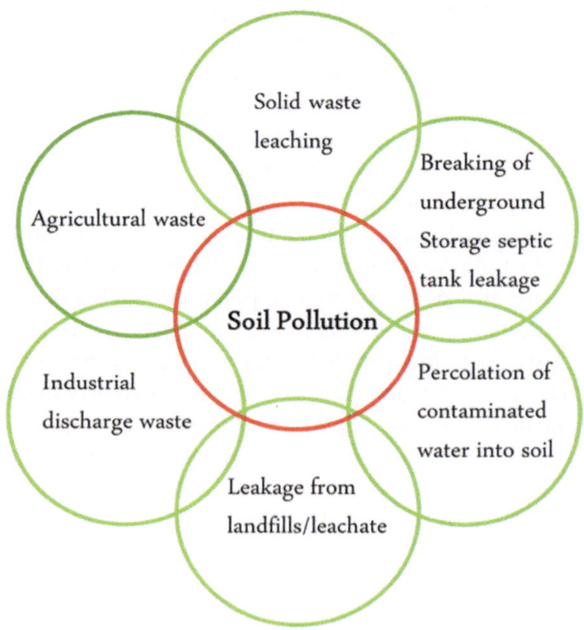

Fig. 2 Schematic pathways of some soil pollutants from different sources

Table 1 Major pollutants in soil

Group	Individual compounds	Type of soil
Trace and heavy metals (2003b; Meijer et al., 2003a)	Titanium (Ti), Vanadium (V), Chromium (Cr), Manganese (Mn), Ferrum (Fe), Cobaltum (Co), Niccolum (Ni), Zincum (Zn), Cuprum (Cu), Arsenicum (As), strontium (Sr), Argentum (Ag), Cadmium (Cd), Stibium (Sb), Barium (Ba), Thallium (Tl), Plumbum (Pb), Hydrargyrum (Hg)	Mixed soil
PAHs (Bucheli et al., 2004; Desaules et al., 2010)	Acenaphthylene, Acenaphthene, Fluoren, Phenanthrene, Anthracene, Fluoranthene, Pyrene, Benzo-anthracene, Chrysene, Benzo fluoranthene, Benzo fluoranthene, Benzo pyrene, Indeno pyrene, Dibenzo anthracene, Benzo perylene	Mixed soil
PAEs (Alliot et al., 2014)	Dimethyl phthalate (DMP), Di-ethyl phthalate (DEP), Di-iso-butyl phthalate (DiBP), Di-n-butyl phthalate (DnBP), Butyl-benzyl phthalate (BBP), Di-ethyl-hexyl phthalate (DEHP), Di-n-octyl phthalate (DnOP)	Urban soil Sub-urban Park soil
Aps (Pape et al., 2014)	Nonylphenols, Nonylphenol monoethoxylate, Nonylphenol diethoxylate, Octylphenol, Octylphenol monoethoxylate, Octylphenol diethoxylate, and Nonylphenoxy acetic acid	Soils Urban soils Agricultural soils
PFASs (Thévenot et al., 2007)	Perfluorobutanoïc acid, Perfluoropentanoic acid, Perfluorohexanoic acid, Perfluorobutane sulfonic acid, Perfluoroheptanoic acid, Perfluoroheptanoic acid, 6:2 Fluorotelomere sulfonate, Perfluorooctanoc acid, Perfluorohexane sulfonic acid, Perfluoronoanoic acid, Perfluoroheptane sulfonic acid, Perfluorodecanoique acid, N-methyl perfluorooctane sulfonamidoacetic acid	Mixed soils
Others (Cladière et al., 2013)	Bisphenol A (BPA), tetrabromobisphenol A (TBBPA), hexachlorobenzene (HCB), pentachlorobenzene (PeCB)	Agricultural soil

(OCs) and polychlorinated biphenyls (PCBs). The impacts of these substances vary depending on their type and form, including endocrine disruption, carcinogenicity, gastric disorders, and degradation of the nervous system, among others. The ratios PEC/PNEC and MEC/PNEC are commonly utilized to evaluate the risk associated with these substances. Moreover, certain soil parameters can influence the sequestration of contaminants in soil, including the relationships between soil organic matter, which plays a crucial role in the partitioning and fate of pollutants (Chiou et al., 1986; Freed et al., 1977; Howard, 1989; Kenaga, 1975; Ling et al., 2022; Maria et al., 2015; Shi et al., 2019; Weber et al., 1991). In addition to the previously mentioned soil pollutants, other substances like fire-retardants, heavy metals such as cadmium, lead, and mercury, commonly used industrial chemicals like Bisphenol A, and certain pesticides have been demonstrated to affect natural endocrine systems (Caliman & Gavrilescu, 2009; Cook et al., 2012; Gavrilescu, 2009; Preda et al., 2012).

The focus of extensive research on emerging pollutants primarily centers on aquatic environments, including surface water, soil, groundwater, and leachate discharged from landfills. In soil, these pollutants become partially immobilized and can significantly impact microorganisms, worms, and other organisms over extended periods. These microorganisms are synthesized by denaturing proteins (Zavarzin & Kolotilova, 2001). Also, microorganisms are cultivated using antibiotics and drugs. The mechanism of their impact varies. In numerous landfills, leachates not only contain microorganic pollutants but also have elevated levels of heavy metals and metalloids, such as arsenic, which can exhibit varying oxidation states, including As(V) and As(III). Arsenic is a hazard due to its antibiotic change in the biogeochemical cycle. For example, bacteria capable of reducing As(V) show resistance to various types of antibiotics. Conversely, As(III)-oxidizing bacteria, primarily belonging to the Proteobacteria group, exhibit reduced activity when exposed to chloramphenicol (Yamamura et al., 2014). There is evidence that these pollutants affect not only microorganisms. Additionally, these factors contribute to the overall degradation of ecosystems (Diamond et al., 2011; Pal et al., 2010a; Rosi-Marshall et al., 2013). The accumulation of drugs in soil and water poses a significant danger to invertebrates, fishes, and humans. Soil pollutants enter our bodies through the food chain, leading to various illnesses. Furthermore, the proliferation of antibiotics in the environment contributes to the increased resistance of pathogens.

Earthworms are commonly considered soil engineers (Jones et al., 1994). The contamination of soil is shown schematically in Scheme 1. Earthworms and microorganisms constitute the majority of the living biomass in soils. Their significant influence on soil structure and fertility supports crucial soil functions essential for soil health and ecosystem services (Barrios, 2007; Costanza et al., 1997; Delgado-Baquerizo et al., 2020; Gavrilescu, 2005; Hanajík, 2014; Pulleman et al., 2012). Through continuous burrowing and feeding, they alter the physicochemical and biological characteristics of soils by creating galleries, thereby enhancing soil macro porosity. Additionally, their ongoing burrowing activity disperses microorganisms throughout the soil (Yang & Elsas, 2018). Earthworms are highly susceptible to chemical contamination, including the biodegradation of pesticides and other organic pollutants. They have a strong impact on the soil structure and fertility, which support essential soil functions that contribute to soil health and ecosystem services (Barrios, 2007; Costanza et al., 1997; Delgado-Baquerizo et al., 2020; Gavrilescu, 2005; Hanajík, 2014; Pulleman et al., 2012) Their continual burrowing and feeding activities modify the physicochemical and biological properties of soils where they bore galleries and thus increase soil macro porosity. They also disperse microorganisms in the bulk soil through their continual burrowing activity (Yang & Elsas, 2018). Earthworms are very sensitive to chemical contamination, such as the biodegradation of pesticides and other organic pollutants (Bart et al., 2019; Kersant_e et al.,., 2006; Sanchez-Hernandez et al., 2015). Due to their sensitivity to pollutants (Pelosi et al., 2013), soil-dwelling earthworms are a cause of environmental contaminants even though they stimulate soil microbial proliferation (Fründ et al., 2011; Rodríguez-Castellanos & Sanchez-Hernandez, 2007; Solé, 2020). Numerous studies focus on the absorption of pharmaceuticals by earthworms (Carter et al., 2014). From the ecotoxicological studies, earthworms in pharmaceutical dissipation in soil (Brown et al., 2000; Dempsey et al., 2013). In Table 2, crop protection fertilizers (insecticides, fungicides, and herbicides) are listed with residual chemicals that affect the soil, surface, and groundwater. The majority of crops are affected by arthropod pests, leading to a greater emphasis on the production of insecticides compared to herbicides and fungicides. Organophosphate pesticides are becoming more prevalent due to their effectiveness.

Scheme 1 A flow chart showing the effect of soil pollutants on earthworms

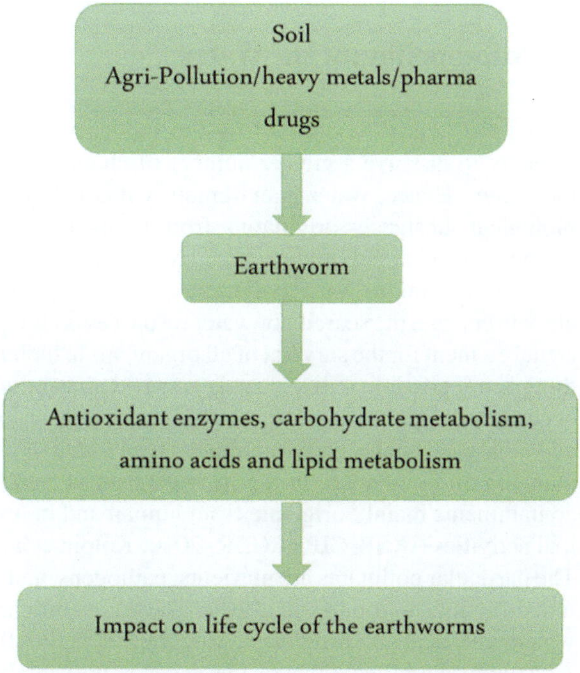

Table 2 Crop protection fertilizers produced along with their residual derivatives in water (Database & of Directorate of Plant Protection, Quarantine Storage, GOI., 2021; Knowledge & Strategic Partner Next Generation Indian Agriculture-Role of Crop Protection Solutions Indian Agrochemical Industry, 2016; State of Agriculture in India, 2017)

Chemical name	Type	Residues expected
Crop protection (insecticide, herbicide and fungicide)		
Cartap	Carbamate	4-(N,N-dimethylamino)-1,2 dithiolane
Acephate	Organophosphate	Methamidophos
Chlorpyriphos	Organophosphate	3,5,6-Trichloro-2-Pyridinol
Glyphosate	Organophosphate	Aminomethylphosphoric acid
Pretilachlor	Chloroacetamide	2-Chloro-N-(2,6-dimethylphenyl)-acetamide
Hexaconazole	–	1,2,4-triazole,1H-1,2,4-teiazol-1yl-aceticacid

Overall, a literature review can be found on the occurrence, environmental behavior, and fate of emerging pollutants (Deblonde et al., 2011; Lapworth & Goodly, 2006; Pal et al., 2010b; Stuart et al., 2012). Limited information is available regarding micropollution, primarily due to challenges posed by the physiochemical properties of target compounds. This deficiency in knowledge extends to environmental and food security concerns, while also encompassing issues such as soil degradation and the health burdens associated with exposure to contaminated soil.

3 Micropollutants in Water

Water, frequently recognized as the universal solvent, possesses the distinctive capacity to dissolve a greater number of chemicals than any other liquid found on the planet. Hence, water is extremely vulnerable to contamination. Chemical and biological substances originating from farms, urban areas, and chemical industries are being discharged into streams, rivers, lakes, oceans, aquifers, and other bodies of water, leading to water pollution. Against the backdrop of industrialization and climate change, the scarcity of water resources is emerging as a critical issue. Water is a vital element for the survival of all organisms in their lives, and without water, everything is not possible to live, just as we need air, sunlight, and food. Increasing water needs have made the water quality worse due to various pollutions. Many chemical and biological agents and contaminants are found in various environmental compartments in different areas during their persistence and long-distance transport. These contaminants mainly originate from human and industrial, agricultural, and municipal activities (IKSR-CIPR-ICBR, 2010; Kolpin et al., 2002; Sørensen et al., 2007). The particular pollutants are nutrients, pathogens, and chemical elements, which turn into different environmental effects. These contaminants change the quality of water, which creates health problems (Alaguraja et al., 2010; Kroeze et al., 2016).

Around the globe, 99% of the water is not suitable for use. Only 1% is usable for human beings. Figure 1 shows the flow of water in the pie chart. Figure 1 shows how important water is in life and different parts of the oceans, seas, ice, saline, and atmospheric water (U.S geological survey, 1998). Water pollutants can be described as the presence of other substances in the water until they cannot be used for a specific purpose. The quality of water is decreasing due to the change in human surroundings caused by several human factors. Biological pollution emerges during the production and distribution of drinking water, influenced by various factors such as shifts in human population and demographic trends. Globally, agriculture stands out as a primary contributor to water deterioration, particularly due to shifts toward intensive farming practices that result in leaching into surface and groundwater, leading to associated health concerns (Goel, 2009; Houtman, 2010). During rainfall, the infiltration of fertilizers, pesticides, and animal waste from farms into our rivers occurs, resulting in the transportation of nutrients and diseases, including bacteria and viruses. Nutrient pollutants, like excess nitrogen and phosphorus, can lead to low-oxygen conditions that harm aquatic life (Vadiraj et al., 2021). Additionally, pollutants have been detected in drinking water, sourced from wastewater originating from domestic, industrial, and pharmaceutical sources. Table 3 outlines the waste from drugs and the adverse effects on living organisms. It's not just water and soil that have gained attention; pollutants in the air are also of concern.

In recent years, there has been a growing focus on researching the removal of micropollutants, providing evidence that these pollutants can accumulate as chemical sinks in the environment. Soil and water pollution resulting from the accumulation of toxic chemicals or biological micropollutants could potentially be mitigated or eliminated through the implementation of various technologies. Furthermore, scientists have devised technologies to control microplastic pollution, garnering significant attention. Certain microorganisms in the environment have demonstrated the ability

Table 3 Pharmaceutical drugs and adverse effects (https://www.business-standard.com/article/companies/hereare-top-20-generic-drugs-in-india-anti-diabetic-medicines-toplist-117051300977_1.html)

Chemical/ Brand name	Type of drugs	Residues expected	Adverse effects (reference)
Mixtard	Anti-diabetic	Soluble insulin	–
Spasmo proxyvon plus	Anticholinergic and analgesic	Dicyclomine, paracetamol, tramadol	
Galvus	Anti-diabetic	Metformin-vildagliptin	
Augmentin	Anti-biotics	Amoxicillin, Clavulanate potassium	Resistance among bacterial pathogens, reduction of efficiency of treatment plants by reducing the microbial number (Daughton & Ternes, 1999; W W, 1998)
Clavam	Anti-biotics	Amoxicillin, Clavulanic acid	
Monocef	Anti-biotics	Ceftriaxone	
Voveran	Anti-biotics	Analgesic and anti-inflammatory	
Novomix	Anti-biotics	Protamine crystallized Insulin	
Betadine	Antiseptic	Povidone-iodine	–
Volini	Analgesic and anti-inflammatory	Diclofenac diethyl ammonium topical, menthol topical methyl salicylate topical, oleum lini topical	Renal lesions and gill alterations of rainbow trout (Schwaiger et al., 2004)
Zephiran	Disinfectant	Benzalkonium chloride	Increase in antibiotic resistance (Tandukar et al., 2013)
Aquasept, Triclotrex-B	Disinfectant/ antiseptic	Triclosan	Resuction of the number of microbial in treatment plants by acting as biocide agents (McMurry et al., 1998; Okumura & Nishikawa, 1996)

to decompose microplastics (Ball, 2017; Bejgarn et al., 2015; Boyle et al., 2020; Browne et al., 2013; Cadogan et al., 1993).

4 Conclusions

Pollution is a global issue due to the increasing ingress of pollutants into the environment every year. This involves organic compounds originating from various sources including industrial, agricultural, and household activities, such as plasticizers, surfactants, personal care items, and pharmaceuticals. These contaminants have attracted the attention of environmental agencies due to their extensive presence, long-lasting nature, harmful effects, and tendency to accumulate in organisms. Major sources of environmental pollution include the use of sewage and sewage sludge for soil fertilization, as well as the discharge of leachate from municipal landfills. These pollutants pollute drinking water sources, rivers, lakes, and soils, impacting soil-dwelling organisms and the plants they support. Moreover, they infiltrate into the soil and groundwater.

These pollutants need disposal with minimum deterioration. From a literature review, it has been proven that my treatment of wastewater or contaminated water is not effective in contradiction of pollutants. However, it does not completely eradicate them. Fortunately, there are some straightforward methods to prevent water contamination or, at the very least, to significantly reduce it. These include: (a) Need to reconsider methods of wastewater disposal and properly dispose of chemical cleaners such as oils and nonbiodegradable items. (b) reduce or maintain your car very often, so control leak oil and antifreeze. (c) Implementing soil-aquifer treatment systems. (d) Promoting the reuse of water for consumption and wastewater for irrigation. (e) Encouraging water recharge activities indirectly through these practices. It remains crucial to enhance research and development investment to create products with improved performance and broader applications. These innovations can effectively reduce water contamination and enhance the capacity for removing hazardous contaminants.

References

Alaguraja, P., Yuvaraj, D., Sekar, M., Muthuveerran, P., & Manivel, M. (2010). Remote sensing and GIS approach for the water pollution and management in Tiruchirappli Taluk, Tamil Nadu, India. *International Journal of Environmental Science and Technology, 1*(1), 66–70.

Alliot, F., Moreau-Guigon, E., Bourges, C., Desportes, A., Teil, M.-J., Blanchard, M., & Chevreuil, M. (2014). A multi-residue method for characterization of endocrine disruptors in gaseous and particulate phases of ambient air. *Atmospheric Environment, 92*, 1–8.

Amy, P., Pei, R., Heather, S., & Carlson, K. H. (2006). Antibiotic resistance genes as emerging contaminants: Studies in Northern Colorado†. *Environmental Science and Technology, 40*(23), 7445–7450.

Ball, P. (2017). Material witness: Plastics on the menu. *Nature Materials, 16*, 606.

Barrios, E. (2007). Soil biota, ecosystem services and land productivity. *Ecological Economics, 64*(2), 269–285.

Bart, S., Pelosi, C., Barraud, A., Péry, A. R. R., Cheviron, N., Grondin, V., Mougin, C., & Crouzet, O. (2019). Earthworms mitigate pesticide effects on soil microbial activities. *Frontiers in Microbiology, 10*, 1–11.

Bejgarn, S., Macleod, M., Bogdal, C., & Breitholtz, M. (2015). Toxicity of leachate from weathering plastics: An exploratory screening study with Nitocra spinipes. *Chemosphere, 132*, 114–119.

Bowman, J. J., & McGettigan, M. (1994). Atmospheric deposition in acid sensitive areas of Ireland—The influence of wind direction and a new coal burning electricity generation station on precipitation quality. *Water, Air and Soil Pollution, 75*, 159–175.

Boyle, D., Catarino, A. I., Clark, N. J., & Henry, T. B. (2020). Polyvinyl chloride (PVC) plastic fragments release Pb additives that are bioavailable in zebrafish. *Environmental Pollution, 263*, 114422.

Brown, G. G., Barois, I., & Lavelle, P. (2000). Regulation of soil organic matter dynamics and microbial activity in the drilosphere and the role of interactions with other edaphic functional domains. *European Journal of Soil Biology, 36*(3–4), 177–198.

Browne, M. A., Niven, S. J., Galloway, T. S., Rowland, S. J., & Thompson, R. C. (2013). Microplastic moves pollutants and additives to worms, reducing functions linked to health and biodiversity. *Current Biology, 23*, 2388–2392.

Bucheli, T. D., Blum, F., Desaules, A., & Gustafsson, O. (2004). Polycyclic aromatic hydrocarbons, black carbon, and molecular markers in soils of Switzerland. *Chemosphere, 56*, 1061–1076.

Byrne, E. (1979). An foras taluntais, Dublin. *Chemical Analysis of Agricultural Material*s, 194

Cadogan, D. F., Papez, M., Poppe, A. C., & Scheubel, J. (1993). An assessment of the release; occurrence and possible effect of plasticizers in the environment. *Progress in Rubber and Plastic Technology, 10*(1), 1–19.

Caliman, F. A., & Gavrilescu, M. (2009). Pharmaceuticals, personal care products and endo-crine disrupting agents in the environment—A review. *Clean, 37*(4–5), 277–303.

Camacho, M., Herrera, A., Gomez, M., Acosta-Dacal, A., Martinez, I., Henriquez- Hernandez, L. A., & Luzardo, O. P. (2019). Organic pollutants in marine plastic debris from Canary Islands beaches. *Science of the Total Environment, 662*, 22–31.

Carter, L. J., Garman, C. D., Ryan, J., Dowle, A., Bergstrom, E., Tomas-Oates, J., & Boxall, A. B. A. (2014). Fate and uptake of pharmaceuticals in soil-earthworm systems. *Environmental Science and Technology, 48*, 5955–5963.

Chiou, C. T., Malcolm, R. L., Brinton T. I., & Kile D. E. (1986). Water solubility enhancement of some organic pollutants and herbicides. Dissolved humic and fulvic acids. *Environmental Science & Technology, 20*(5), 502–508.

Cladière, M., Gaspéri, J., Lorgeoux, C., Bonhomme, C., Rocher, V., & Tassin, B. (2013). Alkylphenolic compounds and bisphenol a contamination within a heavily urbanized area: Case study of Paris. *Environmental Science and Pollution Research, 20*, 2973–2983.

Cook, S. M., VanDuinen, B. J., Love, N. G., & Skerlos, S. J. (2012). Life cycle comparison of environmental emissions from three disposal options for unused pharmaceuticals. *Environmental Science and Technology, 46*, 5535–5541.

Costanza, R., D'Arge, R., de Groot, R., Farber, S., Grasso, M., Hannon, B., Limburg, K., Naeem, S., O'Neill, R. V., Paruelo, J., Raskin, R. G., Sutton, P., & van den Belt, M. (1997). The value of the world's ecosystem services and natural capital. *Nature, 387*, 253–260.

Statistical Database of Directorate of Plant Protection, Quarantine & Storage, GOI. Retrieved July 19, 2021, from http://ppqs.gov.in/statistical-database.

Daughton, C. G., & Ternes, T. A. (1999). Pharmaceuticals and personal care products in the environment: Agents of subtle change? *Environmental Health Perspectives, 107*(Suppl 6), 907.

Deblonde, T., Cossu-Leguille, C., & Hartemann, P. (2011). Emerging pollutants in waste-water: A review of the literature. *International Journal of Hygiene and Environ-Mental Health, 214*, 442–448.

Delgado-Baquerizo, M., Reich, P. B., Trivedi, C., Eldridge, D. J., Abades, S., Alfaro, F. D., Bastida, F., Berhe, A. A., Cutler, N. A., Gallardo, A., García-velázquez, L., Hart, S. C., Hayes, P. E., He, J., Hseu, Z., Hu, H., Kirchmair, M., Neuhauser, S., Perez, C. A., … Singh, B. K. (2020). Multiple elements of soil biodiversity drive ecosystem functions across biomes. *Nature Ecology & Evolution, 4*(2), 210–220.

Dempsey, M. A., Fisk, M. C., Yavitt, J. B., Fahey, T. J., & Balser, T. C. (2013). Exotic earthworms alter soil microbial community composition and function. *Soil Biology & Biochemistry, 67*, 263–270.

Desaules, A., Ammann, S., & Schwab, P. (2010). Advances in long-term soil-pollution monitoring of Switzerland. *Journal of Plant Nutrition and Soil Science, 173*, 525–535.

Diamond, J. M., Latimer, H. A., Mukittick, K. R., Thornton, K. W., Bartell, S. M., & Kidd, K. A. (2011). Prioritizing contaminants of emerging concern for ecological screening assessment Environ. *Toxicol. Chem, 30*(11), 2385–2394.

EUROSTAT. http://epp.eurostat.ec.europa.eu/tgm/table.do?tab=table&init=1&plugin=1&language=en&pcode=ten00011.

Freed, V. H., Chiou, C. T., & Haque, R. (1977). Chemodynamics: Transport and behavior of chemicals in the environment. A problem in environmental health. *Environmental Health Perspectives, 20*, 55–70.

Fründ, H.-C., Graefe, U., & Tischer, S. (2011). *Earthworms as Bioindicators of Soil Quality* (pp. 261–278). Springer.

Gavrilescu, M. (2005). Fate of pesticides in the environment and its bioremediation. *Engineering in Life Sciences, 5*, 497–526.

Gavrilescu, M. (2009). Behaviour of persistent pollutants and risks associated with their presence in the environment–integrated studies. *Environmental Engineering and Management Journal, 8*, 1517–1531.

Gavrilescu, M. (2010). Achievements, opportunities and challenges. *Dynamic Biochemistry Process Biotechnology and Molecular Biology. Environmental Biotechnology, 4*(1):1–36.

Goel, P. K. (2009). *Water Pollution: Causes, Effects and Control.* New Age International.

Hanajík, P. (2014). Role of soil and soil microorganisms in evaluation process of ecosystem services: a review. *Phytopedon, 13*, 9–13.

Here are top 20 generic drugs in India, anti-diabetic medicines top list|Business Standard News. Retrieved July 20, 2021, from https://www.business-standard.com/article/companies/hereare-top-20-generic-drugs-in-india-anti-diabetic-medicines-toplist-117051300977_1.html.

Houtman, C. J. (2010). Emerging contaminants in surface waters and their relevance for the production of drinking water in Europe. *Journal of Integrative Environmental Sciences, 7*, 271–295.

Howard, P. H. (1989). Handbook of Environmental Fate and Exposure Data for Organic Chemicals (Vol. 1, p. 563). Lewis Publishers.

IKSR-CIPR-ICBR. (2010). *Strategy for micro-pollutants—Strategy for municipal and industrial wastewater*. Koblenz, Germany: International Commission for the Protection of the Rhine (ICPR).

Jones, C. G., Lawton, J. H., & Shachak, M. (1994). Organisms as ecosystem engineers. *Oikos, 69*(3), 373.

Kenaga, E. E. (1975). Partitioning and uptake of pesticides in biological systems. In: V.H. Freed (Eds.), Environment Dynamics of Pesticides, Haquer (p. 261). Plenum Press.

Kersanté, A., Martin-Laurent, F., Soulas, G., & Binet, F. (2006). Interactions of earthworms with Atrazine-degrading bacteria in an agricultural soil. *FEMS Microbiology Ecology, 57*, 192–205.

Knowledge and Strategic Partner Next Generation Indian Agriculture-Role of Crop Protection Solutions Indian Agrochemical Industry. Published online 2016.

Kolpin, D. W., Furlong, E. T., Meyer, M. T., Thurman, E. M., Zaugg, S. D., & Barber, L. B. (2002). Pharmaceuticals, hormones, and other organic wastewater contaminants in U.S. streams, 1999–2000: A national reconnaissance. *Environmental Science and Technology, 36*, 1202–1211.

Kroeze, C., Gabbert, S., Hofstra, N., Koelmans, A. A., Li, A., Löhr, A., Ludwig, F., Strokal, M., Verbur, C., Vermeulen, L., van Vliet, M. T. H., de Vries, W., Wang, M., & van Wijnen, J. (2016). Global modelling of surface water quality: a multi-pollutant approach, The Netherlands and faculty of management, science and technology, open university, The Netherlands, current opinion in enviormnet sustainability.

Lapworth, D. J., & Goodly, D. C. (2006). Source and persistence of pesticides in a semi-con-fined chalk aquifer of southeast England. *Environmental Pollution, 144*, 1031–1044.

Le Pape, P., Quantin, C., Morin, G., Jouvin, D., Kieffer, I., Proux, O., Ghanbaja, J., & Ayrault, S. (2014). Zinc speciation in the suspended particulate matter of an Urban River (Orge, France): Influence of seasonality and urbanization gradient. *Environmental Science and Technology, 48*, 11901–11909.

Ling, D., Daofen, H., Zhuozhi, O., & Xuwtao, G. (2022). The effect of microplastics on soil ecosystem a review. *Current Opinion in Environmental Science and Health, 26*, 100344.

Maria, G., Katerina, D., Jens, A., Spiros, A., & Fabio, F. (2015). Emerging pollutants in the environment: Present and future challenges in biomonitoring, ecological risks and bioremediation. *New Biotechnology, 32*(1), 147–156.

McMurry, L. M., Oethinger, M., & Levy, S. B. (1998). Triclosan targets lipid synthesis. *Nature, 394*(6693), 531–532.

Meijer, S. N., Ockenden, W. A., Steinnes, E., Corrigan, B. P., & Jones, K. C. (2003a). Spatial and temporal trends of POPs in Norwegian and UK background air: Implications for global cycling. *Environmental Science and Technology, 37*, 454–461.

Meijer, S. N., Ockenden, W. A., Sweetman, A., Breivik, K., Grimalt, J. O., & Jones, K. C. (2003b). Global distribution and budget of PCBs and HCB in background surface soils: Implications or sources and environmental processes. *Environmental Science and Technology, 37*, 667–672.

Naveen, B., & Tsai, T.-Y. (2020). *Advances and Applications in Environmentally Friendly Carbon Nanomaterials, Emerging Carbon-Based Nanocomposites for Environmental Applications. Green Nanocomposites* (pp. 55–70). John Wiley & Sons, Inc.

Naveen, B., Tsai, T.-Y., & Hwang, S.S. (2023) Nano-engineered Materials for textile waste remediation. In *Environmental Hazards on Textile Waste* (pp. 153–161). Springer Nature.

Okumura, T., & Nishikawa, Y. (1996). Gas chromatography—mass spectrometry determination of triclosans in water, sediment and fish samples via methylation with diazomethane. *Analytica Chimica Acta, 325*(3), 175–184.

Pal, A., Gin, K.Y.-H., Lin, A.Y.-G., & Reinhard, M. (2010a). Impacts of emerging organic contaminants in freshwater resources: Review of recent occurrences, sources, fate, and effects. *Science of the Total Environment, 408*, 6062–6069.

Pal, A., Gin, K.Y.-H., Lin, A.Y.-C., & Reinhard, M. (2010b). Impacts of emerging organic contaminants on freshwater resources: Review of recent occurrences, sources, fate and effects. *Science of the Total Environment, 408*, 6062–6069.

Pelosi, C., Joimel, S., & Makowski, D. (2013). Searching for a more sensitive earthworm species to be used in pesticide homologation tests—A meta-analysis. *Chemosphere, 90*(3), 895–900.

Preda, C., Ungureanu, M. C., & Vulpoi, C. (2012). Endocrine disruptors in the environment and their impact on human health. *Environmental Engineering and Management Journal, 11*, 1697–1706.

Pulleman, M., Creamer, R., Hamer, U., Helder, J., Pelosi, C., Pérès, G., & Rutgers, M. (2012). Soil biodiversity, biological indicators and soil ecosystem services-an overview of European approaches. *Current Opinion in Environment Sustainability, 4*(5), 529–538.

Rodríguez-Castellanos, L., & Sanchez-Hernandez, J. C. (2007). Earthworm biomarkers of pesticide contamination. Current status and perspectives. *Journal of Pesticide Science, 32*, 360–371.

Rosi-Marshall, E. J., Kincaid, D. W., Bechtold, H. A., Royer, T. V., Rojas, M., & Kelly, J. J. (2013). Pharmaceuticals suppress algal growth and microbial respiration and alter bacterial communities in stream biofilms. *Ecological Applications, 23*, 583–593.

Sanchez-Hernandez, J. C., Notario del Pino, J., & Domínguez, J. (2015). Earth worm induced carboxylesterase activity in soil, Assessing the potential for detoxification and monitoring organophosphorus pesticides. *Ecotoxicology and Environmental Safety, 122*, 303–312.

Schwaiger, J., Ferling, H., Mallow, U., Wintermayr, H., & Negele, R. D. (2004). Toxic effects of the non-steroidal anti- inflammatory drug diclofenac. Part I: Histopathological alterations and bioaccumulation in rainbow trout. *Aquatic Toxicology, 68(2), 141–150.*

Shi, M., Sun, Y. Y., Wang, Z. H., He, G., & Hanxiang, Q. (2019) Plastic film mulching increased the accumulation and human health risks of phthalate esters in in wheat grains. *Environmental Pollution, 250*, 1–7.

Solé, M. (2020). Biomarkers in earthworms. In *Interaction and Fate of Pharmaceuticals in Soil-Crop Systems. The Handbook of Environmental Chemistry* (Vol. 103, pp. 1–27). Springer.

Sørensen, S. R., Holtze, M. S., Simonsen, A., & Aamand, J. (2007). Degradation and mineralization of nanomolar concentrations of the herbicide dichlobenil and its persistent metabolite 2,6-dichlorobenzamide by Aminobacter spp. isolated from Dichlobe-nil-treated soils. *Applied Environmental Microbiology, 73*, 399–406.

State of Agriculture in India. Published online 2017.

Stuart, M., Lapworth, D., Crane, E., & Hart, A. (2012). Review of risk from potential emerging contaminants in UK groundwaters. *Science of the Total Environment, 446*, 1–21.

Tandukar, M., Oh, S., Tezel, U., Konstantinidis, K. T., & Pavlostathis, S. G. (2013). Long-term exposure to benzalkonium chloride disinfectants results in change of microbial community structure and increased antimicrobial resistance. *Environmental Science and Technology, 47*(17), 9730–9738.

Thévenot, D. R., Moilleron, R., Lestel, L., Gromaire, M.-C., Rocher, V., Cambier, P., Bonté, P., Colin, J.-L., de Ponteves, C., & Meybeck, M. (2007). Critical budget of metal sources and pathways in the seine river basin (1994–2003) for Cd, Cr, Cu, Hg, Ni, Pb and Zn. *Science of the Total Environment, Human Activity and Material Fluxes in a Regional River Basin: THe Seine River Watershed—Seine Special Issue, 375*, 180–203.

Thomaidis, N. S., Asimakopoulos, A. G., & Bletsou, A. A. (2012). A tutorial mini-review. *Emerging Contaminants Global NEST Journal, 14*(1), 72–79.

U.S geological survey 1998. Retrieved October 01, 2017, from https://www.usgs.gov/media/files/survey-methods-1998-pdf. U.S department of the interior.

Vadiraj K. T., Achar, R. R., & Sirigeri, S. (2021). A review on emerging micropollutants. *Environmental Concentration and Toxicity, 6*(4). http://www.revistabionatura.com

W W. (1998). Medical consequence.

Weber, W. J., Mcgley P., & Katz L. E. (1991). Sorption phenomena in subsurface systems concepts, models and effects on contaminant fate and transport. *Water Research, 25*(5), 499–528.

WWF, TOXIC CHEMICALS. http://wwf.panda.org/about_our_earth/teacher_resources/webfieldtrips/toxics/.

Yamamura, S., Watanabe, K., Suda, W., Tsuboi, S., & Watanabe, M. (2014). Effect of antibiotics on redox transformation of arsenic and diversity of arsenite-oxidizing bacteria in sediment microbial. *Environmental Science and Technology, 48*, 350–357.

Yang, P., & Van Elsas, J. D. (2018). Mechanisms and ecological implications of the movement of bacteria in soil. *Applied Soil Ecology, 129*, 112–120.

Yingfei, S., Xueying, Y., Ying, Z., & Li, R. (2021). Microplastics (MPs) act as sources and vector of pollutants-impact hazards and preventive measures. *Bulletin of Environmental Contamination and Toxicology, 107*, 722–729.

Zavarzin, G. A., & Kolotilova, N. N. (2001). *Introduction to environmental microbiology* (p. 256) (Universitet, Moscow, [in Russian]).

Chapter 5
Microplastics Interactions with PFAS and Co-transport in the Soil

Alaa El Din Mahmoud, Dina M. El-Sherif, and Ghada Abdallah

Abstract Microplastics and Per- and Polyfluoroalkyl Substances (PFAS) are two emerging environmental pollutants that have gained significant considerable interest due to their long-term existence, mobility, and possible negative effects on human health safety and the environment. Both microplastics and PFAS have been detected in different ecosystems that include soil, and their potential interactions and co-transport in soil can have significant implications for their fate and behavior. This chapter presents an overview of the interactions between microplastics and PFAS in soil, including their co-transport, sorption, and desorption. The chapter highlights the need for an improved understanding of the interactions between microplastics and PFAS in soil and developing successful and effective mitigation methods and techniques to minimize their adverse effects on environment. The results highlight the critical need for a deeper comprehension of the interactions between microplastics and PFAS, especially in soil environments. To minimize their negative environmental effects, appropriate mitigation techniques must be developed, which requires a greater understanding. In addition to highlighting the present level of study on MPs

A. E. D. Mahmoud (✉)
Environmental Sciences Department, Faculty of Science, Alexandria University, Alexandria 21511, Egypt
e-mail: alaa-mahmoud@alexu.edu.eg

Green Technology Group, Faculty of Science, Alexandria University, Alexandria 21511, Egypt

Civil and Environmental Engineering Department, College of Engineering, Temple University, Philadelphia, PA 19122, USA

Water and Environmental Technology Center (WET), Temple University, Philadelphia, PA 19122, USA

D. M. El-Sherif
Collaborative Innovation Center of Food Safety and Quality Control in Jiangsu Province, School of Food Science and Technology, Jiangnan University, Wuxi 214122, China

National Institute of Oceanography and Fisheries (NIOF), Cairo, Egypt

G. Abdallah
Environmental Sciences Department, Faculty of Postgraduate Studies for Advanced Sciences, Beni-Suef, Egypt

A. K. Mishra et al. (eds.), *Microplastics*, https://doi.org/10.1007/978-981-97-6461-7_5

and PFAS in soil, this synopsis of the literature also stresses the necessity of taking preventative action in order to mitigate potential environmental dangers.

Keywords Per- and polyfluoroalkyl substances · Plastics · Soils · Food chain · Climate change · Mitigation

1 Introduction

Microplastics (MPs) are tiny particles of plastic smaller than 5 mm (Isangedighi et al., 2020; Sridharan et al., 2021; El-Sherif et al., 2022). They might originate from a wide range of sources, including industrial operations, the decomposition of bigger plastic pieces, microbeads in cosmetic products, synthetic fibers from textiles, and textile degradation (Khan et al., 2022). Primary MPs are those that are directly released into the environment, while secondary MPs are produced when it is bigger pieces of plastic break down (El-Sherif et al., 2022).

MPs are commonly found in oceans, rivers, and other bodies of water (Wu et al., 2020), additionally found in soil (Sajjad et al., 2022). MPs in soil can come from a variety of sources (Fig. 1), including plastic mulch used in agriculture, sewage sludge used as fertilizer, compost, street runoff, and microplastic pollution in the air (Sajjad et al., 2022; Zhang et al., 2020). Deep and regional variations in the distribution of microplastics in soil are substantial. Additionally, microplastics were found in soil samples from the majority of countries and remote regions. Microplastics exhibit complex migratory behavior in soil environments (Zhao et al., 2022). The source of soil microplastics (Fig. 1) (Yang et al., 2021) was correlated with their properties (polymer type, size, and shape). Soil microplastics commonly contain polymers polyethylene (PE), poly-propylene (PP), and poly-styrene (PS). Most frequent MP shapes observed in various areas include fibers, films, pieces or sheets, pellets or spheres, and foams.

MPs have the potential to harm soil health by modifying soil structure, decreasing water retention, and influencing nutrient availability. They can also disrupt soil microbial populations, which are necessary for optimal soil function (Sajjad et al., 2022; Wu et al., 2020). MPs can additionally have an negative impact on plant health by slowing growth and interfering with nutrient intake. MPs have been reported to accumulate in plant tissues in several studies, possibly impacting the safety and quality of food crops (Roy et al., 2023). MPs in the soil's environment can potentially contaminate the ecosystem. They can transfer pollutants to adjacent ecosystems and perhaps damage wildlife since they can be transported by the wind or the water (Shafea et al., 2023).

Plastics have accumulated in several environmental compartments as a result of their extensive usage and inappropriate disposal (Evode et al., 2021). MPs were discovered in air, soil, rivers, lakes, seas, and even far-off places like Arctic ice (Hale et al., 2020). Due to their small dimension, they may be readily carried across great distances by wind and water currents (Kye et al., 2023).

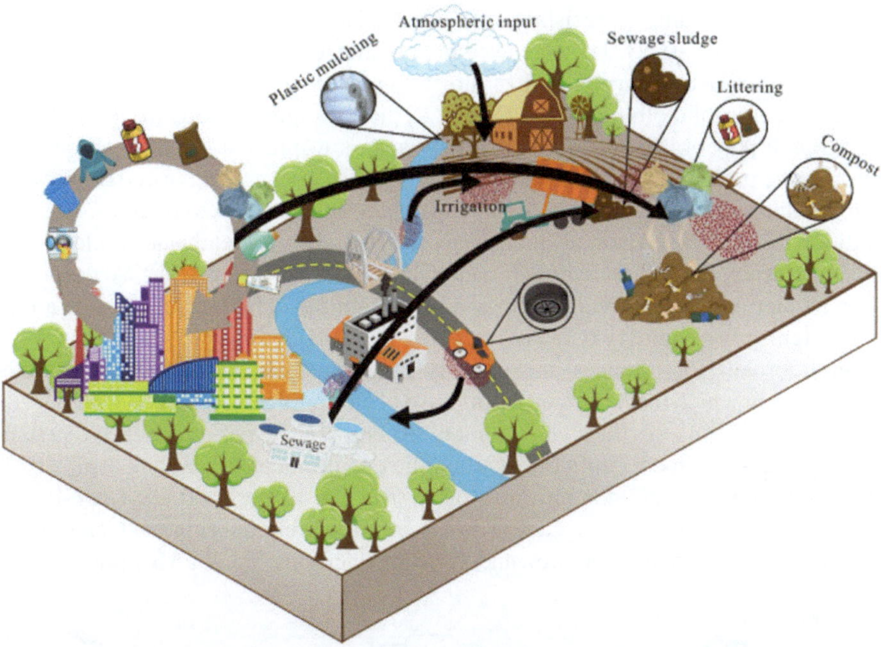

Fig. 1 The most common sources of soil's microplastics

MPs' impacts on ecosystems and human wellness are currently being studied. Marine species can consume MPs directly or indirectly by eating polluted and infected prey. This consumption can cause physical injury, such as obstruction or damage to the digestive system. Furthermore, MPs have the ability to absorb or interact or affect hazardous compounds found in the environment, such as pesticides (Mahmoud et al., 2022a; Chaukura et al., 2022), heavy metals (Badr et al., 2020; Mahmoud et al., 2021, 2022b), and organics (Mahmoud, 2020; Mahmoud et al., 2020, 2022c, d). When creatures consume these substances, they can go up the food chain, potentially affecting human health (Bhuyan, 2022; Chaukura et al., 2022).

In the last few years, there has been a rising in worry about this prevalence and impact of per- and polyfluoroalkyl substances (PFAS) and microplastics in our environment (Navarathna et al., 2023). Because of their vast dispersion, durability, and possible detrimental impacts on human health and environment, all of those types of pollutants have attracted a lot of attention (Meng et al., 2023; Navarathna et al., 2023).

PFAS are a group of man-made chemicals that have been widely used in various industrial and consumer products (Fig. 2) since the 1950s (Dauchy et al., 2017; Sunderland et al., 2018; Fair et al., 2019; Bonato et al., 2020; Meegoda et al., 2020; Roth et al., 2020; Seshasayee et al., 2021; Vorst et al., 2021; Mahmoudnia et al., 2022). PFAS are a group of fluorine-containing chemical compounds with outstanding qualities that are mostly used in the manufacture of stain-resistant materials such as being water-resistant, oil-resistant, and heat-resistant (Glüge et al., 2020a, 2020b; Baker & Knappe, 2022). They are used in non-stick cookware such as Teflon (Teng, 2012), stain-resistant carpets and furniture to repel water and stains (Young et al., 2022), waterproof clothing (Xia et al., 2022), the packaging of food like pizza boxes, microwave popcorn bags, and fast-food wrappers to prevent grease from soaking through the packaging (Curtzwiler et al., 2021). PFAS are present in certain personal care products, including shampoo, shaving cream, and dental floss (Gaines, 2022). In addition PFAS are used in many industrial processes such as metal plating, electronics manufacturing, and oil recovery (Juliane Glüge et al., 2020a, 2020b). According to Clean Water Action, PFAS are often used in firefighting foams because of their ability to extinguish fires quickly (Clean Water Action, 2023).

Perfluoroalkyl substances are typically fluorinated aliphatic carbon-containing substances in which the hydrogen atoms have been completely replaced with fluorine atoms, resulting in a perfluoroalkyl moiety structure (C_nF_2n+1-) attached to a functional group head, most commonly carboxylic or sulfonic acids (Zhang et al., 2022). Polyfluoroalkyl compounds have the same structure as perfluoroalkyl substances, with the distinction that some carbon atoms aren't entirely fluorinated (Gaines et al., 2023). Based on their structures and functional groups, the PFAS family can be

Fig. 2 Sources of PFAS

Fig. 3 PFAS common types classification

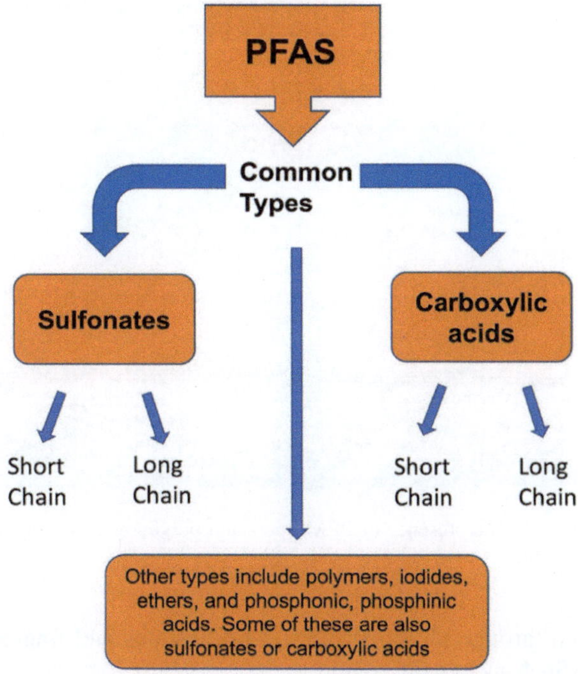

divided into several subclasses (Fig. 3) (Baldwin et al., 2023), including perfluorinated carboxylic acids, perfluorinated sulfonic acids, perfluorinated phosphonic acids, and perfluoroalkyl phosphinic acids (Wang et al., 2017a; Zhang et al., 2022).

Because of their strong chemical linkages (C-F), PFAS are resistant to environmental degradation (Mahinroosta & Senevirathna, 2020). Numerous PFAS are anticipated to endure in the environment as a result (Fig. 4). Landfills can be a source of PFAS contamination if products containing these chemicals are disposed of improperly (Hepburn et al., 2019). PFAS can enter drinking water sources through contaminated groundwater or surface water runoff from areas where these chemicals have been used or disposed of improperly (Bonato et al., 2020; Dauchy et al., 2017). Biosolids, wastewater effluent, landfill leachate, and aqueous film-forming firefighting foam are the main sources of PFAS entry into soil (Bolan et al., 2021).

These organic chemicals are extremely persistent in the environment and unable to simply break down, which has led to their extensive contamination of soil, water, and air (Bayabil et al., 2022; Mahmoud, 2022; Mahmoud & Fawzy, 2023). Their exceptional chemical and thermal stability are a result of their strong and robust CF bonds (Wen et al., 2020; Xu et al., 2021). Additionally, several biological and physicochemical processes can transport per- and polyfluoroalkyl compounds across the world, even into far-off and pristine environment (Zhao et al., 2020). Under typical conditions, PFAS do not break down in soil, however it is thought that they may

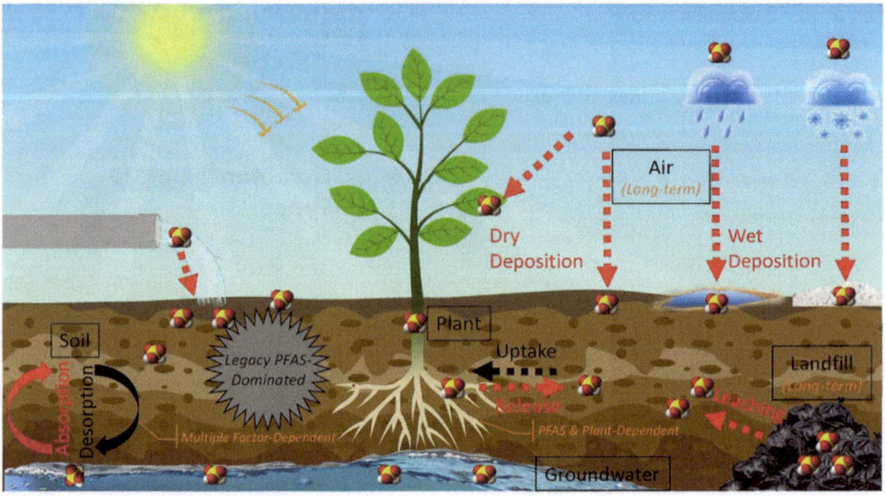

Fig. 4 The presence of PFAS (per- and polyfluoroalkyl compounds) in soil (Wang et al., 2023a, 2023b)

go through physiochemical modifications and fragment into simpler alkyl groups (Shahsavari et al., 2021).

PFAS have a variety of negative effects, possible hazards, and possible long-term impacts including health concerns (Fenton et al., 2021; Sunderland et al., 2019), and environmental damage because of their persistence and bioaccumulation (Cousins et al., 2020). The human body's main source of PFAS exposure is contaminated food and water. Unsurprisingly, these fluorinated chemical compounds have been found in meat, fish tissues, plant grains, and dairy products (Stahl et al., 2011).

According to (Brusseau et al., 2020), PFAS may swiftly migrate from polluted soil to plants. In a different investigation, Stahl and his research team found that plants might store PFAS chemicals in large quantities, transferring them into the human food chain (Stahl et al., 2009).

This chapter is focused on providing an overview of the interactions between microplastics and PFAS in soil and the development of appropriate mitigation methods to reduce their adverse effects on the environment. It will discuss the co-transport, sorption, and desorption of microplastics and PFAS in soil, as well as the potential implications of these interactions for the fate and behavior of these contaminants. The chapter will conclude by summarizing the key findings and highlighting the need for further research.

2 Pathways of Co-transport of Microplastics in Soil

There is a shortage of information and understanding regarding MP distribution and movement in soil. According to the results of several investigations, MP concentration varied with soil depth, suggesting that the MPs migrated down by unidentified means. It is difficult to predict how MPs would migrate in soil, and a variety of factors could impact both their vertical and horizontal distributions (Zhou et al., 2020).

Both topsoil and deep soil have been shown to contain microplastics, according to recent studies. These findings suggest that soil microplastics may be able to migrate vertically. It is possible for dissolved chemicals or small particles to move through soil because it is a permeable medium with macro- and mesopores in the m range. Numerous investigations have shown that the leaching process can transport tiny particles along soil pores (Li et al., 2020).

Microplastics in soil may age as a result of various chemical, physical and biological variables, the most prominent of which being ultraviolet radiation in topsoil (Ren et al., 2021a). Also, the full impact of geographical, meteorological, and other factors on the distribution of MPs is taken into consideration (Feng et al., 2023). With soil features, pore water conditions, soil biota, flora, and human activities, as well as polymer particle characteristics (Gao et al., 2021).

Microplastics (MPs) provide a covert threat to human health by entering ecosystems through river drainages, sewage outfalls, and runoff after a lot of rain, furthermore, through air input and ocean currents (Hao et al., 2023). Recent media attention has focused on the possible harm and toxic effects of MPs to ecosystems, and it is predicted that by 2050, 12 billion metric tonnes of waste including plastic will be dumped in landfills or other areas (Xu et al., 2022b). Microplastics can enter soil through a variety of means, including landfills, agricultural mulching films, sewage irrigation, flooding, bioturbation, and diffuse air deposition. The agroecosystem is a primary entry point for microplastics into soils due to inputs of plastic mulches and the application of sewage sludge (Ren et al., 2021b).

There are numerous ways that microplastics might permeate the terrestrial ecosystem, including landfills, air fallout (0.343 million tonnes globally each year), and various approaches to agriculture, such as the application of plastic mulch. Microplastics enter the terrestrial environment mostly through agroecosystems, especially when applying wastewater treatment sludge (biosolids) to the soil (Baho et al., 2021).

MPs' tiny particle size and high specific area of surface make them ideal carriers for adsorption and coexistence with developing micropollutants like PFAS (Mejías et al., 2023). The passage of PFAS combined with the MPs carrier may increase the risk of their aggregation in animal's body organs and tissues and the food chain (Sobhani et al., 2021). And by reviewing the results of some previous studies, including the study of Scott et al. (2021). The presence of inorganic and/or organic matter linked with microplastics in the environment significantly increases PFAS adsorption, which may pose an environmental danger to biota in soil or aquatic environments.

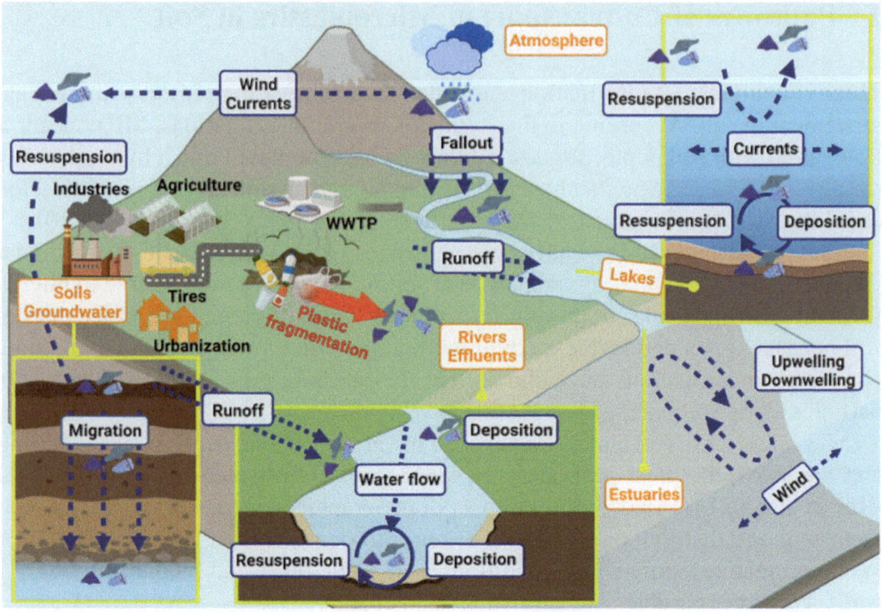

Fig. 5 The scheme depicts the continental environmental channels connecting the continent's many microplastics reservoirs (Alfonso et al., 2021)

In Fig. 5, diagram showing the routes by which microplastics travel through an environment on a continent. An outline of a system or process is depicted graphically in a diagram, chart, or other representation when the term "scheme" is used in this context. In this instance, the scheme is utilized to represent the environmental paths or channels that microplastics use to migrate. The fragmentation of plastic debris, microbeads in personal care items, and textile fibers are a few of the sources from which these microplastics enter the environment.

About PFAS we can say that like many other toxins pollutants, it can enter the environment directly or indirectly. Direct sources include PFAS contamination is released from a variety of organizations, including sludge disposal sites, wastewater treatment plants, and landfills (Abunada et al., 2020). Because of its special physical and chemical characteristics, PFAS have multiple avenues for long-term stability in soils, which is a drawback (Jha et al., 2021a, 2021b, 2021c).

PFAS chemicals' bioaccumulation in plants and animals occurs because they are soluble in soil which can then be transmitted to surface water or leak into groundwater. Plant roots can absorb PFAS by sorption and diffusion from the soil (Jha et al., 2021a, 2021b, 2021c).

2.1 Biota

Biota contributes to the spread of MPs by incorporating it in one location and excreting it in another (Waldschläger et al., 2020). Earthworms and other soil biota, such as soil collembolans, ingested microplastics and significantly contributed to their transportation in the terrestrial environment. (Surendran et al., 2023).

2.2 Compost

Compost soil amendment can potentially provide an entry point for microplastics to make their way into the soil. Organic waste is typically placed on farming land as nutrients after it is composted and fermented in order to repurpose nutrients, trace elements, and humus (Yang et al., 2021). In addition to PFAS pollutants, there is a relationship between their presence or transmission path in the soil and the compost, PFAS in many kinds of commercially available, non-biosolid organic fertilizers produced from food, yard waste, manure, and mushroom-based materials (Sivaram et al., 2022).

2.3 Wastewater

Urban runoff and wastewater treatment plant effluent (WWTP) effluent are examples of land-based microplastics pathways (Ziajahromi et al., 2017). In Fig. 6, microplastics enter the WWTP and reach the soil in several pathways. Where the sewage networks that collect both home and commercial wastewater allow microplastics to enter WWTPs, stormwater runoff is another way that microplastics can get into WWTPs. Microplastics are carried into storm drains, which eventually connect to the wastewater treatment system.

Average MPs concentration in wastewater effluents and sewage sludge was 14,750 $\pm$ 8612.9 particles.kg^{-1}and 276.3 $\pm$ 137.3 particles.L^{-1}, consequently, with 90% of MPs being smaller than 0.5 mm. In terms of sewage sludge and wastewater effluents, WWTP had the highest proportion of MPs (Ragoobur et al., 2021). The spreading of sludge appears to be a crucial mechanism for the dispersion of MPs in the environment (Collivignarelli et al., 2021).

By looking at a Caniglia et al. (2022) study, The Biosolid-PFAS interaction provides a different path for their movement inside a wastewater treatment plant (WWTP) and biosolids from municipal WWTPs are frequently applied to the soil as a fertilizer and soil conditioner.

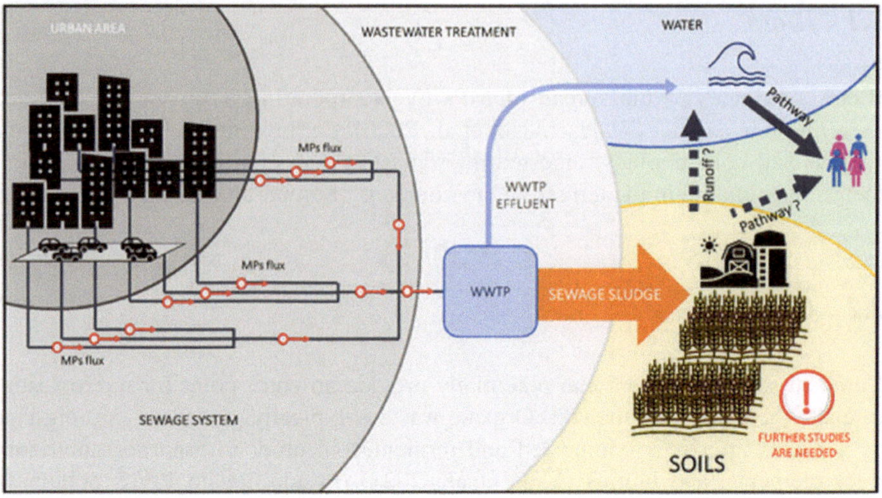

Fig. 6 MPs' routes into and out of WWTPs and their accumulation in soil (Collivignarelli et al., 2021)

3 Microplastic Interactions with PFAS and Co-transport in the Soil

3.1 PFAS and MPs Interaction in the Food Supply Chain

PFAS and MPs can interact in many ways, contributing to food chain contamination (Cheng et al., 2021). They can be emitted into the environment from industrial sites or landfills and contaminate soil and water (Jha et al., 2021a, 2021b, 2021c; Meegoda et al., 2020). PFAS are occasionally discharged into the environment through using of firefighting foam, which includes significant quantities of these chemicals (Reinikainen et al., 2022). PFAS can bind to the surface of MPs, making them more mobile and potentially more bioavailable to species in the food chain. MPs can transport PFAS over far distances, possibly spreading pollution to new places and ecosystems (Fenton et al., 2021).

When PFAS-containing water penetrates the soil, the chemicals can adsorb on soil particles or dissolve in the water. Some PFAS may be more susceptible than others to migrate with water through the soil, depending on their physicochemical characteristics. PFAS that are more mobile in water may be more easily transferred through the soil, while those that are highly adsorbed to soil particles may remain in place (Jha et al., 2021a, 2021b, 2021c).

Co-transport of PFAS in soil has the potential to pollute surface and groundwater, and potentially drinking water sources. The degree of PFAS mobility in soil is determined by several parameters, including soil characteristics, PFAS properties, and the quantity and duration of rainfall (Wang et al., 2023a, 2023b).

MPs have the ability to accumulate in the environment and have an impact on people's health, aquatic life, and soil quality. The co-transport of MPs in the soil can occur through several mechanisms; when it rains, water can move across the surface of the soil, carrying MPs with it. Infiltration of water into the soil can also carry MPs deeper into the soil profile. Erosion of soil particles can also transport MPs attached to the surface of the soil particles (Pérez-Reverón et al., 2022; Yu et al., 2022). Wind erosion can sometimes cause MPs to move from one position to another (Ren et al., 2021a, 2021b, 2021c; Sajjad et al., 2022). MP particle distribution in soils has been influenced by MPs' characteristics such as size, shape, density, and surface charge (Table 1). MPs could also affect soil physiological factors and undergo a variety of declining and transport behaviors (Sajjad et al., 2022). Herein, we summarize the effect of each parameter as follows:

A. **Effect of MPs Particle Size**

MPs' characteristics, including density, shape, and size, had an impact on how they transferred through porous media. Differences in maximum depth and MPs' particle amounts in columns show that the size of MPs' particles had a substantial impact on the depth profiles' permeation curves (Gao et al., 2021).

(Gao et al., 2021) study found that medium-sized sand (0.21–0.60 mm) was used as the porous medium for the column transport studies, and PE and PET MPs particles of three sizes—10–20, 100–150, and 300–450 m, respectively were used. And the

Table 1 MPs properties affecting their transport with other pollutants via soil environment

MPs properties	Explanation	References
Size	MPs that have smaller sizes travel farther than ones with greater sizes. A smaller PS particles (0.1–0.6 m) in seawater-saturated sand move faster than larger plastic particles (0.8–1.5 m)	O'Connor et al. (2019)
Density	High-density PET (about 1.3–1.4 g cm^3) can accelerate sedimentation by gravity, resulting in increased soil retention	Dong et al. (2021)
Shape	MPs with asymmetrical forms may be more likely to cling to other particles, which may affect how well they move. Round or spherical particles are more easily carried by wind and water currents and are less prone to cling to other particles	Ulusoy (2023)
Surface charge	Negatively charged MPs resist other negatively charged particles, whereas positive-charged MPs attract negatively charged particles. This can have an effect on the aggregation and deposition of MPs in the environment	Yasir et al. (2022)
Hydrophobicity	MPs that are hydrophobic (repellent to water) are more prone to adsorb other compounds, such as persistent organic pollutants (POPs), which might affect their distribution and bioavailability in the environment	Martín et al. (2022)

results showed them that after wet–dry processes the tiny PE MPs particles (10–20 m) penetrated to an absolute maximum depth of 9.5 cm, the medium-sized ones (100–150 m) to 5.5 cm, and the largest-sized ones (300–450 m) to 1.5 cm. The effect of the size of microplastic granules on the soil after penetration into it, MPs lower the bulk density of soil. Smaller MP particles have a more negative impact on microbial community dynamics and soil fertility than bigger microplastic particles have on soil porosity and aeration. Additionally, MPs weaken crucial soil processes like water penetration and change the distribution of soil's water-stable aggregates, which reduces soil stability. Reduced soil permeability also prevents crop roots from growing vertically, which has an adverse effect on yields (Guo et al., 2022).

B. **Effect of MPs Shape**

Due to their various interactions with soil aggregation, the shapes of microplastics may affect how they flow through soil and may even have a stumbling block effect. For instance, it has been observed that plastic microspheres and microparticles can migrate more easily to deep layers of soil than microfibers since the latter are able to tangle soil particles to create clods (Ren et al., 2021b).

C. **Effect of MPs Polymer Types**

As manufacturers work to create plastics with particular qualities (such as flexibility, roughness, resistance, and durability), each microplastic structure can be represented by various types of polymer. These polymer varieties are made up of several monomers, some of which may be environmentally harmful (Lozano et al., 2021).

The most common polymer forms in soils are PE and PP (Schenkel, 2022). Earlier studies revealed that the primary properties of a polymer's type that affect mobility through soil are density and hydrophobicity. Polyamide (PA) will be the material in the transport process that is retained the least, followed by polyethylene (PE) and polypropylene (PP), with polyester (PEST) being the material that is held the most.

D. **Effect of Soil Biota and Flora**

The process of MPs migration and transformation can be significantly impacted by soil microorganisms. These are transported from the surface of the soil downward by soil biota, for instance, earthworms and local fungi, altering the different soil horizons (O'Kelly et al., 2021).

E. **Effect of PFAS Presence in Soil with Microplastics**

Soil pH increased considerably, most probably due to enhanced waste decomposition impacted by PFAS. PFAS severely reduced soil breathing. PFOS had a deleterious impact on water-stable aggregates, which could be attributed to microbial changes. PFAS the number of bacteria and fungi in the soil but not microbial or enzyme activity (Xu et al., 2023). In soil, at a concentration of 1 $\mu g\ g^{-1}$, PFAS reduced overall bacterial diversity while increasing the number of specific bacteria (Wang et al., 2023a, 2023b). In the study of (Xu et al., 2022a), it was noted that TFA, an ultra-short-chain PFAS, seemed to affect soil functions and characteristics differently than long- and short-chain PFAS.

The co-transport of MPs and sorbed pollutants in the soil can aid in understanding the dangers associated with heavy metals, organic pollutants, and manufactured nanomaterials (El-Kady et al., 2023; Ren et al., 2021a, 2021b, 2021c). (Wan et al., 2019) founded that adding 1% w/w polyethylene film in two clay soils accelerated water evaporation and desiccation cracking.

Also, PFAS and MPs can then be taken up by plants or animals that live in or near contaminated areas (Jha et al., 2021a, 2021b, 2021c a). Consumers can be receiving PFAS and MPs from their diet (Cheng et al., 2021; Roth et al., 2020). Foods such as fish, shellfish, meat (especially beef), dairy products, and processed foods that use plastic and PFAS containing packaging materials are known to contain high levels of PFAS and MPs as illustrated in Fig. 7 (Cheng et al., 2021; Vorst et al., 2021). Studies have shown that people who consume a lot of these foods may have elevated traces of PFAS and MPs in their bloodstreams (Bonato et al., 2020; Seshasayee et al., 2021).

Both PFAS and MPs have the ability to accumulate in the bodies of species in the food chain, resulting in increased exposure and potentially hazardous consequences. This is especially dangerous for top predators like marine animals and birds, which can collect significant quantities of these contaminants in their tissues (Ma et al., 2022). There is some indication that PFAS and MPs may have synergistic effects, meaning that their combined impact on species in the food chain may be higher than the sum of their separate impacts (Peritore et al., 2023). PFAS and MPs have both been related to negative health effects in organisms, including reproductive and developmental effects, immune system dysfunction, and endocrine disruption. The cumulative effects of these contaminants on organisms in the food chain are still unclear, but there is growing fear that they may constitute a major danger to human health (Fenton et al., 2021).

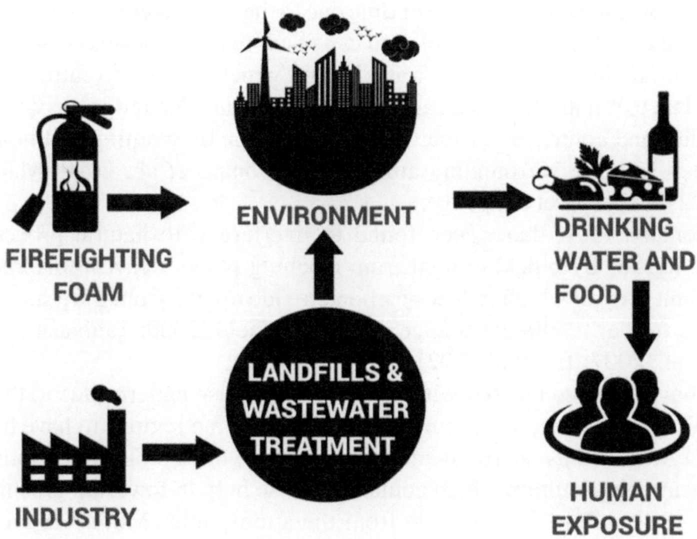

Fig. 7 PFAS and MPs transport in the environment

However, despite these efforts to regulate PFAS use and contamination levels in the environment, they continue to be identified at high amounts in multiple regions of the world (Brennan et al., 2021). This is due to their longevity in the environment as well as their extensive use over several decades (Ali & Khan, 2017).

4 The Relationship Between PFAS, MPs, and Climate Change

PFAS have been linked to global warming through their contribution to greenhouse gas emissions (Bonato et al., 2020). PFAS are known to be emitted into the atmosphere from industrial sources, such as manufacturing plants and wastewater treatment facilities. These emissions can help cause climate change by storing heat in the atmosphere, resulting in an increase in global temperatures (Jha et al., 2021a, 2021b, 2021c). Additionally, PFAS can be discharged into the environment via runoff from landfills or agricultural activities, where they can enter rivers and lakes and eventually make movements into the ocean (Bonato et al., 2020; Taniyasu et al., 2003). Once in the ocean, PFAS can accumulate in marine organisms such as shellfish and fish, which can then be eaten by humans. This process is known as bioaccumulation, and it has been linked to increased levels of greenhouse gases in the atmosphere due to its ability to trap heat (Bonato et al., 2020; Dauchy et al., 2017; Meegoda et al., 2020).

In addition to contributing directly to climate change through emissions, PFAS have also been linked to other environmental issues that can further exacerbate global warming (Mahmoudnia et al., 2022). For example, PFAS have been observed to accumulate into soil and water bodies over time due to their persistence in the environment (Bonato et al., 2020). This accumulation can result in higher levels of nutrient runoff from agricultural lands into nearby water bodies, which can lead to eutrophication (an overabundance of nutrients) that can cause algal blooms that reduce oxygen levels in water bodies and contribute further to climate change by emitting additional greenhouse gases into the surrounding atmosphere (Bonato et al., 2020; Mahmoudnia et al., 2022; Meegoda et al., 2020).

Additionally, PFAS have been found to interfere with natural processes such as photosynthesis by blocking light from reaching plants' leaves; this interference reduces plant growth which reduces carbon dioxide uptake from the atmosphere and contributes further to climate change (Moody & Field, 2000; Taniyasu et al., 2003; Haukås et al., 2007; Jha et al., 2021a, 2021b, 2021c).

PFAS have also been linked with changes in land use patterns due to their ability to accumulate over time; this accumulation has led some regions to have high quantities of PFAS (such as nearby manufacturing sites) unsuitable for certain types of development or agriculture which could otherwise help in lowering greenhouse gas emissions or absorb carbon dioxide from the atmosphere (Moody & Field, 2000; Parolini et al., 2022).

Finally, there is growing evidence linking PFAS with climate change through direct emissions as well as indirect pathways such as bioaccumulation and land use changes (Borgå et al., 2022). As such it is important for governments around the world to take steps toward reducing or eliminating sources of these chemicals so that their impacts on global warming can be minimized or eliminated altogether (Kumar et al., 2022).

MPs may contribute to climate change in a variety of ways. One approach is through their influence on marine ecosystems, which play a crucial role in managing the earth's carbon cycle. When marine ecosystems are disrupted or harmed, they may become less efficient in regulating the carbon cycle, resulting in higher atmospheric CO_2 levels and contributing to global warming (Lincoln et al., 2022; Mahmoud & Krasucka, 2023).

MPs can accumulate in marine organisms' tissues, resulting in slower growth, reproduction, and survival rates. This can have an impact on the overall productivity and biodiversity of marine ecosystems, altering the Earth's carbon cycle and climate (Tursi et al., 2022).

MPs can also contribute to climate change through their manufacture and disposal. Plastics, particularly MPs, require the usage of fossil fuels, which are a major contributor significantly to the release of greenhouse gases and global warming. Plastics, especially MPs, can potentially contribute to climate change by emitting greenhouse gases during burning or decomposition in landfills (Shen et al., 2020). Consequently, MPs do not directly cause global warming, their effects on marine ecosystems as well as the manufacturing and disposal of plastics may have a negative influence on the environment. Plastic consumption should be reduced, and more sustainable methods of manufacturing and disposing of plastic should be supported (Ford et al., 2022). Furthermore, the movement of MPs in ocean currents can aid in the spread of invasive species, altering marine ecosystems and altering the Earth's carbon cycle (Kumar et al., 2021).

5 Current Knowledge and Awareness and Comprehension of PFAS and MPs Pollution

The current knowledge and awareness of PFAS pollution is growing rapidly due to recent studies that have highlighted the probable health consequences of exposure to these chemicals (Renfrew & Pearson, 2021; Rogers et al., 2021). In particular, studies indicate a link between PFAS exposure and a number of negative health impacts, including cancer, thyroid alteration, reproductive damage, immune system suppression, and endocrine disruption. (Ducatman et al., 2022; Rogers et al., 2021). Additionally, a study has shown that PFAS may accumulate in the food chain and be passed from mother to kid during pregnancy or breastfeeding (Marks et al., 2019; Mitro et al., 2020).

Given the potential negative health effects linked with PFAS exposure, there is an increasing need for public awareness about this issue (Marks et al., 2019; Mitro et al., 2020; Renfrew & Pearson, 2021; Rogers et al., 2021; Ducatman et al., 2022). In the past few years, there has been an increase in coverage by the media about PFAS pollution as well as increased public concern about this issue. This has led to increased pressure on governments at all levels to take action on this issue (Blum et al., 2015; Tian et al., 2022). As a result of this, many countries worldwide have implemented regulations or policies aimed at reducing or eliminating PFAS pollution (Cornelsen et al., 2021). For example, the European Union recently adopted legislation that bans certain uses of PFAS chemicals and requires companies to phase out other uses by 2030 (Commission, n.d.). Additionally, several US states have passed laws restricting the use of certain types of PFAS chemicals or requiring companies to disclose their use of these substances (Brennan et al., 2021).

Despite these efforts to reduce or eliminate PFAS pollution, it remains an ongoing problem due to its continued existence in the environment and its ability to accumulate in living organisms over time (Langenbach et al., 2021). As such, persons must be aware of potential sources of exposure so that they can take steps to reduce their own risk from these chemicals (EPA, 2022). This includes avoiding products containing PFAS when possible (such as non-stick cookware), limiting consumption of fish caught from contaminated waters (such as those near industrial sites), avoiding contact with contaminated soils or dust (such as those near airports), and using water filters certified for removal of PFAS compounds when drinking tap water from potentially contaminated sources. Additionally, individuals should be aware that some foods may contain traces of these compounds due to their widespread use in the packaging of food materials such as paperboard cartons and microwaved popcorn bags (Sun et al., 2016; Wang et al., 2017; Langenbach et al., 2021; EPA, 2022).

In conclusion, current knowledge, and awareness about PFAS pollution is growing rapidly due to recent studies highlighting its potential health risks as well as increased media coverage about this issue (Kudo & Kawashima, 2003; Rogers et al., 2021). This has prompted numerous countries around the world to implement regulations aimed at reducing or eliminating these pollutants from our environment; however, it persists in the environment and accumulates over time, making it a continuing problem. (Kudo & Kawashima, 2003). Therefore, individuals should be informed of potential sources of exposure, so that they can take steps toward reducing their own risk from these chemicals when possible (Fuller et al., 2022; Landrigan et al., 2018).

MPs contamination in the environment has raised public awareness and concern in recent years. While MPs have been prevalent in the environment for several decades, their presence and possible implications have only just become publicly recognized. Research on MPs' pollution has advanced significantly during the last ten years, as has public awareness of the problem (Lamichhane et al., 2023). Numerous studies have demonstrated that MPs can be found in various types of environmental media, including water, sediment, soil, and air (Barnes et al., 2009; Isangedighi et al., 2020; Lamichhane et al., 2023; Rios Mendoza et al., 2018; Sridharan et al., 2021). Almost everywhere on the planet, from distant mountain peaks to the bottom of the ocean,

MPs have been discovered. The fact that MPs are so prevalent suggests that they have the ability to seriously impair both the environment and the wellness of persons (Lamichhane et al., 2023).

Some initiatives to solve the problem globally have been made in response to the rising concern about MPs pollution. The 2030 Agenda for the Development of Sustainability, which was approved by the UN in 2015, contains a specific objective to "prevent and significantly reduce marine pollution of all kinds, in particular from land-based activities, including marine debris and nutrient pollution (Mahmoud et al., 2023; Martín-Lara et al., 2021)." There have also been initiatives to decrease plastic waste and promote more environmentally friendly methods of producing and discarding plastic, as well as prohibitions on specific single-use plastic products in various nations and areas (Pandey et al., 2023).

More action is required to solve the problem of MPs pollution despite these initiatives (Kibria et al., 2023). This entails creating more accurate techniques for identifying and quantifying MPs in the environment, as well as putting into practice more successful methods for lowering plastic waste and supporting sustainable methods for the manufacture and disposal of plastic (Abu Jadayil et al., 2022). In addition, the integration of nanosensors to monitor the pollutants is crucial (Mahmoud & Fawzy, 2021) and using nanomembranes could mitigate the effect of MPs and other pollutants (Mahmoud & Mostafa, 2023).

6 Conclusion

In the subject of the transfer of microplastics and other pollutants like PFAS to the soil, and their co-transport dynamics. Through an exploration of relevant studies and empirical evidence, we have gained a nuanced understanding of the multifaced interactions that occur between these environmental contaminates. We found that a number of elements were involved in the production of microplastics in the environment throughout our examination. These included human actions like weathering-induced plastic fragmentation, industrial operations, and inappropriate disposal techniques. Natural processes such as the decomposition of bigger plastic items also contributed to the formation of microplastics. Afterwards, our analysis identified a number of critical elements that promote these pollutants' spread throughout various habitats, with an emphasis on their movement into the soil. Because of their small size and low weight, microplastics can be carried by the wind and water. For instance, rainwater runoff from metropolitan areas may carry plastic particles into the surrounding soil, bringing microplastics into terrestrial ecosystems. Moreover our study focused on the complex interactions between soil and microplastics. Notably, MPs' intrinsic physical and chemical characteristics allow for an effective transfer to soil matrices. As we focus on and mentioned during their passage through the environment, persistent organic pollutants like PFAS transported via MPs. A secondary degree of complication is added by this co-transport mechanism, since the combination of PFAS and MPs affects the composition and quality of the soil.

After examining the latest research and the data and information it contains on the spread of microplastics and their transfer in the soil to other pollutants, the direction of scientific research will be directed toward evaluating the risks in a broader and more interesting way, working on removal techniques, researching more about the path of microplastics in the environment, starting from the source and ending with the recipient, passing through the carrier or mediator that helps spread.

References

Abu Jadayil, W., Qureshi, M. R. N. M., Ajaj, R., Aqil, E., Shawahin, G., Anver, H., & Aljeawi, S. (2022). An empirical investigation on plastic waste issues and plastic disposal strategies to protect the environment: A UAE perspective. *Sustainability, 14*(24), 16719.

Abunada, Z., Alazaiza, M. Y., & Bashir, M. J. (2020). An overview of per-and polyfluoroalkyl substances (PFAS) in the environment: Source, fate, risk and regulations. *Water, 12*, 3590.

Alfonso, M. B., Arias, A. H., Ronda, A. C., & Piccolo, M. C. (2021). Continental microplastics: Presence, features, and environmental transport pathways. *Science of the Total Environment, 799*, 149447.

Ali, H., & Khan, E. (2017). Environmental chemistry in the twenty-first century. *Environmental Chemistry Letters, 15*(2), 329–346.

Baho, D. L., Bundschuh, M., & Futter, M. N. (2021). Microplastics in terrestrial ecosystems: Moving beyond the state of the art to minimize the risk of ecological surprise. *Global Change Biology, 27*, 3969–3986.

Badr, N. B., Al-Qahtani, K. M., & Mahmoud, A. E. D. (2020). Factorial experimental design for optimizing selenium sorption on Cyperus laevigatus biomass and green-synthesized nano-silver. *Alexandria Engineering Journal, 59*(6), 5219–5229.

Baker, E. S., & Knappe, D. R. (2022). Per-and polyfluoroalkyl substances (PFAS)—contaminants of emerging concern. *Analytical and Bioanalytical Chemistry, 414*(3), 1187–1188.

Baldwin, W., Davis, T. T., & Eccles, J. A. (2023). Per-and polyfluoroalkyl substances (PFAS) and their toxicology as evidenced through disease and biomarkers. In *Biomarkers in Toxicology* (pp. 989–1016). Springer.

Barnes, D. K., Galgani, F., Thompson, R. C., & Barlaz, M. (2009). Accumulation and fragmentation of plastic debris in global environments. *Philosophical Transactions of the Royal Society B: Biological Sciences, 364*(1526), 1985–1998.

Bayabil, H. K., Teshome, F. T., & Li, Y. C. (2022). Emerging contaminants in soil and water. *Frontiers in Environmental Science, 10*, 873499.

Bhuyan, M. S. (2022). Effects of microplastics on fish and in human health. *Frontiers in Environmental Science, 10*, 827289.

Blum, A., Balan, S. A., Scheringer, M., Trier, X., Goldenman, G., Cousins, I. T., Diamond, M., Fletcher, T., Higgins, C., Lindeman, A. E., Peaslee, G., Voogt, P. D., Wang, Z., & Weber, R. (2015). The madrid statement on poly- and perfluoroalkyl substances (PFASs). *Environmental Health Perspectives, 123*, A107–A111.

Bolan, N., Sarkar, B., Yan, Y., Li, Q., Wijesekara, H., Kannan, K., Tsang, D. C., Schauerte, M., Bosch, J., & Noll, H. (2021). Remediation of poly-and perfluoroalkyl substances (PFAS) contaminated soils–to mobilize or to immobilize or to degrade? *Journal of Hazardous Materials, 401*, 123892.

Bonato, M., Corrà, F., Bellio, M., Guidolin, L., Tallandini, L., Irato, P., & Santovito, G. (2020). PFAS environmental pollution and antioxidant responses: An overview of the impact on human field. *International Journal of Environmental Research and Public Health, 17*, 8020.

Borgå, K., McKinney, M. A., Routti, H., Fernie, K. J., Giebichenstein, J., Hallanger, I., & Muir, D. C. G. (2022). The influence of global climate change on accumulation and toxicity of persistent organic pollutants and chemicals of emerging concern in Arctic food webs. *Environmental Science: Processes & Impacts, 24*, 1544–1576.

Brennan, N. M., Evans, A. T., Fritz, M. K., Peak, S. A., & von Holst, H. E. (2021). Trends in the regulation of per-and polyfluoroalkyl substances (PFAS): A scoping review. *International Journal of Environmental Research and Public Health, 18*, 10900.

Brusseau, M. L., Anderson, R. H., & Guo, B. (2020). PFAS concentrations in soils: Background levels versus contaminated sites. *Science of the Total Environment, 740*, 140017.

Caniglia, J., Snow, D. D., Messer, T., & Bartelt-Hunt, S. (2022). Extraction, analysis, and occurrence of per-and polyfluoroalkyl substances (PFAS) in wastewater and after municipal biosolids land application to determine agricultural loading. *Frontiers in Water, 4*, 892451.

Chaukura, N., Muzawazi, E. S., Katengeza, G., & Mahmoud, A. E. D. (2022). Remediation technologies for contaminated soil systems. In W. Gwenzi (Ed.), *Emerging Contaminants in the Terrestrial-Aquatic-Atmosphere Continuum* (pp. 353–365). Elsevier.

Cheng, Y., Mai, L., Lu, X., Li, Z., Guo, Y., Chen, D., & Wang, F. (2021). Occurrence and abundance of poly- and perfluoroalkyl substances (PFASs) on microplastics (MPs) in Pearl River Estuary (PRE) region: Spatial and temporal variations. *Environmental Pollution, 281*, 117025.

Clean Water Action. (2023). PFAS-Containing Firefighting Foams|Clean Water Action. Commission E, (n.d.) Chemicals strategy.

Collivignarelli, M. C., Carnevale Miino, M., Caccamo, F. M., & Milanese, C. (2021). Microplastics in sewage sludge: A known but underrated pathway in wastewater treatment plants. *Sustainability, 13*, 12591.

Cornelsen, M., Weber, R., & Panglisch, S. (2021). Minimizing the environmental impact of PFAS by using specialized coagulants for the treatment of PFAS polluted waters and for the decontamination of firefighting equipment. *Emerging Contaminants, 7*, 63–76.

Cousins, I. T., Dewitt, J. C., Glüge, J., Goldenman, G., Herzke, D., Lohmann, R., Miller, M., Ng, C. A., Scheringer, M., Vierke, L., & Wang, Z. (2020). Strategies for grouping per- and polyfluoroalkyl substances (PFAS) to protect human and environmental health. *Environmental Science: Processes & Impacts, 22*, 1444–1460.

Curtzwiler, G. W., Silva, P., Hall, A., Ivey, A., & Vorst, K. (2021). Significance of Perfluoroalkyl Substances (PFAS) in food packaging. *Integrated Environmental Assessment and Management, 17*, 7–12.

Dauchy, X., Boiteux, V., Bach, C., Rosin, C., & Munoz, J. F. (2017). Per- and polyfluoroalkyl substances in firefighting foam concentrates and water samples collected near sites impacted by the use of these foams. *Chemosphere, 183*, 53–61.

Dong, S., Xia, J., Sheng, L., Wang, W., Liu, H., & Gao, B. (2021). Transport characteristics of fragmental polyethylene glycol terephthalate (PET) microplastics in porous media under various chemical conditions. *Chemosphere, 276*, 130214.

Ducatman, A., LaPier, J., Fuoco, R., & DeWitt, J. C. (2022). Official health communications are failing PFAS-contaminated communities. *Environmental Health: A Global Access Science Source, 21*, 1–18.

El-Kady, M. M., Ansari, I., Arora, C., Rai, N., Soni, S., Verma, D. K., Singh, P., & Mahmoud, A. E. D. (2023). Nanomaterials: A comprehensive review of applications, toxicity, impact, and fate to environment. *Journal of Molecular Liquids, 370*, 121046.

El-Sherif, D. M., Eloffy, M. G., Elmesery, A., Abouzid, M., Gad, M., El-Seedi, H. R., Brinkmann, M., Wang, K., & Al Naggar, Y. (2022). Environmental risk, toxicity, and biodegradation of polyethylene: a review. *Environmental Science and Pollution Research, 29*, 81166–81182.

EPA US. (2022). Protect Your Family from Sources of Lead|US EPA.

Evode, N., Qamar, S. A., Bilal, M., Barceló, D., & Iqbal, H. M. N. (2021). Plastic waste and its management strategies for environmental sustainability. *Case Studies in Chemical and Environmental Engineering., 4*, 100142.

Fair, P. A., Wolf, B., White, N. D., Arnott, S. A., Kannan, K., Karthikraj, R., & Vena, J. E. (2019). Perfluoroalkyl substances (PFASs) in edible fish species from Charleston Harbor and tributaries, South Carolina, United States: Exposure and risk assessment. *Environmental Research, 171*, 266–277.

Feng, S., Lu, H., & Yao, T. (2023). Effects of multiple environmental factors and land use patterns on microplastic distribution in the topsoil of the Qinghai and Sichuan provinces of China. *Journal of Environmental Chemical Engineering, 11*, 109657.

Fenton, S. E., Ducatman, A., Boobis, A., DeWitt, J. C., Lau, C., Ng, C., Smith, J. S., & Roberts, S. M. (2021). Per- and polyfluoroalkyl substance toxicity and human health review: Current state of knowledge and strategies for informing future research. *Environmental Toxicology and Chemistry, 40*, 606.

Ford, H. V., Jones, N. H., Davies, A. J., Godley, B. J., Jambeck, J. R., Napper, I. E., Suckling, C. C., Williams, G. J., Woodall, L. C., & Koldewey, H. J. (2022). The fundamental links between climate change and marine plastic pollution. *Science of the Total Environment, 806*, 150392.

Fuller, R., Landrigan, P. J., Balakrishnan, K., Bathan, G., Bose-O'Reilly, S., Brauer, M., Caravanos, J., Chiles, T., Cohen, A., Corra, L., Cropper, M., Ferraro, G., Hanna, J., Hanrahan, D., Hu, H., Hunter, D., Janata, G., Kupka, R., Lanphear, B., et al. (2022). Pollution and health: A progress update. *The Lancet Planetary Health, 6*, e535–e547.

Gaines, L. G. T. (2022). Historical and current usage of per- and polyfluoroalkyl substances (PFAS): A literature review. *American Journal of Industrial Medicine.*

Gaines, L. G. T., Sinclair, G., & Williams, A. J. (2023). A proposed approach to defining per- and polyfluoroalkyl substances (PFAS) based on molecular structure and formula. *Integrated Environmental Assessment and Management* 1.

Gao, J., Pan, S., Li, P., Wang, L., Hou, R., Wu, W.-M., Luo, J., & Hou, D. (2021). Vertical migration of microplastics in porous media: Multiple controlling factors under wet-dry cycling. *Journal of Hazardous Materials, 419*, 126413.

Glüge, J., Scheringer, M., Cousins, I. T., Dewitt, J. C., Goldenman, G., Herzke, D., Lohmann, R., Ng, C. A., Trier, X., & Wang, Z. (2020a). An overview of the uses of per- and polyfluoroalkyl substances (PFAS). *Environmental Science: Processes & Impacts, 22*, 2345–2373.

Guo, Z., Li, P., Yang, X., Wang, Z., Lu, B., Chen, W., Wu, Y., Li, G., Zhao, Z., & Liu, G. (2022). Soil texture is an important factor determining how microplastics affect soil hydraulic characteristics. *Environment International, 165*, 107293.

Hale, R. C., Seeley, M. E., La, Guardia, M. J., Mai, L., Zeng, E. Y. (2020). A global perspective on microplastics. *Journal of Geophysical Research: Oceans,* 125, e2018JC014719.

Hao, X., Sun, H., Zhang, Y., Li, S., & Yu, Z. (2023). Co-transport of arsenic and micro/nano-plastics in saturated soil. *Environmental Research, 228*, 115871.

Haukås, M., Berger, U., Hop, H., Gulliksen, B., & Gabrielsen, G. W. (2007). Bioaccumulation of per- and polyfluorinated alkyl substances (PFAS) in selected species from the Barents Sea food web. *Environmental Pollution, 148*, 360–371.

Hepburn, E., Madden, C., Szabo, D., Coggan, T. L., Clarke, B., & Currell, M. (2019). Contamination of groundwater with per- and polyfluoroalkyl substances (PFAS) from legacy landfills in an urban re-development precinct. *Environmental Pollution, 248*, 101–113.

Isangedighi, I. A., David, G. S., & Obot, O. I. (2020). Plastic waste in the aquatic environment: impacts and management. *Analysis of Nanoplastics and Microplastics in Food* (pp. 15–43). CRC Press.

Jha, G., Kankarla, V., McLennon, E., Pal, S., Sihi, D., Dari, B., Diaz, D., & Nocco, M. (2021a) Per- and polyfluoroalkyl substances (PFAS) in integrated crop–livestock systems: environmental exposure and human health risks. *International Journal of Environmental Research and Public Health, 18*, 12550.

Jha, G., Kankarla, V., McLennon, E., Pal, S., Sihi, D., Dari, B., Diaz, D., & Nocco, M. (2021b). Per- and polyfluoroalkyl substances (PFAS) in integrated crop-livestock systems: Environmental exposure and human health risks. *International Journal of Environmental Research and Public Health, 18*, 12550.

Jha, G., Kankarla, V., McLennon, E., Pal, S., Sihi, D., Dari, B., Diaz, D., & Nocco, M. (2021c). Per-and polyfluoroalkyl substances (PFAS) in integrated crop–livestock systems: Environmental exposure and human health risks. *International Journal of Environmental Research and Public Health, 18*, 12550.

Glüge, J., Martin Scheringer, T., Cousins, I. C., DeWitt, J., Goldenman, G., Herzke, D., Rainer Lohmann, A., Ng, C., Trier, X., & Wang, Z. (2020b). An overview of the uses of per- and polyfluoroalkyl substances (PFAS). *Environmental Science: Processes & Impacts, 22*, 2345–2373.

Khan, N. A., Khan, A. H., López-Maldonado, E. A., Alam, S. S., López López, J. R., Méndez Herrera, P. F., Mohamed, B. A., Mahmoud, A. E. D., Abutaleb, A., & Singh, L. (2022). Microplastics: Occurrences, treatment methods, regulations and foreseen environmental impacts. *Environmental Research, 215*, 114224.

Kibria, M. G., Masuk, N. I., Safayet, R., Nguyen, H. Q., & Mourshed, M. (2023). Plastic waste: Challenges and opportunities to mitigate pollution and effective management. *International Journal of Environmental Research, 17*, 1–37.

Kudo, N., & Kawashima, Y. (2003). Toxicity and toxicokinetics of perfluorooctanoic acid in humans and animals. *Journal of Toxicological Sciences, 28*, 49–57.

Kumar, A., Singh, P., Raizada, P., & Hussain, C. M. (2022). Impact of COVID-19 on greenhouse gases emissions: A critical review. *Science of the Total Environment, 806*, 150349.

Kumar, R., Verma, A., Shome, A., Sinha, R., Sinha, S., Jha, P. K., Kumar, R., Kumar, P., Shubham, Das, S., Sharma, P., & Prasad, P. V. V. (2021). Impacts of plastic pollution on ecosystem services, sustainable development goals, and need to focus on circular economy and policy interventions. *Sustainability, 13*, 9963.

Kye, H., Kim, J., Ju, S., Lee, J., Lim, C., & Yoon, Y. (2023). Microplastics in water systems: A review of their impacts on the environment and their potential hazards. *Heliyon, 9*, e14359.

Lamichhane, G., Acharya, A., Marahatha, R., Modi, B., Paudel, R., Adhikari, A., Raut, B. K., Aryal, S., & Parajuli, N. (2023). Microplastics in environment: Global concern, challenges, and controlling measures. *International Journal of Environmental Science and Technology, 20*, 4673.

Landrigan, P. J., Fuller, R., Acosta, N. J., Adeyi, O., Arnold, R., Baldé, A. B., Bertollini, R., Bose-O'Reilly, S., Boufford, J. I., Breysse, P. N., & Chiles, T. (2018). The lancet commission on pollution and health. *The Lancet, 391*(10119), 462–512.

Langenbach, B., Wilson, M., Zhang, T., Kim, U. J., & Pilar Martinez Moral, M. (2021). Per- and polyfluoroalkyl substances (PFAS): Significance and considerations within the regulatory framework of the USA. *International Journal of Environmental Research and Public Health, 18*, 11142.

Li, J., Song, Y., & Cai, Y. (2020). Focus topics on microplastics in soil: Analytical methods, occurrence, transport, and ecological risks. *Environmental Pollution, 257*, 113570.

Lincoln, S., Andrews, B., Birchenough, S. N. R., Chowdhury, P., Engelhard, G. H., Harrod, O., Pinnegar, J. K., & Townhill, B. L. (2022). Marine litter and climate change: Inextricably connected threats to the world's oceans. *Science of the Total Environment, 837*, 155709.

Lozano, Y. M., Lehnert, T., Linck, L. T., Lehmann, A., & Rillig, M. C. (2021). Microplastic shape, polymer type, and concentration affect soil properties and plant biomass. *Frontiers in Plant Science, 12*, 616645.

Ma, T., Ye, C., Wang, T., Li, X., & Luo, Y. (2022). Toxicity of per- and polyfluoroalkyl substances to aquatic invertebrates, planktons, and microorganisms. *International Journal of Environmental Research and Public Health, 19*, 16729.

Mahinroosta, R., & Senevirathna, L. (2020). A review of the emerging treatment technologies for PFAS contaminated soils. *Journal of Environmental Management, 255*, 109896.

Mahmoud, A. E. D. (2020). Graphene-based nanomaterials for the removal of organic pollutants: Insights into linear versus nonlinear mathematical models. *Journal of Environmental Management, 270*, 110911.

Mahmoud, A. E. D., & Fawzy, M. (2021). Nanosensors and nanobiosensors for monitoring the environmental pollutants. In A. S. H. Makhlouf & G. A. M. Ali (Eds.), *Waste Recycling Technologies for Nanomaterials Manufacturing* (pp. 229–246). Springer International Publishing.

Mahmoud, A. E. D., & Krasucka, P. (2023). *Global Water Challenges and Sustainability* (pp. 1–12). CRC Press.

Mahmoud, A. E. D., Al-Qahtani, K. M., Alflaij, S. O., Al-Qahtani, S. F., & Alsamhan, F. A. (2021). Green copper oxide nanoparticles for lead, nickel, and cadmium removal from contaminated water. *Scientific Reports, 11*, 12547.

Mahmoud, A. E. D., El-Maghrabi, N., Hosny, M., & Fawzy, M. (2022c). Biogenic synthesis of reduced graphene oxide from Ziziphus spina-christi (Christ's thorn jujube) extracts for catalytic, antimicrobial, and antioxidant potentialities. *Environmental Science and Pollution Research.*

Mahmoud, A. E. D., Fawzy, M., Khairy, H., & Sorour, A. (2022b). Environmental bioremediation as an eco-sustainable approach for pesticides: A case study of MENA region. In S. Siddiqui, M. K. Meghvansi, & K. K. Chaudhary (Eds.), *Pesticides Bioremediation* (pp. 479–494). Springer International Publishing.

Mahmoud, A. E. D., Franke, M., & Braeutigam, P. (2022c). Experimental and modeling of fixed-bed column study for phenolic compounds removal by graphite oxide. *Journal of Water Process Engineering, 49*, 103085.

Mahmoud, A. E. D., Franke, M., Stelter, M., & Braeutigam, P. (2020). Mechanochemical versus chemical routes for graphitic precursors and their performance in micropollutants removal in water. *Powder Technology, 366*, 629–640.

Mahmoud, A. E. D., Hosny, M., El-Maghrabi, N., & Fawzy, M. (2022d). Facile synthesis of reduced graphene oxide by Tecoma stans extracts for efficient removal of Ni (II) from water: Batch experiments and response surface methodology. *Sustainable Environment Research, 32*, 22.

Mahmoud, A. E. D. (2022). Recent advances of TiO_2 nanocomposites for photocatalytic degradation of water contaminants and rechargeable sodium ion batteries. In A. E. Shalan, A. S. Hamdy Makhlouf, & S. Lanceros-Méndez (Eds.), *Advances in Nanocomposite Materials for Environmental and Energy Harvesting Applications* (pp. 757–770). Springer International Publishing.

Mahmoud, A. E. D., Fawzy, M., & Khan, N. (2023). *Artificial Intelligence and Modeling for Water Sustainability: Global Challenges* (1st ed.). CRC Press.

Mahmoud, A. E. D., & Mostafa, E. (2023). Nanofiltration membranes for the removal of heavy metals from aqueous solutions: Preparations and applications. *Membranes, 13*(9), 789.

Mahmoud, A. E. D., & Fawzy, M. (2023). Decontamination of levofloxacin from water using a novel chitosan–walnut shells composite: Linear, nonlinear, and optimization modeling. *Applied Water Science, 13*(12), 1–19.

Mahmoudnia, A., Mehrdadi, N., Baghdadi, M., & Moussavi, G. (2022). Change in global PFAS cycling as a response of permafrost degradation to climate change. *Journal of Hazardous Materials Advances, 5*, 100039.

Marks, K. J., Jeddy, Z., Flanders, W. D., Northstone, K., Fraser, A., Calafat, A. M., Kato, K., & Hartman, T. J. (2019). Maternal serum concentrations of perfluoroalkyl substances during pregnancy and gestational weight gain: The Avon Longitudinal Study of Parents and Children. *Reproductive Toxicology, 90*, 8–14.

Martín, J., Santos, J. L., Aparicio, I., & Alonso, E. (2022). Microplastics and associated emerging contaminants in the environment: Analysis, sorption mechanisms and effects of co-exposure. *Trends in Environmental Analytical Chemistry, 35*, e00170.

Martín-Lara, M. A., Godoy, V., Quesada, L., Lozano, E. J., & Calero, M. (2021). Environmental status of marine plastic pollution in Spain. *Marine Pollution Bulletin, 170*, 112677.

Meegoda, J. N., Kewalramani, J. A., Li, B., Marsh, R. W. (2020). A review of the applications, environmental release, and remediation technologies of per- and polyfluoroalkyl substances. *International Journal of Environmental Research and Public Health, 17*, 8117.

Mejías, C., Martín, J., Santos, J. L., Aparicio, I., & Alonso, E. (2023). Adsorption of perfluoroalkyl substances on polyamide microplastics: Effect of sorbent and influence of environmental factors. *Environmental Research, 216*, 114834.

Meng, L., Tian, H., Lv, J., Wang, Y., & Jiang, G. (2023). Influence of microplastics on the photodegradation of perfluorooctane sulfonamide (FOSA). *Journal of Environmental Sciences, 127*, 791–798.

Mitro, S. D., Sagiv, S. K., Rifas-Shiman, S. L., Calafat, A. M., Fleisch, A. F., Jaacks, L. M., Williams, P. L., Oken, E., & James-Todd, T. M. (2020). Per- and polyfluoroalkyl substance exposure, gestational weight gain, and postpartum weight changes in project viva. *Obesity, 28*, 1984–1992.

Moody, C. A., & Field, J. A. (2000). Perfluorinated surfactants and the environmental implications of their use in fire-fighting foams. *Environmental Science and Technology, 34*, 3864–3870.

Navarathna, C. M., Pray, H., Rodrigo, P. M., Arwenyo, B., McNeely, C., Reynolds, H., Hampton, N., Lape, K., Roman, K., Heath, M., & Stokes, S. (2023). Microplastics and per-and polyfluoroalkyl substances (PFAS) analysis in sea turtles and bottlenose dolphins along Mississippi's coast. *Analytica, 4*(1), 12–26.

O'Connor, D., Pan, S., Shen, Z., Song, Y., Jin, Y., Wu, W. M., & Hou, D. (2019). Microplastics undergo accelerated vertical migration in sand soil due to small size and wet-dry cycles. *Environmental Pollution, 249*, 527–534.

O'Kelly, B. C., El-Zein, A., Liu, X., Patel, A., Fei, X., Sharma, S., Mohammad, A., Goli, V. S. N. S., Wang, J. J., & Li, D. (2021). Microplastics in soils: An environmental geotechnics perspective. *Environmental Geotechnics, 8*, 586–618.

Pandey, P., Dhiman, M., Kansal, A., & Subudhi, S. P. (2023). Plastic waste management for sustainable environment: Techniques and approaches. *Waste Disposal & Sustainable Energy, 5*, 205–222.

Parolini, M., De Felice, B., Rusconi, M., Morganti, M., Polesello, S., & Valsecchi, S. (2022). A review of the bioaccumulation and adverse effects of PFAS in free-living organisms from contaminated sites nearby fluorochemical production plants. *Water Emerging Contaminants & Nanoplastics, 1*, 18.

Pérez-Reverón, R., Álvarez-Méndez, S. J., Kropp, R. M., Perdomo-González, A., Hernández-Borges, J., & Díaz-Peña, F. J. (2022). Microplastics in agricultural systems: Analytical methodologies and effects on soil quality and crop yield. *Agriculture (Switzerland), 12*, 1162.

Peritore, A. F., Gugliandolo, E., Cuzzocrea, S., Crupi, R., & Britti, D. (2023). Current review of increasing animal health threat of per- and polyfluoroalkyl substances (PFAS): Harms, limitations, and alternatives to manage their toxicity. *International Journal of Molecular Sciences, 24*, 11707.

Ragoobur, D., Huerta-Lwanga, E., & Somaroo, G. D. (2021). Microplastics in agricultural soils, wastewater effluents and sewage sludge in Mauritius. *Science of the Total Environment, 798*, 149326.

Reinikainen, J., Perkola, N., Äystö, L., & Sorvari, J. (2022). The occurrence, distribution, and risks of PFAS at AFFF-impacted sites in Finland. *Science of the Total Environment, 829*, 154237.

Ren, Z., Gui, X., Xu, X., Zhao, L., Qiu, H., & Cao, X. (2021a). Microplastics in the soil-groundwater environment: Aging, migration, and co-transport of contaminants—A critical review. *Journal of Hazardous Materials, 419*, 126455.

Ren, Z., Gui, X., Wei, Y., Chen, X., Xu, X., Zhao, L., Qiu, H., & Cao, X. (2021b). Chemical and photo-initiated aging enhances transport risk of microplastics in saturated soils: Key factors, mechanisms, and modeling. *Water Research, 202*, 117407.

Ren, Z., Gui, X., Xu, X., Zhao, L., Qiu, H., & Cao, X. (2021c). Microplastics in the soil-groundwater environment: Aging, migration, and co-transport of contaminants–a critical review. *Journal of Hazardous Materials, 419*, 126455.

Renfrew, D., & Pearson, T. W. (2021). The social life of the "forever chemical": PFAS pollution legacies and toxic events. *Environment and Society, 12*, 146–163.

Rios Mendoza, L. M., Karapanagioti, H., & Álvarez, N. R. (2018). Micro(nanoplastics) in the marine environment: Current knowledge and gaps. *Current Opinion in Environmental Science & Health, 1*, 47–51.

Rogers, R. D., Reh, C. M., Breysse, P. (2021). Advancing per- and polyfluoroalkyl substances (PFAS) research: An overview of ATSDR and NCEH activities and recommendations. *Journal of Exposure Science & Environmental Epidemiology, 31*, 961–971.

Roth, K., Imran, Z., Liu, W., & Petriello, M. C. (2020). Diet as an exposure source and mediator of per- and polyfluoroalkyl substance (PFAS) toxicity. *Frontiers in Toxicology, 2*, 6.

Roy, T., Dey, T. K., & Jamal, M. (2023). Microplastic/nanoplastic toxicity in plants: an imminent concern. *Environmental Monitoring and Assessment* 195.

Sajjad, M., Huang, Q., Khan, S., Khan, M. A., Liu, Y., Wang, J., Lian, F., Wang, Q., & Guo, G. (2022). Microplastics in the soil environment: A critical review. *Environmental Technology & Innovation, 27*, 102408.

Schenkel, C. A. (2022). Impact of type and shape on the transport of microplastics in column experiments. Northern Illinois University.

Scott, J. W., Gunderson, K. G., Green, L. A., Rediske, R. R., & Steinman, A. D. (2021). Perfluoroalkylated substances (Pfas) associated with microplastics in a lake environment. *Toxics, 9*, 106.

Seshasayee, S. M., Rifas-Shiman, S. L., Chavarro, J. E., Carwile, J. L., Lin, P. I. D., Calafat, A. M., Sagiv, S. K., Oken, E., & Fleisch, A. F. (2021). Dietary patterns and PFAS plasma concentrations in childhood: Project Viva, USA. *Environment International, 151*, 106415.

Shafea, L., Yap, J., Beriot, N., Felde, V. J. M. N. L., Okoffo, E. D., Enyoh, C. E., & Peth, S. (2023). Microplastics in agroecosystems: A review of effects on soil biota and key soil functions. *Journal of Plant Nutrition and Soil Science, 186*, 5–22.

Shahsavari, E., Rouch, D., Khudur, L. S., Thomas, D., Aburto-Medina, A., & Ball, A. S. (2021). Challenges and current status of the biological treatment of PFAS-contaminated soils. *Frontiers in Bioengineering and Biotechnology, 8*, 602040.

Shen, M., Huang, W., Chen, M., Song, B., Zeng, G., & Zhang, Y. (2020). (Micro)plastic crisis: Un-ignorable contribution to global greenhouse gas emissions and climate change. *Journal of Cleaner Production., 254*, 120138.

Sivaram, A. K., Panneerselvan, L., Surapaneni, A., Lee, E., Kannan, K., & Megharaj, M. (2022). Per-and polyfluoroalkyl substances (PFAS) in commercial composts, garden soils, and potting mixes of Australia. *Environmental Advances, 7*, 100174.

Sobhani, Z., Fang, C., Naidu, R., & Megharaj, M. (2021). Microplastics as a vector of toxic chemicals in soil: Enhanced uptake of perfluorooctane sulfonate and perfluorooctanoic acid by earthworms through sorption and reproductive toxicity. *Environmental Technology & Innovation, 22*, 101476.

Sridharan, S., Kumar, M., Singh, L., Bolan, N. S., & Saha, M. (2021). Microplastics as an emerging source of particulate air pollution: A critical review. *Journal of Hazardous Materials, 418*, 126245.

Stahl, T., Heyn, J., Thiele, H., Hüther, J., Failing, K., Georgii, S., & Brunn, H. (2009). Carryover of perfluorooctanoic acid (PFOA) and perfluorooctane sulfonate (PFOS) from soil to plants. *Archives of Environmental Contamination and Toxicology, 57*, 289–298.

Stahl, T., Mattern, D., & Brunn, H. (2011). Toxicology of perfluorinated compounds. *Environmental Sciences Europe, 23*, 1–52.

Sun, M., Arevalo, E., Strynar, M., Lindstrom, A., Richardson, M., Kearns, B., Pickett, A., Smith, C., & Knappe, D. R. U. (2016). Legacy and emerging perfluoroalkyl substances are important drinking water contaminants in the cape fear river watershed of North Carolina. *Environmental Science and Technology Letters, 3*, 415–419.

Sunderland, E. M., Hu, X. C., Dassuncao, C., Tokranov, A. K., Wagner, C. C., & Allen, J. G. (2018). A review of the pathways of human exposure to poly- and perfluoroalkyl substances (PFASs) and present understanding of health effects. *Journal of Exposure Science & Environmental Epidemiology, 29*, 131–147.

Sunderland, E. M., Hu, X. C., Dassuncao, C., Tokranov, A. K., Wagner, C. C., & Allen, J. G. (2019). A review of the pathways of human exposure to poly- and perfluoroalkyl substances (PFASs) and present understanding of health effects. *Journal of Exposure Science & Environmental Epidemiology, 29*, 131.

Surendran, U., Jayakumar, M., Raja, P., Gopinath, G., & Chellam, P.V., 2023. Microplastics in terrestrial ecosystem: Sources and migration in soil environment. *Chemosphere*, 137946.

Taniyasu, S., Kannan, K., Horii, Y., Hanari, N., & Yamashita, N. (2003). A survey of perfluorooctane sulfonate and related perfluorinated organic compounds in water, fish, birds, and humans from Japan. *Environmental Science and Technology, 37*, 2634–2639.

Teng, H. (2012). Overview of the development of the fluoropolymer industry. *Applied Sciences, 2*, 496–512.

Tian, H., Gaines, C., Launi, L., Pomales, A., Vazquez, G., Goharian, A., Goodnight, B., Haney, E., Reh, C.M., & Rogers, R.D. (2022). Understanding public perceptions of per- and polyfluoroalkyl substances: Infodemiology study of social media. *Journal of Medical Internet Research, 24*(3), e25614. https://www.jmir.org/2022/3/e25614.

Tursi, A., Baratta, M., Easton, T., Chatzisymeon, E., Chidichimo, F., De Biase, M., & De Filpo, G. (2022). Microplastics in aquatic systems, a comprehensive review: Origination, accumulation, impact, and removal technologies. *RSC Advances, 12*, 28318–28340.

Ulusoy, U. (2023). A review of particle shape effects on material properties for various engineering applications: From macro to nanoscale. *Minerals, 13*, 91.

Vorst, K. L., Saab, N., Silva, P., Curtzwiler, G., & Steketee, A. (2021). Risk assessment of per- and polyfluoroalkyl substances (PFAS) in food: Symposium proceedings. *Trends in Food Science & Technology., 116*, 1203–1211.

Waldschläger, K., Lechthaler, S., Stauch, G., & Schüttrumpf, H. (2020). The way of microplastic through the environment–Application of the source-pathway-receptor model. *Science of the Total Environment, 713*, 136584.

Wan, Y., Wu, C., Xue, Q., & Hui, X. (2019). Effects of plastic contamination on water evaporation and desiccation cracking in soil. *Science of the Total Environment, 654*, 576–582.

Wang, Y., Munir, U., & Huang, Q. (2023a). Occurrence of per- and polyfluoroalkyl substances (PFAS) in soil: Sources, fate, and remediation. *Soil & Environmental Health, 1*, 100004.

Wang, Z., Dewitt, J. C., Higgins, C. P., & Cousins, I. T. (2017). A never-ending story of per- and polyfluoroalkyl substances (PFASs)? *Environmental Science and Technology, 51*, 2508–2518.

Wang, Y., Munir, U., & Huang, Q. (2023b). Occurrence of per-and polyfluoroalkyl substances (PFAS) in soil: Sources, fate, and remediation. *Soil & Environmental Health*, 100004.

Wen, Y., Mirji, N., & Irudayaraj, J. (2020). Epigenetic toxicity of PFOA and GenX in HepG2 cells and their role in lipid metabolism. *Toxicology in Vitro, 65*, 104797.

Wu, M., Yang, C., Du, C., & Liu, H. (2020). Microplastics in waters and soils: Occurrence, analytical methods and ecotoxicological effects. *Ecotoxicology and Environmental Safety, 202*, 110910.

Xia, C., Diamond, M. L., Peaslee, G. F., Peng, H., Blum, A., Wang, Z., Shalin, A., Whitehead, H. D., Green, M., Schwartz-Narbonne, H., Yang, D., & Venier, M. (2022). Per- and polyfluoroalkyl substances in North American school uniforms. *Environmental Science and Technology, 56*, 13845–13857.

Xu, B., Liu, S., Zhou, J. L., Zheng, C., Weifeng, J., Chen, B., Zhang, T., & Qiu, W. (2021). PFAS and their substitutes in groundwater: Occurrence, transformation and remediation. *Journal of Hazardous Materials, 412*, 125159.

Xu, B., Alizray, R., Lammel, D. R., Riedel, S., & Rillig, M. C. (2022a). Concentration dependent response of soil parameters and functions to trifluoroacetic acid. *European Journal of Soil Science, 73*, e13266.

Xu, B., Yang, G., Lehmann, A., Riedel, S., & Rillig, M. C. (2023). Effects of perfluoroalkyl and polyfluoroalkyl substances (PFAS) on soil structure and function. *Soil Ecology Letters, 5*, 108–117.

Xu, L., Liang, Y., Zhang, R., Xu, B., Liao, C., Xie, T., & Wang, D. (2022b). Facilitated transport of microplastics and nonylphenol in porous media with variations in physicochemical heterogeneity. *Environmental Pollution, 315*, 120297.

Yang, L., Zhang, Y., Kang, S., Wang, Z., & Wu, C. (2021). Microplastics in soil: A review on methods, occurrence, sources, and potential risk. *Science of the Total Environment, 780*, 146546.

Yasir, A. M., Ma, J., Ouyang, X., Zhao, J., Zhao, Y., Weng, L., Islam, M. S., Chen, Y., & Li, Y. (2022). Effects of selected functional groups on nanoplastics transport in saturated media under diethylhexyl phthalate co-contamination conditions. *Chemosphere, 286*, 131965.

Young, A. S., Pickard, H. M., Sunderland, E. M., & Allen, J. G. (2022). Organic fluorine as an indicator of per- and polyfluoroalkyl substances in dust from buildings with healthier versus conventional materials. *Environmental Science and Technology, 56*, 17090–17099.

Yu, H., Zhang, Y., Tan, W., & Zhang, Z. (2022). Microplastics as an Emerging environmental pollutant in agricultural soils: Effects on ecosystems and human health. *Frontiers in Environmental Science., 10*, 855292.

Zhang, B., Yang, X., Chen, L., Chao, J., Teng, J., & Wang, Q. (2020). Microplastics in soils: A review of possible sources, analytical methods and ecological impacts. *Journal of Chemical Technology & Biotechnology, 95*(8), 2052–2068.

Zhang, Z., Sarkar, D., Biswas, J. K., & Datta, R. (2022). Biodegradation of per- and polyfluoroalkyl substances (PFAS): A review. *Bioresource Technology., 344*, 126223.

Zhao, Z., Cheng, X., Hua, X., Jiang, B., Tian, C., Tang, J., Li, Q., Sun, H., Lin, T., Liao, Y., & Zhang, G. (2020). Emerging and legacy per- and polyfluoroalkyl substances in water, sediment, and air of the Bohai Sea and its surrounding rivers. *Environmental Pollution, 263*, 114391.

Zhao, S., Zhang, Z., Chen, L., Cui, Q., Cui, Y., Song, D., & Fang, L. (2022). Review on migration, transformation and ecological impacts of microplastics in soil. *Applied Soil Ecology, 176*, 104486.

Zhou, Y., Wang, J., Zou, M., Jia, Z., Zhou, S., & Li, Y. (2020). Microplastics in soils: A review of methods, occurrence, fate, transport, ecological and environmental risks. *Science of the Total Environment, 748*, 141368.

Ziajahromi, S., Neale, P. A., Rintoul, L., & Leusch, F. D. (2017). Wastewater treatment plants as a pathway for microplastics: Development of a new approach to sample wastewater-based microplastics. *Water Research, 112*, 93–99.

Chapter 6
Personal Protective Equipment Associated with COVID-19 Pandemic and Microplastic

Sameeksha Rawat and Madhuben Sharma

Abstract The COVID-19 pandemic has led to a significant increase in the use of personal protective equipment (PPE), which has proven essential in reducing viral transmission. Microplastic contamination, on the other hand, is a new environmental hazard brought about by the extensive dependence on PPE. This chapter investigates the relationship between the use of PPE and microplastic pollution, looking at the materials used in PPE manufacture, the processes by which PPE breaks down into microplastics, and their environmental effects. PPE is mostly made of synthetic polymers, such as polypropylene and polyethylene, which break down easily and release microplastics into the environment. This paper explains the effects of PPE-related microplastic contamination on ecosystems and human health by thoroughly reviewing the existing studies and the measures taken. The strategies for reducing single-use PPE, waste management procedures, recycling programs, and other mitigating measures are also assessed. The findings highlight the pressing need for sustainable options for conventional PPE materials as well as the implementation of appropriate disposal procedures. By tackling these issues, we could provide the possibility of a PPE usage strategy that is more ecologically friendly while still preserving ecological integrity and human health in the event of future health emergencies.

Keywords Personal protective equipment · COVID-19 · Microplastics · Environmental consequences

---

S. Rawat · M. Sharma (✉)
Sustainability Cluster, School of Advanced Engineering, University of Petroleum and Energy Studies, Bidholi, Energy Acres, Dehradun 248007, Uttarakhand, India
e-mail: madhubensharma06@gmail.com; mbsharma@ddn.upes.ac.in

A. K. Mishra et al. (eds.), *Microplastics*, https://doi.org/10.1007/978-981-97-6461-7_6

1 Introduction

Synthetic polymers, such as plastics, are frequently sourced from petroleum. These extensive carbon chain polymers have special qualities that may be used to their advantage, such as strength, flexibility, or both (Al-Tohamy et al., 2023). Petroleum-based polymers, both biodegradable and non-biodegradable, account for 90% of the world's plastic output. Polypropylene (PP), Polyurethane (PU), Polyethylene Terephthalate (PET), Polyethylene (PE), Polystyrene (PS), and Polyvinyl chloride (PVC) are among the petroleum-based biodegradable polymers. Polylactic acid (PLA), polyhydroxybutyrate (PHB), poly(butylene succinate adipate) (PBSA), polycaprolactone (PCL), poly (3-hydroxybutyrateco-3-hydroxyvalerate) (PHBV), and polybutylene succinate (PBS) are examples of petroleum-based polymers that are not biodegradable (Al-Tohamy et al., 2023; Urbanek et al., 2018).

Plastics, which are being employed to make an extensive variety of items because of their low cost, great resilience, and lightweight, are emerging as the primary material generating marine pollution. However, issues like different types of environmental pollution are developing because of poor disposal of plastics, in contrast to the increased demand in different products (Ray et al., 2022). Microplastics (MP) can arise from the fragmentation and degradation of plastics in the environment due to physicochemical or biological weathering processes (Al-Tohamy et al., 2023). Plastic trash (less than 5 mm in diameter) is widely present and is becoming an increasing environmental problem (Horton & Dixon, 2018). MPs have a high specific surface area and are tiny, thus they can collect on the surfaces of different contaminants. MPs are easily consumed by animals and serve as a vehicle for several hazardous substances, including antibiotics, polycyclic aromatic hydrocarbons (PAH), heavy metals, and organo-chorine insecticides. As a result, they accumulate in the tissues of animals, leading to pathological stress and hindering reproduction (Al-Tohamy et al., 2023; Elgarahy et al., 2021). There are two sorts of MPs: primary MPs, which are generated at this microscopic size and are often found in cosmetic items, and other one is the degradation products of bigger plastic objects known as secondary MPs (Picó & Barceló Apr., 2019).

Plastic garbage rapidly fragments due to photo-oxidation caused by ultraviolet (UV) radiation. The degradation of plastics can proceed extremely slowly in a marine environment due to its cold and anoxic conditions. Numerous origins of microplastics (MPs) and the distinct fragmentation processes they go through account are based on the differences in their morphologies, which can include pellets, fibres, and pieces (Akdogan & Guven, 2019; Klein et al., 2015). After the degradation, MPs are widely distributed, and they have a potential to either sink to the seafloor or become a part of the food chain. It has been shown that pregnant women's placentas contain plastic fragments that range in diameter from 5 to 10 μm (Ragusa et al., 2021).

In one of the studies, researchers reported that in the selected 36 countries, the total estimated amount of plastic waste and MPs generated were 4.2 million tonnes and 9774 thousand tonnes, respectively. This poses a serious risk to international initiatives to reduce the use of plastic (Shukla et al., 2022). Before the COVID-19 epidemic, it was estimated that despite many international government initiatives to decrease single-use items, the amount of plastic waste would quadruple by 2030 (Patrício Silva et al., 2021).

The COVID-19 pandemic initially surfaced at Wuhan, China's Hunan seafood market in December 2019 (Rowan & Laffey, 2021). The World Health Organisation (WHO) classified the illness brought on by the severe acute respiratory syndrome coronavirus 2 (SARS-CoV-2) as a global public health emergency (Khan et al., 2023). The sickness soon became widespread when it first appeared, eventually endangering the health of everyone on the planet (Wu et al., 2020). One preventative measure used during the pandemic to stop viral transmission was personal protective equipment (PPE). One-use PPE such as surgical gowns is frequently composed of nonwoven materials comprising polypropylene, polyethylene, and polyethylene terephthalate, and is mostly used for single-use applications (Prata et al., 2021; Uddin et al., 2022). The world's ecology has been negatively impacted by the plastic garbage produced by PPE kits, gloves, single-use face masks, and other gear.

The environment is endangered by the inappropriate use of PPE kits, the dumping of biological waste, and rise in the volume of plastic trash from residential homes. According to certain studies, there may be long-term effects on both human health and the ecosystem (Ray et al., 2022). Numerous studies have demonstrated ways to address the growing waste produced by PPE kits and lessen their prolonged consequences on the environment and public health. An efficient pyrolysis technique to convert PPE kit trash into liquid fuels was proposed by Jain et al. (2022). Pyrolysis is one energy conversion method that suggests using COVID-19 polymeric wastes to produce bio-crude oil and char (Aragaw & Mekonnen, 2021). One of the researchers has shown how to decontaminate facemasks and N95 respirators while maintaining their filtering capacity using a dry heat pasteurisation process (Xiang et al., 2020). This chapter looks into the association between the COVID-19 epidemic's widespread use of PPE and the production of microplastics. It also clarifies the types and materials used in PPE production, analyses the mechanisms of releasing microplastics, evaluates the consequences of the contamination on ecosystems and human health, and suggests mitigating measures like recycling and sustainable materials.

2 PPE: Types and Materials

Wearing clothes or other protective gear to guard against fire, hazardous chemicals, or physical harm is known as PPE (Sutton, 2017). Viral infection risk and airborne virus load are significantly decreased by N95 respirators, which can remove 95% of the airborne particles in circulation (Xiang et al., 2020). Globally, there were 163,738,674 confirmed COVID-19 cases as of May 20, 2021, and 3,384,750 fatalities (Ray et al., 2022). In many nations, the COVID-19 wave recurrence has left healthcare systems depleted.

PPE became a life-saving tool for any member of the medical staff who is either directly or indirectly involved in hospital setup during and after COVID-19 (Derraik et al., 2020). The majority of authorities advise that wearing specific PPE and masks help in reducing the spread of viruses among the public, while each country has its methods depending on the socioeconomic situation (Nguyen et al., 2020).

3 Materials Used in PPE Production

A summary of the raw materials utilised in the production of PPE components is provided in Table 1 (Jain et al., 2022). PP, a non-woven material with a single usage, is the polymer that is primarily utilised in the production of PPE. PP's fundamental structure is a saturated carbon chain with a methyl group joined to a different carbon atom. Methyl groups give PP its toughness and set it apart from polyethylene. Its intrinsic, long-lasting electrostatic charge, attracts and holds onto particles to shield the wearer from infections of any type. Its hardness, high tensile strength, and low weight further qualify it for use in industrial production (Mondal, 2022).

The physical properties of PP have certain values such as the notched impact strength and thermal coefficient of expansion should be 3.0–30.0 kJ/m^2 and 100–150 $\times$ 10^{-6} respectively. Similarly, the glass transition temperature and maximum service temperature should be −10 and 140 °C, respectively. The specific heat capacity must be 1520 J/ (kg K) at 20 °C.

The melting point is 160–168 °C, tensile strength should be 0.95–1.30 N/mm^2, and density must be 0.905 g/cm^3. The elastic modulus and impact strength should be 1.5–3 GPa and 2–6 kJ/m^2 at 20 °C, respectively. The values of flammability, thermal conductivity and electric resistivity should be UL 94 HB, 0.41 W/(m K) at 20 °C, and 1*1013–1*1014°Ω.m at 20 °C, respectively ("Polypropylene: Properties, Processing, & Applications", 2020; Jain et al., 2022).

There are three distinct stereo-specific configurations in which PP can be synthesised (Maddah, 2016; Malpass & Band, 2012):

a. *Atactic*: There is no consistency in the arrangement of the chiral carbons, and the methyl groups are sporadically found on both sides.
b. *Isotactic*: Every chiral carbon with a methyl group has the same configuration. Either above or below the plane, all of the methyl groups are present.
c. *Syndiotactic*: The methyl groups alternately exist, one above and one below the plane, and the chiral carbon configurations are inversely similar to one another.

Out of the three, the isotactic PP has an advantage due to its more crystalline form. Unlike other forms, the crystalline nature results from the regular array of the polymer's chains. PP has a melting temperature of 171 °C that is fully isotactic (Shubhra et al., 2013). This enables the items to be steam sterilised, making the use of PP in medicine possible. The utilisation of polypropylene (PP) for surgical tools and PPE is favoured due to its high thermal stability, high dimensional stability, bacterial resistance and the response to different chemicals, bases, acid alcohols,

Table 1 Raw materials utilised in the manufacture of PPE components ("Polypropylene: Properties, Processing, & Applications", 2020; Mondal, 2022)

Component of PPE	Image	Raw material used	Chemical formula
Single use protective gowns	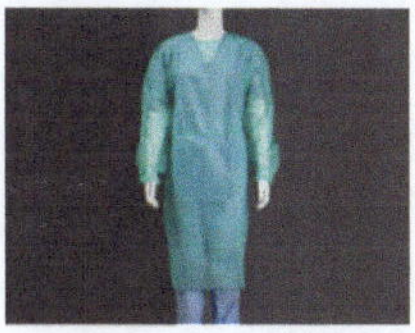 (*source* Isolation Gown, 2024)	Normally polypropylene	$(C_3H_6)n$
Normal surgical masks	 (*source* "Face mask" & rawpixel., 2024)	Polypropylene	$(C_3H_6)n$
Face shields	 (*source* "Qatar Scientific Club Manufactures 6 Thousand Face Shields for HMC & Oryx Staff, 2020)	Polycarbonate, Acetate, Propionate, Polyethylene terephthalate glycol, and Polyvinyl chloride	$C_{15}H_{16}O_2$, $C_2H_3O_2^-$, $C_3H_5O_2$, $-(C_2H_4)n-$, $(C_{10}H_8O_4)n$, C_2H_3Cl
Goggles	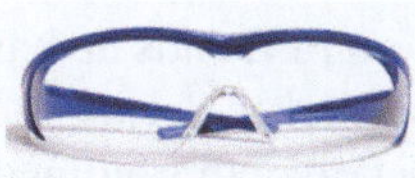 (*source* Lilly, 2024)	High quality polycarbonates	$C_{15}H_{16}O_2$
N95 respirators	 (*source* Face Mask, 2024)	Polypropylene	$(C_3H_6)n$

(continued)

Table 1 (continued)

Component of PPE	Image	Raw material used	Chemical formula
Coveralls	 (*source* "Coveralls" & BooksChatter. Accessed, 2024)	High density polyethylene	–(C2H4)n–,
Powered air purifying respirators	 (*source* Al-Mallah, 2024)	Rubber or silicone	(C_5H_8)n or (R_2SiO)x

esters, aldehydes, ketones, and aliphatic hydrocarbons, as well as its flexibility for moulding (Jain et al., 2022).

4 Durability and Disposal of PPE

PPE worn just once during emergencies like COVID-19 was difficult to be broken down at room temperature. The plastic material from PPE ends up harming the environment by winding up in the water and landfills. It takes decades for the microorganism to decompose such biomedical waste. To get rid of this biomedical trash there is an option of recycling polymers by physical and chemical processes. this will stop the discharge of biomedical surplus into the environment. This technique incorporates the three pillars of sustainable development goals: reduce, reuse, and recycle (Jain et al., 2022; Sarker et al., 2012). The Indian Central Board of Pollution Control (CPCB) has released rules for the collection, disposal, and segregation of COVID-19 waste. The CPCB has provided guidelines on the administration of used PPE kits in its fourth amended guideline for managing, treating, and discarding waste produced during the diagnosis, course of therapy, and isolation of COVID-19 victims. These guidelines state that (CPCB) (Gupta, 2021):

a. PPEs that the public discards at businesses, organisations, malls, offices, etc., shall be kept in a separate container for three days after which they should be chopped or shredded and disposed of as dry general solid trash.
b. To prevent reuse, gloves and waste masks from homes should be stored in paper bags for at least 72 h before being disposed of as dry general solid trash after being cut.
c. Used PPEs, including goggles, a splash-proof apron, face shields, hazmat suits, plastic coveralls, and nitrile gloves, must be gathered from COVID-19 isolation wards and placed in a red bag.
d. Used PPE kits at hospitals are required to be packaged in yellow, double-layered bags marked as COVID-19 trash.

Additionally, all garbage linked to COVID-19 must be transported in separate vehicles, with each vehicle being cleaned after each journey. The Biomedical Trash Management Act's Schedule 2 specifies how to dispose of this kind of trash: deep burial or incineration for yellow-coloured waste, and microwave, autoclaving, or chemical treatment for red-coloured waste. PPE disposal is one of the causes of microplastic pollution and is responsible for the disintegrating the environmental sustainability.

The subsequent sections discuss what is microplastic pollution, how it is distributed in the environment, what are its environmental consequences, how is PPE usage related to microplastic pollution, and lastly, what are the mitigating strategies to reduce this pollution.

5 Microplastics Pollution

Plastics are thought to be among the most persistent and common contaminants. Plastic fragmentation and buildup on the surface of the planet are recent concerns related to its use. Plastics accumulate because they are long-lasting and have a low recycling rate. These large plastic fragments eventually break down into smaller particles called microplastics (Gregory, 1999). Plastic pollutants were divided into three categories: mega-debris, which has a diameter of more than 100 mm, meso-debris, which has a diameter of 5 to 20 mm, and macro-debris, which has a diameter of more than 20 mm. Most macroplastics are tiny particles, and they are easily seen with the help of the eye (Loganathan & Kizhakedathil, 2023). MPs have been regarded as pollutants since the year 2000. MP is linked to fibres, granules, and tiny pieces of plastic (Cole et al., 2011). Since there are discrepancies in the research, it is challenging to create a scientific guideline for the definition and classification of microplastics. MPs are described by Graham and Thompson (Graham & Thompson, 2009) as elements with a diameter of less than 10 mm. Moore (2008) separated the plastic debris into two types according to size: MPs are smaller than 5 mm while macroplastics are greater than 5 mm. According to Betts (2008) and Barnes et al. (2009), they have less than 5 mm. Derraik (2002) stated the size of these particles should be between 2 and 6 mm, whereas Ryan et al. (2009) stated that MP diameters less than 2 mm, and other literary works advised the size of these particles should be less than 1 mm. MPs are categorised as primary, secondary, and nano plastics depending on where they come from van Cauwenberghe et al. (2015).

Primary MPs: Primary MPs are polymers that are minor in size. They originate from consumer-based products like cosmetics directly or indirectly (Wang et al., 2019). Primary MPs also originate from the wear and tear of sizable plastic substances during production, usage, or upkeep—for example, the eroding of tyres when driving or the wearing and tearing of synthetic fabrics while being washed (Boucher & Friot, 2017).

Secondary MPs: The combined effects of physical, chemical, thermal, biological, and photochemical processes result in the formation of secondary MPs from bigger plastic debris (Dawson et al., 2018; Loganathan & Kizhakedathil, 2023). This occurs because of fishing net losses or other inadvertent losses caused by weathering processes and photodegradation of improperly managed garbage, such as abandoned plastic bags (Boucher & Friot, 2017). Secondary MPs are characterised as irregularly sized and shaped pieces of worn, embrittled plastic. Many of the fractured MPs are angular, subrounded, rounded, and sub-angular in shape. Rough, sharp, broken edges, and deteriorated are other ways that they might be depicted (Loganathan & Kizhakedathil, 2023).

Nano-plastics: Tiny plastics that are less than 100 nm are known as nano-plastics. In addition to coming from manufactured material, nano-plastics are produced as the MP detritus breaks down and weathered (Abu-Hilal & Al-Najjar, 2009). When plastic waste is exposed to sunlight, the pace at which photo-oxidation occurs is accelerated, leading to increased brittleness. Plastic fragmentation is further broken down by abrasion and wave action into nano- and micro-sized ($\leq 0.1\ \mu m$) and 0.1–$1000\ \mu m$) particles (da Costa et al., 2016).

6 MPs Distribution in the Environment

There are several ways that MPs may get into the environment, for instance, soil, surface water, air, subsurface water, etc. (Shukla et al., 2022).

Few research studies address the complicated and varied pathways that primary MPs from various sources take before entering the aquatic environment. The quantity of MPs entering the aquatic environment is therefore difficult to measure. To make this easier, the channels from various sources that infiltrate aquatic ecosystems have been categorised into four groups: wind, road runoff, neighbouring waters (mostly marine coating), and residential sewage (Wang et al., 2019).

7 Environmental Consequences of MPs Pollution

Plastics generated from disposable face masks and used PPE soak the toxic and organic pollutants present in the aquatic ecosystem, which results in a perilous film in the system. When aquatic creatures consume plastic-based particles, this process may result in poisoning (Williams-Wynn & Naidoo, 2020). Table 2 illustrates how the fragmentation of MPs may cause bioaccumulation in the food chain and perhaps detrimental consequences. Consuming MPs, even in tiny doses, can have a significant influence on seabirds' blood calcium levels, morphometrics, cholesterol, amylase levels, and uric acid levels (e.g., Ardenna carneipes) (Lavers et al., 2019).

8 Relationship Between PPE Usage and Microplastic Pollution

The broad use of PPE in the wake of the COVID-19 pandemic has been crucial in preserving public health and slowing the virus's spread. Microplastic contamination is an unexpected environmental consequence of the increased use of PPE, despite its crucial function in infection prevention. To shed light on how the materials used in PPE manufacture, their degradation processes, and the ensuing environmental repercussions contribute to this critical issue, this section investigates the complex relationship between the use of PPE and the spread of microplastics in the environment.

Materials Utilised in the Production of PPE: PPE is made from a wide range of diverse materials and polymers, including synthetic polymers, vinyl, latex, nitrile disposable gloves, polyurethane and/or polypropylene face masks, and other synthetic fibres (e.g. disinfection wipes) (Ammendolia et al., 2021). However, MP contamination has emerged as a new threat to both terrestrial and aquatic environments as a result of the increasing use of PPE during the COVID-19 epidemic. More than 10 million face masks are reportedly discarded improperly each month, releasing them into the environment (Khan et al., 2023). Global MP pollution has gotten worse due to massive amounts of plastic waste released into the environment.

PPE Breakdown into Microplastics: The environment may need several hundred years for commercial and conventional plastics to break down since they are not biodegradable. Plastics are subject to both biotic and abiotic weathering in the environment. There are alterations to MPs' mechanical, functional group, hydrophobicity, and crystallinity characteristics throughout the degradation process (Pizarro-Ortega et al., 2022). Several variables might affect the degradation process, including mechanical stress, microbial action, UV radiation, and many more (Ray et al., 2022).

Environmental Effects of PPE-related Microplastic Pollution: MPs can have ecotoxicological effects on a variety of creatures depending on their form, size, and concentration (Rochman et al., 2019). Plastic made from waste PPE and single-use face masks can absorb different organic and poisonous contaminants, resulting in the

Table 2 The effects of aquatic plastic contamination on the environment

Sources	Problems	Consequences	References
Bags made of plastic for medical waste	– Improper handling of solid waste in the medical industry – Mishandling and inappropriate garbage bag disposal from medical waste	– Organic pollution bioaccumulation – Toxin bioaccumulation – Heavy toxic metal and persistent organic pollutant (POP) bioaccumulation	Yang et al., (2020)
Plastic waste in day-to-day living	Tangle brought on by plastic strands	Loss of aquatic life, including marine wildlife (e.g., seagulls and turtles) tangle brought on by plastic strands	Haque et al., (2021)
Polymers found in PPE kits	– Improper handling of solid waste – Improper disposal of medical gloves and PPE kits	– Hazardous compounds bioaccumulating and eventually making their way into the food chain – COVID-19 outbreak medium	Haque et al., (2021), Klemeš et al., (2020)
Face mask-derived macroplastics	Misuse of face shields and masks intended for one usage	Serves as a conduit for COVID-19 pandemics due to the ability of MPs to accumulate in the food chain	Klemeš et al., (2020)
Other plastic wastes	– Adsorption of POPs and a variety of hazardous substances – Comprise a range of hazardous materials and compounds	– Negatively impact marine life and worsen the ecosystem – As a chemical or toxic material degrades, it may leak into aquatic habitats	Schnurr et al., (2018)

formation of a hazardous film. When aquatic species consume plastic-based particles, this process may result in toxicity. After their fragmentation, MPs bioaccumulate in the food chain, which may have a negative influence on the ecosystem (Williams-Wynn & Naidoo, 2020).

9 Mitigating Strategies to Reduce MP and PPE Pollution

Environmental management techniques must be put into practice to protect human health and the stability of aquatic ecosystems. MP generation can be greatly decreased by creating and using sustainable substitutes for conventional PPE materials, such as biodegradable polymers or natural fibres. Due to their accelerated environmental degradation, these compounds are less persistent and provide less risk of damage (Ray et al., 2022).

Prevention of contamination requires the use of efficient waste management techniques, such as the appropriate disposal and recycling of PPE. The introduction of MPs into the environment can be reduced by initiating recycling programs, encouraging the segregation of PPE waste, and building specialised collection and recycling facilities (Singh et al., 2020).

The amount of PPE waste generated overall and the ensuing MP pollution can be reduced by putting pollution control measures into place, such as encouraging responsible consumption habits or decreasing the usage of single-use PPE by switching to reusable choices (Eckelman et al., 2012).

10 Conclusion

An analysis of the relationship between the use of PPE and MP pollution sheds light on the environmental consequences of our response to the COVID-19 pandemic. Examining PPE components, breakdown mechanisms, and environmental effects highlights a concerning fact: the extensive use of synthetic polymers in PPE production has increased microplastic pollution. PPE products emit microplastics into ecosystems when they break down, frequently because of incorrect disposal techniques, which causes ecological degradation and endangers biodiversity. Microplastics absorb and transfer harmful contaminants, making their way into the food chain, posing health hazards to humans.

The chapter also points out viable approaches for mitigating these issues. Pathways to lessen PPE and microplastic pollution include adopting sustainable material alternatives, improving waste management procedures, and putting pollution control measures into place. Furthermore, governmental interventions are crucial because they drive the creation and implementation of laws controlling the manufacture, use, and disposal of PPE. We can protect ecosystems and human health for present and future generations by reducing the negative consequences of microplastic contamination via coordinated actions.

References

Abu-Hilal, A. H., & Al-Najjar, T. H. (2009). Plastic pellets on the beaches of the Northern Gulf of Aqaba, Red Sea. *Aquatic Ecosystem Health and Management, 12*(4), 461–470. https://doi.org/10.1080/14634980903361200

Akdogan, Z., & Guven, B. (2019). "Microplastics in the environment: A critical review of current understanding and identification of future research needs. *Environmental Pollution, 254.* Elsevier Ltd. https://doi.org/10.1016/j.envpol.2019.113011.

Al-Mallah, Z. (2024) Everything you need to know about medical masks. Noon Post. Retrieved April 07, 2024, from https://www.noonpost.com/content/36489.

Al-Tohamy, R., et al. (2023). Can wood-feeding termites solve the environmental bottleneck caused by plastics? A critical state-of-the-art review. *Journal of Environmental Management, 326.* Academic. https://doi.org/10.1016/j.jenvman.2022.116606.

Ammendolia, J., Saturno, J., Brooks, A. L., Jacobs, S., & Jambeck, J. R. (2021). An emerging source of plastic pollution: Environmental presence of plastic personal protective equipment (PPE) debris related to COVID-19 in a metropolitan city. *Environmental Pollution, 269*, 116160. https://doi.org/10.1016/j.envpol.2020.116160

Aragaw, T. A., & Mekonnen, B. A. (2021). Current plastics pollution threats due to COVID-19 and its possible mitigation techniques: A waste-to-energy conversion via Pyrolysis. *Environmental Systems Research, 10*(1). https://doi.org/10.1186/s40068-020-00217-x.

Barnes, D. K. A., Galgani, F., Thompson, R. C., & Barlaz, M. (2009). Accumulation and fragmentation of plastic debris in global environments. *Philosophical Transactions of the Royal Society B, 364*(1526), 1985–1998. https://doi.org/10.1098/rstb.2008.0205

Betts, K. (2008). Why small plastic particles may pose a big problem in the oceans. *Environmental Science and Technology, 42*(24), 8995. https://doi.org/10.1021/es802970v

Boucher, J., & Friot, D. (2017) *Primary Microplastics in the Oceans: a Global Evolution of Sources* (Vol. 10). IUCN. https://doi.org/10.2305/IUCN.CH.2017.01.en.

Cole, M., Lindeque, P., Halsband, C., & Galloway, T. S. (2011). Microplastics as contaminants in the marine environment: A review. *Marine Pollution Bulletin, 62*(12), 2588–2597. https://doi.org/10.1016/j.marpolbul.2011.09.025

da Costa, J. P., Santos, P. S. M., Duarte, A. C., & Rocha-Santos, T. (2016). (Nano)plastics in the environment—Sources, fates and effects. *Science of the Total Environment, 566–567*, 15–26. https://doi.org/10.1016/j.scitotenv.2016.05.041

Coveralls, Books Chatter. Retrieved April 07, 2024, from https://bookschatter.blogspot.com/2016/04/Q-A-fibers.html.

Dawson, A. L., et al. (2018). Turning microplastics into nanoplastics through digestive fragmentation by Antarctic krill. *Nature Communications, 9*(1). https://doi.org/10.1038/s41467-018-034 65-9.

Derraik, J. G. B. (2002). The pollution of the marine environment by plastic debris: a review. *Marine Pollution Bulletin, 44*, 842–852. 2002. www.elsevier.com/locate/marpolbul.

Derraik, J. G. B., Anderson, W. A., Connelly, E. A., & Anderson, Y. C. (2020). Rapid review of SARS-CoV-1 and SARS-CoV-2 viability, susceptibility to treatment, and the disinfection and reuse of ppe, particularly filtering facepiece respirators. *International Journal of Environmental Research and Public Health, 17*(17), 1–31. https://doi.org/10.3390/ijerph17176117

Eckelman, M., Mosher, M., Gonzalez, A., & Sherman, J. (2012). Comparative life cycle assessment of disposable and reusable laryngeal mask airways. *Anesthesia and Analgesia, 114*(5), 1067–1072. https://doi.org/10.1213/ANE.0b013e31824f6959

Elgarahy, A. M., Akhdhar, A., & Elwakeel, K. Z. (2021). Microplastics prevalence, interactions, and remediation in the aquatic environment: A critical review. *Journal of Environmental Chemical Engineering, 9*(5). Elsevier Ltd. https://doi.org/10.1016/j.jece.2021.106224.

Face mask. Rawpixel. Retrieved April 07, 2024, from https://www.rawpixel.com/search/facemask.

Graham, E. R., & Thompson, J. T. (2009). Deposit- and suspension-feeding sea cucumbers (Echinodermata) ingest plastic fragments. *Journal of Experimental Marine Biology and Ecology, 368*(1), 22–29. https://doi.org/10.1016/j.jembe.2008.09.007

Gregory, M. R. (1999). Plastics and South Pacific Island shores: Environmental implications. *Ocean and Coastal Management, 42*(6–7), 603–615. https://doi.org/10.1016/S0964-5691(99)00036-8

Gupta, H. (2021) Introduction of PPE kit as a new pollutant in environment. *International Journal of Law Management & Humanities, 4*. https://doi.org/10.10000/IJLMH.11969.

Haque, M. S., Uddin, S., Sayem, S. M., & Mohib, K. M. (2021). Coronavirus disease 2019 (COVID-19) induced waste scenario: A short overview. *Journal of Environmental Chemical Engineering, 9*, 104660. https://doi.org/10.1016/j.jece.2020.104660

Horton, A. A., & Dixon, S. J. (2018). Microplastics: An introduction to environmental transport processes. *Wiley Interdisciplinary Reviews: Water, 5*(2). https://doi.org/10.1002/WAT2.1268.

Isolation Gown. (2024). *CVprotection.com*. Retrieved April 07, 2024, from https://cvprotection.com/product/isolation-gown/.

Jain, S., Yadav Lamba, B., Kumar, S., & Singh, D. (2022). Strategy for repurposing of disposed PPE kits by production of biofuel: Pressing priority amidst COVID-19 pandemic. *Biofuels, 13*(5), 545–549. https://doi.org/10.1080/17597269.2020.1797350.

Khan, M. T., et al. (2023). Personal protective equipment (PPE) disposal during COVID-19: An emerging source of microplastic and microfiber pollution in the environment. *Science of the Total Environment, 860*, 160322. https://doi.org/10.1016/j.scitotenv.2022.160322

Klein, S., Worch, E., & Knepper, T. P. (2015). Occurrence and spatial distribution of microplastics in river shore sediments of the rhine-main area in Germany. *Environmental Science and Technology, 49*(10), 6070–6076. https://doi.org/10.1021/acs.est.5b00492

Klemeš, J. J., Van Fan, Y., Tan, R. R., & Jiang, P. (2020). Minimising the present and future plastic waste, energy and environmental footprints related to COVID-19. *Renewable and Sustainable Energy Reviews, 127*, 109883. https://doi.org/10.1016/j.rser.2020.109883

Lavers, J. L., Hutton, I., & Bond, A. L. (2019). Clinical pathology of plastic ingestion in marine birds and relationships with blood chemistry. *Environmental Science and Technology, 53*(15), 9224–9231. https://doi.org/10.1021/acs.est.9b02098

Lilly, M., (2024). Laboratory protection goggles. Wikipedia Commons. Retrieved April 07, 2024, from https://commons.wikimedia.org/wiki/File:Laboratory_protection_goggles-blue.jpg.

Loganathan, Y., & Kizhakedathil, M. P. J. (2023). A review on microplastics—An indelible ubiquitous pollutant. *Biointerface Research in Applied Chemistry, 13*(2). https://doi.org/10.33263/BRIAC132.126.

Maddah, H. A. (2016). Polypropylene as a promising plastic: A review. *American Journal of Polymer Science, 6*(1), 1–11. https://doi.org/10.5923/j.ajps.20160601.01

Malpass, D. B., & Band, E. I. (2012). *Introduction to Industrial Polypropylene*. John Wiley & Sons.

Mondal, S. (2022). Environmental catastrophe amidst COVID-19 pandemic: Disposal and management of PPE kits for the production of biofuel with the sustainable approach in solar thermal energy. *Materials Today Proceedings, 64*, 1266–1271. https://doi.org/10.1016/j.matpr.2022.03.721

Moore, C. J. (2008). Synthetic polymers in the marine environment: A rapidly increasing, long-term threat. *Environmental Research, 108*(2), 131–139. https://doi.org/10.1016/j.envres.2008.07.025

N95 Face Mask, PNG ALL. Retrieved April 07, 2024, from https://www.pngall.com/n95-mask-png/download/44156.

Nguyen, L. H., et al. (2020). Risk of COVID-19 among front-line health-care workers and the general community: A prospective cohort study. *Lancet Public Health, 5*(9), e475–e483. https://doi.org/10.1016/S2468-2667(20)30164-X

Patrício Silva, A. L., et al. (2021). Increased plastic pollution due to COVID-19 pandemic: Challenges and recommendations. *Chemical Engineering Journal, 405*. Elsevier B.V. https://doi.org/10.1016/j.cej.2020.126683.

Picó, Y., & Barceló, D. (2019). Analysis and prevention of microplastics pollution in water: Current perspectives and future directions. *ACS Omega, 4*(4), 6709–6719. https://doi.org/10.1021/acsomega.9b00222

Pizarro-Ortega, C. I., et al. (2022). Degradation of plastics associated with the COVID-19 pandemic. *Marine Pollution Bulletin, 176*, 113474. https://doi.org/10.1016/j.marpolbul.2022.113474

Polypropylene: Properties, Processing, and Applications. (2020, May 05) *Matmatch.* Retrieved April 07, 2024, from https://matmatch.com/learn/material/polypropylene

Prata, J. C., Patrício Silva, A. L., Duarte, A. C., & Rocha-Santos, T. (2021). Environments disposable over reusable face masks: Public safety or environmental disaster? https://doi.org/10.3390/environments.

Qatar Scientific Club Manufactures 6 Thousand Face Shields for HMC and Oryx Staff (2020, October 07) *What's Goin On Qatar*. Retrieved April 07, 2024. https://wgoqatar.com/65215.

Ragusa, A., et al. (2021). Plasticenta: First evidence of microplastics in human placenta. *Environment International, 146*. Elsevier Ltd. https://doi.org/10.1016/j.envint.2020.106274.

Ray, S. S., Lee, H. K., Huyen, D. T. T., Chen, S. S., & Kwon, Y. N. (2022). Microplastics waste in environment: A perspective on recycling issues from PPE kits and face masks during the COVID-19 pandemic. *Environmental Technology and Innovation, 26*. Elsevier B.V. https://doi.org/10.1016/j.eti.2022.102290.

Rochman, C. M., et al. (2019). Rethinking microplastics as a diverse contaminant suite. *Environmental Toxicology and Chemistry, 38*(4), 703–711. https://doi.org/10.1002/etc.4371

Rowan, N. J., & Laffey, J. G. (2021). Unlocking the surge in demand for personal and protective equipment (PPE) and improvised face coverings arising from coronavirus disease (COVID-19) pandemic—Implications for efficacy, re-use and sustainable waste management. *Science of the Total Environment, 752*. Elsevier B.V. https://doi.org/10.1016/j.scitotenv.2020.142259.

Ryan, P. G., Moore, C. J., Van Franeker, J. A., & Moloney, C. L. (2009). Monitoring the abundance of plastic debris in the marine environment. *Philosophical Transactions of the Royal Society B, 364*(1526), 1999–2012. https://doi.org/10.1098/rstb.2008.0207

Sarker, M., Rashid, M. M., Molla, M., & Rahman, M. S. (2012) A new technology proposed to recycle waste plastics into hydrocarbon fuel in USA *3*(5), 2076–2909. Journal homepage: www.IJEE.IEEFoundation.org ISSN. www.IJEE.IEEFoundation.org.

Schnurr, R. E. J., et al. (2018). Reducing marine pollution from single-use plastics (SUPs): A review. *Marine Pollution Bulletin, 137*, 157–171. https://doi.org/10.1016/j.marpolbul.2018.10.001

Shubhra, Q. T. H., Alam, A. K. M. M., & Quaiyyum, M. A. (2013). Mechanical properties of polypropylene composites: A review. *Journal of Thermoplastic Composite Materials, 26*(3), 362–391. https://doi.org/10.1177/0892705711428659

Shukla, S., Khan, R., Saxena, A., & Sekar, S. (2022). Microplastics from face masks: A potential hazard post Covid-19 pandemic. *Chemosphere, 302*, 134805. https://doi.org/10.1016/j.chemosphere.2022.134805

Singh, N., Tang, Y., & Ogunseitan, O. A. (2020). Environmentally sustainable management of used personal protective equipment. *Environmental Science and Technology, 54*(14), 8500–8502. https://doi.org/10.1021/acs.est.0c03022

Sutton, I. (2017). Personal protective equipment. In *Plant Design and Operations* (pp. 401–415). Elsevier. https://doi.org/10.1016/B978-0-12-812883-1.00014-0.

Uddin, M. A., Afroj, S., Hasan, T., Carr, C., Novoselov, K. S., & Karim, N. (2022). Environmental impacts of personal protective clothing used to combat COVID-19. *Advanced Sustainable Systems, 6*(1). John Wiley and Sons Inc. https://doi.org/10.1002/adsu.202100176.

Urbanek, A. K., Rymowicz, W., & Mirończuk, A. M. (2018). Degradation of plastics and plastic-degrading bacteria in cold marine habitats. *Applied Microbiology and Biotechnology, 102*(18), 7669–7678. Springer. https://doi.org/10.1007/s00253-018-9195-y.

Van Cauwenberghe, L., Devriese, L., Galgani, F., Robbens, J., & Janssen, C. R. (2015). Microplastics in sediments: A review of techniques, occurrence and effects. *Marine Environment Research, 111*, 5–17. https://doi.org/10.1016/j.marenvres.2015.06.007

Wang, T., et al. (2019). Emission of primary microplastics in mainland China: Invisible but not negligible. *Water Research, 162*, 214–224. https://doi.org/10.1016/j.watres.2019.06.042

Williams-Wynn, M. D., & Naidoo, P. (2020). A review of the treatment options for marine plastic waste in South Africa. *Marine Pollution Bulletin, 161*, 111785. https://doi.org/10.1016/j.marpolbul.2020.111785

Wu, J. T., Leung, K., & Leung, G. M. (2020). Nowcasting and forecasting the potential domestic and international spread of the 2019-nCoV outbreak originating in Wuhan, China: A modelling study. *The Lancet, 395*(10225), 689–697. https://doi.org/10.1016/S0140-6736(20)30260-9

Xiang, Y., Song, Q., & Gu, W. (2020). Decontamination of surgical face masks and N95 respirators by dry heat pasteurization for one hour at 70°C. *American Journal of Infection Control, 48*(8), 880–882. https://doi.org/10.1016/j.ajic.2020.05.026

Yang, Y., Liu, W., Zhang, Z., Grossart, H.-P., & Gadd, G. M. (2020). Microplastics provide new microbial niches in aquatic environments. *Applied Microbiology and Biotechnology, 104*, 6501–6511. https://doi.org/10.1007/s00253-020-10704-x/Published

Chapter 7
Microplastics in the Marine Environment

A. D. Forero López, C. V. Colombo, G. E. De la Torre, and M. D. Fernandez-Severini

Abstract In recent years, microplastics (MPs) have emerged as a significant concern due to their widespread presence, non-biodegradable nature, and ability to adsorb, leach, and transport other pollutants in the marine environment. In addition, they pose a potential risk to human health through the consumption of seafood. MPs can enter the oceans by different pathways and can be transported over long trajectories, disseminating various types of pathogens, and organic and inorganic contaminants, in marine ecosystems. Once introduced into aquatic environments, MPs undergo various biotic and abiotic processes, including weathering and biodegradation. These processes alter their physicochemical characteristics, thereby influencing their toxicological impact on organisms. This chapter discusses how these plastic particles enter marine ecosystems, their dynamics once inside, and the degradation processes to which they are exposed, as well as advances in MPs sampling and analyzing techniques.

Keywords Microplastics · Weathering · Oceans · Sources · Marine organisms

1 Introduction

Microplastic (MPs) pollution has become a rapidly growing threat in aquatic environments worldwide. These plastic particles, called MPs (plastic pieces between 1 μm and 5 mm), are ubiquitous, and depending on their source, they can be divided into primary MPs, which are manufactured with this size range for commercial purposes (Andrady, 2011; Boucher & Friot, 2017; Cole et al., 2011; Wang et al., 2019) and secondary MPs, which come from the fragmentation of large pieces due to their weathering in the environment (Andrady, 2011; Cole et al., 2011; Shim &

A. D. F. López (✉) · C. V. Colombo · M. D. Fernandez-Severini
Instituto Argentino de Oceanografía (IADO), CONICET/UNS, CCT-Bahía Blanca, Camino La Carrindanga, km 7.5, Edificio E1, Bahía Blanca, B8000FWB Buenos Aires, Argentina
e-mail: aforero@iado-conicet.gob.ar

G. E. De la Torre
Universidad San Ignacio de Loyola, Av. La Fontana 501, Lima 12, Lima, Perú

A. K. Mishra et al. (eds.), *Microplastics*, https://doi.org/10.1007/978-981-97-6461-7_7

Thompson, 2015; Efimova et al., 2018). There is no exact data on the amount of plastic waste (including MPs) discharged into the oceans because it arrives through river basins, coastal water currents, offshore activities, commercial fishery, landfill in coastal areas, strong winds, and natural events such as heavy rains and tsunamis, among others (Ita-Nagy et al., 2022). However, it has been estimated that around 11 million tons of plastic waste reach the ocean annually (Geyer et al., 2017), generating billions of MPs as a result of the fragmentation of these plastic pieces over time. Biotic and abiotic degradation of plastic particles is one of the most important processes since they contribute to the MPs reservoir, being <1 mm the most abundant size range reported in the marine environment (Tang et al., 2023). MPs manifest in diverse shapes, including fibers, fragments, pellets, and foams (Bosker et al., 2017). They consist of various synthetic and semisynthetic plastics (e.g., polypropylene, polyethylene, nylon, polyester, polystyrene) and can or not contain additives such as pigments, fillers, and flame retardants, among others (Du et al., 2022). These micro-material pollutants have been commonly detected in all environmental compartments, accumulating and biomagnifying throughout the seafood chain (Goswami et al., 2020; Ma et al., 2016; Wang et al., 2020a, b). Additionally, MPs possess the capacity to adsorb and leach chemical pollutants, act as carriers for contaminants, pathogens, and other organisms (e.g., benthic, pelagic), modify the biogeochemical cycles (e.g., N, P), and alter carbon sequestration in marine ecosystems (Rocha-Santos et al., 2022). Furthermore, MPs can generate direct physical effects on marine wildlife since plastic particles can cause the death of the organisms by constriction or by ingestion (Gall & Thompson, 2015; Jovanovic, 2017; Kurtela & Antolović, 2019). Finally, MPs can exhibit a single or double ecotoxicological effect on organisms depending on their nature (e.g., particle size, shape, surface area, crystal structure, surface charge) and their combination with other chemical pollutants present in the aquatic environment (O'Donovan et al., 2018; Mammo et al., 2020; Gola et al., 2021; Tang, 2021).

Research on MPs pollution in coastal and marine environments has grown exponentially since 2015, after these emerging pollutants were included in the United Nations Environment Program (UNEP) list of top 10 global environmental problems in 2014 (UNEP, 2014) (Fig. 1a), attracting growing public awareness and concern due to the multiple potential dangers they pose to health (Deng et al., 2022; Hale et al., 2020; Issac & Kandasubramanian, 2021; Kramm et al., 2022; Vethaak & Legler, 2021; Vivekanand et al., 2021). China, the United States, and the United Kingdom present the highest number of publications worldwide (Ahmad et al., 2020; Garcés-Ordóñez et al., 2022; Tang et al., 2021; Zhou et al., 2022). However, most of the research has been carried out in coastal areas and beaches, and to a lesser extent in remote ocean regions or continental shelves (Ahmad et al., 2020; Garcés-Ordóñez et al., 2022; Tang et al., 2021; Zhou et al., 2022) (Fig. 1b). The abundance of MPs is closely related to geographical location, population density, and economic development (Tang et al., 2023).

When MPs enter seawater, they can be ingested by marine organisms such as zooplankton (Botterell et al., 2022; Lima et al., 2023), crustaceans (Colombo et al., 2023; Ogunola et al., 2022), fish (Alberghini et al., 2022; Gao et al., 2022), birds

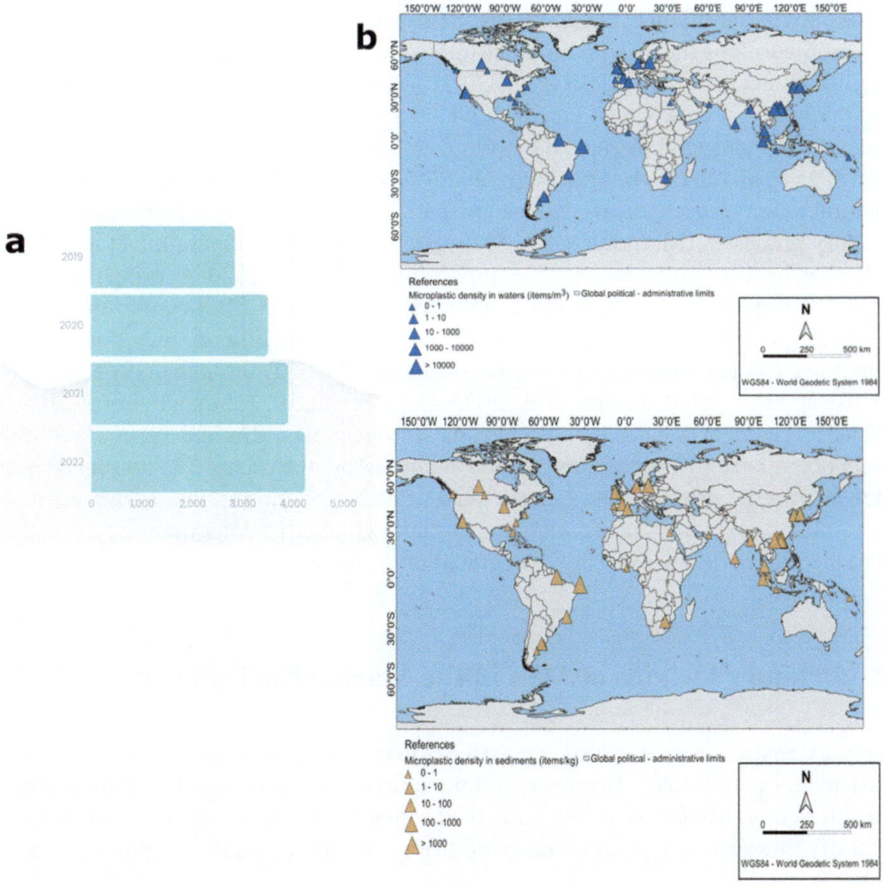

Fig. 1 **a** Estimated number of publications addressing microplastics (MPs) in the marine environment from 2010 to 2022, and **b** Spatial distribution of the abundance of MPs in environmental matrices (seawater and sediment) published between 2020 and 2022

(Liu et al., 2023; Navarro et al., 2023) and even whales (Harlacher et al., 2023; López-Martínez et al., 2023). Once these emerging contaminants enter organisms, they accumulate in tissues, translocating and affecting health in several ways. They can (i) interfere with the feeding process by reducing their ability to detect and catch their natural food, since these particles can be mistaken for prey (Akhbarizadeh et al., 2019; Corinaldesi et al., 2021); (ii) harbor hazardous elements such as phthalates, bisphenols, heavy metals, and pathogenic microorganisms, which may be harmful to health (Amesho et al., 2023; Jinhui et al., 2019; Vieira et al., 2021); (iii) block the digestive tract, causing internal damage and affecting the ability to digest food (Nelms et al., 2018; Fackelmann & Sommer, 2019; Gao et al., 2021); (iv) alter the hormonal functions and development of the organism, and reduce its ability to reproduce because MPs have the potential to function as endocrine disruptors (Hahladakis et al., 2018; Sharifinia et al., 2020; dos-Santos-Coentro, 2022; Kim et al., 2023); (v) accumulate along the food chain, since they are difficult particles to digest and excrete (Hoffmann et al., 2020; Xie et al., 2020); (vi) induce oxidative stress, impacting the functionality of antioxidant enzymes which causes oxidative damage to biological macromolecules, and disrupts the normal activities of the organism (Barboza et al., 2018; Li et al., 2023a, b; Mao et al., 2022); (vii) affect the immune system and the natural defense mechanisms against any pathogen, in addition to increasing inflammatory processes (Mattsson et al., 2015; Hu & Palić, 2020; Sharifinia et al., 2020). MPs can have a significant negative impact on marine biota, underscoring the importance of implementing measures to mitigate the influx of plastics into the oceans and safeguard the well-being of marine ecosystems. In addition, they pose a great unknown risk to human health since they can enter the body through the consumption of fish and shellfish (Colombo et al., 2023; Daniel et al., 2021; Lundebye et al., 2022). Thus, it is evident that oceans have become a large pool of MPs and a prominent sink, with potentially adverse effects on food chains across the ocean. The present chapter delves into the progress and current state of MPs pollution in the marine environment. It also explores the mechanisms through which MPs enter marine ecosystems and the processes involved in the degradation of these plastic particles.

2 Primary Origins of MPs in the Marine Environment

The extensive use of plastic materials makes contamination sources widely distributed and diverse. However, certain sources are more significant than others in different ecosystems. In the case of the marine environment, it is pertinent to classify these sources as either deriving from activities on land or within the sea.

2.1 *Land-Based Sources of Pollution*

Rivers are recognized as a primary conduit through which plastic waste originating from land finds its way into the oceans. It is estimated that 1.5% of the total mismanaged plastic waste generated globally reaches the oceans each year, mainly attributed to riverine plastic emissions (Meijer et al., 2021). Riparian contamination gradients have shown that litter pollutant is significantly higher in the lower course, mainly dominated by plastic debris discharged into the ocean (Ribeiro et al., 2022). River-driven plastic debris is then fragmented into smaller pieces after intensive weathering conditions (see Sect. 4), eventually becoming MPs in the oceans (ter-Halle et al., 2016). Rivers may also directly transport MPs emitted at any section of the river basin by focalized anthropogenic activities or urbanization. Kunz et al. (2023) investigated riverine MPs pollution along an urban–rural gradient in Taiwan. The findings demonstrated a positive correlation between MPs concentration and population density. Additionally, storm sewers are identified as crucial pathways for transporting land-originated plastic into rivers.

On the other hand, studies on floating MPs off the coasts of Tanzania and Mozambique show higher MPs concentrations close to the coast and major river mouths at the start of the rainy seasons (Weideman et al., 2023), suggesting that most MPs are generated on land and transported by the river. The review by Sun et al. (2023) integrates studies on MPs pollution in the silts and surface water of the Yellow River Basin. One of their main findings is that MPs' concentration increased from upstream to downstream. All this evidence confirms that rivers play a pivotal role as major conveyors of land-based plastic debris and MPs into the oceans (Lestari et al., 2023).

Wastewater treatment plants (WWTPs) and urban ditches are other land-based plastic debris sources that contribute to pollution by MPs in the oceans because they release plastic particles into the aquatic environment and soils. WWTPs receive or collect wastewater from households, industries, commercial establishments, and rainwater runoff from urban areas (Okoffo et al., 2019), which can contain MPs of different sizes, shapes, colors, and chemical compositions from diverse primary and secondary sources. During the treatment procedures of WWTPs, MPs may also be adsorbed and act as chemical vectors or pathogens since these plastic particles are released from WWTPs effluents into various environmental matrices with potentially ecotoxicological effects on organisms (Sadia et al., 2022). In this way, some authors have reported the presence of MPs in both samples of influents and effluents, with a concentration range between 1 and 7,000 particles L^{-1} (influents) and 0.0009 and 81 particles L^{-1} (effluents) (Okoffo et al., 2019), evidencing that in primary wastewater treatment procedures, MPs can be removed between 78 and 98% while during secondary treatment between 7 and 20% (Murphy et al., 2016; Talvitie et al., 2017). However, this percentage of remotion/removal efficiency of MPs from influences depends on the advancement of technology and the processes implemented in WWTPs to eliminate the polluting load in wastewater.

Atmospheric deposition of MPs also contributes to microplastic pollution in the marine environment. Some authors have reported concentrations of MPs between 1 particle m^{-3} and 35 particles m^{-3} in the urban atmosphere (Sun et al., 2022); however, there are some differences between different regions worldwide since some researchers have reported high abundances in the urban areas and mountains of UK, France, and the USA with values that oscillate between 771 n m^{-2} day^{-1} (UK) (Wright et al., 2020) and 365 n m^{-2} day^{-1} (Pyrenees Mountains, France) (Allen et al., 2019) in comparison with areas surrounding the ocean, such as the South China Sea (0.01 n m^{-3}) (Wang et al., 2020a, b). Nevertheless, these plastic particles can be transported long distances in the air and then deposited on the waterways (Sun et al., 2022). According to meteorological conditions, MPs depositions can be divided into two categories: dry and wet (Sun et al., 2022). However, a thorough investigation of the distribution gradient and chemical composition of MPs in Earth's atmosphere is needed due to the lack of data in areas located in the middle latitudes, such as South American regions (Yu et al., 2023a, b).

Another new source contributing to MPs (mostly microfibers) pollution in the coastal and marine environments has been Personal Protective Equipment (PPE) waste generated during the COVID-19 pandemic, which has become a new type of waste that threatens the environment (Oliveira et al., 2023; Silva et al., 2021). It has been found on coastal beaches, rivers, and oceans (Oliveira et al., 2023), and is either ingested by wildlife or their bodies become entangled in it, causing them injuries or even death (Hiemstra et al., 2021; Silva et al., 2021). Although there is no exact data on the amount of PPE generated and released during the pandemic, it has been estimated that around 1.56 billion disposable masks, which are the most common type of PPE, were released into the environment (Oceans Aragaw et al., 2022; Asia, 2020).

2.2 *Sea-Based Sources of Pollution*

Among the offshore activities that contribute to MPs pollution, the commercial, aquaculture and industrial fishery are the most important ones. Damaged, abandoned or lost fishing nets enter the sea during fishing operations and end up on coastal beaches or on the ocean floor (Thushari & Senevirathna, 2020; Wu et al., 2023). During their aging processes, they can release thousands of MPs (mainly microfibers) or they can be used by some species (for example, algae, mollusks) to grow and live on them. The accumulation of MPs has become a potential problem due to the release or leaching of various organic additives from the polymeric matrix with potential effects on organisms (Do et al., 2022). The acidification of aquatic environments caused by plastic debris containing chlorine can release HCl into water in their degradation process (Gewert et al., 2015). The bioaccumulation of MPs in organisms can affect their behavior, growth, and reproduction, as well as reduce their commercial quality (Wu et al., 2023). For example, high concentrations of MPs (11.49 particles m^{-3}) were reported in the aquaculture areas in Weihai, China, and according to their characterization, they came from fishing nets, ropes, and foam ball floats employed in the agriculture area (Zhang et al., 2012). Finally, shipping transport and marine structures are other sources of direct discharge of MPs into the oceans due to the degradation or fragmentation of antifouling paint coating into paint particles (PPs); although, in general, PPs consist of a polymeric resin combined with one or more additives and with compositional similarities to MPs (Turner, 2021).

3 MPs Transport and Sinks in the Ocean

As mentioned above, oceans exhibit a heterogeneous abundance of MPs in both horizontal and vertical distribution profiles, with high concentrations reported in coastal areas and even estuaries, and low concentrations in open sea (Vázquez-Rowe et al., 2021; Walkinshaw et al., 2020). Even though the physicochemical properties of MPs (e.g., shape, size, and density) determine their initial distribution in the marine environment, other factors are also influential, such as ocean currents, winds, natural degradation, and the growth of attached marine organisms, which allows MPs to move vertically in the water column (Li et al., 2023a, b). According to various researchers, the number of MPs in marine environments varies with depth, evidencing high concentrations in oceanic surface and subsurface layers, while their concentration decreases in the deeper layers (Li et al., 2023a, b; Liu et al., 2020a, b). In this way, Kukulka et al. (2012) reported that turbulent mixing induced by winds at the sea surface generates a vertical distribution of plastic debris on top of the water, while Liu et al., (2020a, b) suggested that the structure of the water mass or the global thermohaline circulation may also affect the vertical distribution profile of MPs. In particular, the buoyancy of MPs can also affect their distribution in the water column, since MPs less dense than seawater tend to float on the ocean surface, while denser MPs sink in marine sediment (Haave et al., 2019; Lorenz et al., 2019; Maghsodian et al., 2022; Niu et al., 2021). However, the buoyancy of MPs capacity is associated with the plastic density that can generally vary with temperature changes, loss of plasticizers, and change in crystallinity, which takes place during their weathering processes (Jalón-Rojas, et al., 2022; Laursen et al., 2022; Shamskhany & Karimpour, 2022).

Moreover, the colonization of microorganisms and the development of a microcosm called plastisphere on MPs surface (Zettler et al., 2013) can also decrease their buoyancy and modify their distribution in the water column (Li et al., 2023a, b). The plastisphere changes not only the buoyancy of MPs but also its hydrophobicity, roughness, surface charge, and surface functional groups, modifying its sorption behavior in the water column (Okeke et al., 2022; Forero-López et al., 2022b). Moreover, it has been evidenced that the plastisphere acts as a reservoir of pollutants (e.g., heavy metals) and pathogenic microorganisms (e.g., *Vibrio, Escherichia coli,* and *Pseudomonas*), as well as a dispersal vector of these pathogens in moving estuarine and ocean waters (Forero-López et al., 2022b; Liu et al., 2022a, b; Okeke et al., 2022). Other factors such as hetero-aggregation of specific types of MPs with phytoplankton (Wang et al., 2021), or geophysical processes, such as subsurface ocean currents in their transporting from subtropical to remote areas of the ocean, may also affect abundance and distribution (Wichmann et al., 2019). Finally, MPs in surface waters can be accumulated by aquatic plants (e.g., seagrasses) and algae (Li et al., 2023a, b), because they get trapped or entangled through electrostatic interaction caused by the presence of polysaccharides on the plant, as well as by the surface charge of MPs (Ng et al., 2022; Peller et al., 2021). Likewise, aquatic plants also serve as animal feed and human food, generating bioaccumulation and biomagnification of these micropollutants in distinct levels of the trophic chain (Ng et al., 2022). The bioturbation generated in surface sediments by small aquatic organisms such as crabs and other benthic organisms is another external biological action on the settlement of MPs in seawater and can influence their distribution in sediments (Frank et al., 2023; Peller et al., 2021). Some filter-feeders can also accelerate MPs deposition by aggregating them in sinking particulate feces and pseudo-feces; once they are on surface sediments, benthic fauna can rapidly bury these particles during the bioturbation process (Colen et al., 2021; Li et al., 2023a, b). However, plastic particles also can adhere to the carapace and gills of crabs such as *Neohelice granulata* (Truchet et al., 2022;

Villagran et al., 2020). Some authors have already reported a high prevalence of MPs in the respiratory organs of this crab species due to its burrowing behavior, removing sediments and organic matter, including MPs (Truchet et al., 2022).

4 Methods of Analysis of MPs in Marine Environments

Depending on the marine environmental matrix, there are various procedures to assess the quantity and variety of MPs, as shown in Fig. 2. There is no standardized method that sets clear guidelines when it comes to sample collection and preparation, eliminating potential interferents, preventing external contamination, or performing analyses for these microcontaminants; however, the main general steps to follow for the analysis of MPs are (i) sampling, (ii) sample preparation, and (ii) analysis methods. It is important to mention that depending on hydrodynamic and geographic conditions, and the depth of marine waters, samples can be divided into (i) bulk-water sampling and (ii) net sampling (Liu et al., 2020a, b). Bulk-water sampling is a type of sampling that aims to collect a defined volume of water at specific water layers for a limited duration and employs sample devices such as the Rosette sampler system (CTD sampler) (Botterell et al., 2022; Cordova & Hernawan, 2018; Dai et al., 2018), the plexiglass water sampler (Ding et al., 2019), and the lander system (Liu et al., 2019). However, contemporary studies generally opt for net sampling, which primarily includes manta nets (Karlsson et al., 2020; Pan et al., 2021; Tošić et al., 2020), followed by neuston (Athapaththu et al., 2020; Sagawa et al., 2018), and plankton nets (Forero-López et al., 2021; Tsang et al., 2017). Nevertheless, some authors have reported that the abundance of MPs could be affected by (i) the manta/neuston net sizes employed (Lindeque et al., 2020; Bertoldi et al., 2021; Forero-López et al., 2021), and (ii) the volume of the sample taken (De-la-Torre et al., 2022; Lusher et al., 2014), since small volumes of water can bring about the overestimation of the abundance of MPs, whereas large sampling volumes decrease in the abundance of MPs per unit volume. Concerning manta/neuston net sizes, some authors have demonstrated that sampling with smaller mesh sizes (60 μm) collects significantly higher MPs levels than traditional mesh sizes (301 - 400 μm) (Lindeque et al., 2020; Bertoldi et al., 2021; Forero-López et al., 2021). Other sampling methods are also employed such as submersible pumps and *in-situ* collection, where sample devices such as sediment traps and *in-situ* filtration devices are employed among others (Liu et al., 2020a, b). On the other hand, similar to marine waters, sediment sample collection protocols also exhibit a large diversity of methods considering the type of sample (e.g., seafloor, open ocean, coastal, or beach) (Phuong et al., 2021). Depending on the sample collection areas, and access to different types of steel or glass utensils (e.g., box corers, shovels, coring tubes, Van Veen tongs, metal spoons, and glass jars), the most appropriate protocol shall be chosen (Phuong et al., 2021). Once the samples are in the laboratory, methodologies that comprise various steps are employed for the extraction of MPs, such as filtration through stainless steel sieves or glass equipment, depending on the nature of the sample. This is followed by the oxidative digestion to remove organic matter with or without heat (e.g., H_2O_2, Fenton reagent, enzymatic) or separation density with NaCl or $ZnCl_2$, among others (Lee et al., 2023; Mattsson et al., 2022). In the case of organic samples, chemical solutions such as HNO_3, H_2O_2, KOH, or NaOH are used for the digestion of tissues and organs of vertebrates, or invertebrate bodies, being KOH the most employed (Bai et al., 2022; Dellisanti et al., 2023). Once the MPs are extracted from the samples, these anthropogenic particles are identified and quantified by visual equipment combined or not, with stain techniques such as lipophilic Nile red (Erni-Cassola et al., 2017). Followed by the morphological and chemical characterization through a set of instrumental techniques such as Fourier transform infrared spectroscopy (FTIR), Raman spectroscopy, scanning electron microscope (SEM) coupled with an energy-dispersive X-ray analyzer (EDX), and X-ray diffraction (XRD) (Priya et al., 2022). In particular, polyethylene (PE), polypropylene (PP), and polyethylene terephthalate (PET) are the most common types of MPs reported in both marine environments (water and sediments), being the fibers, fragments, and films MPs shapes reported with predominance in the colors transparent, white, and black (Tang et al., 2023).

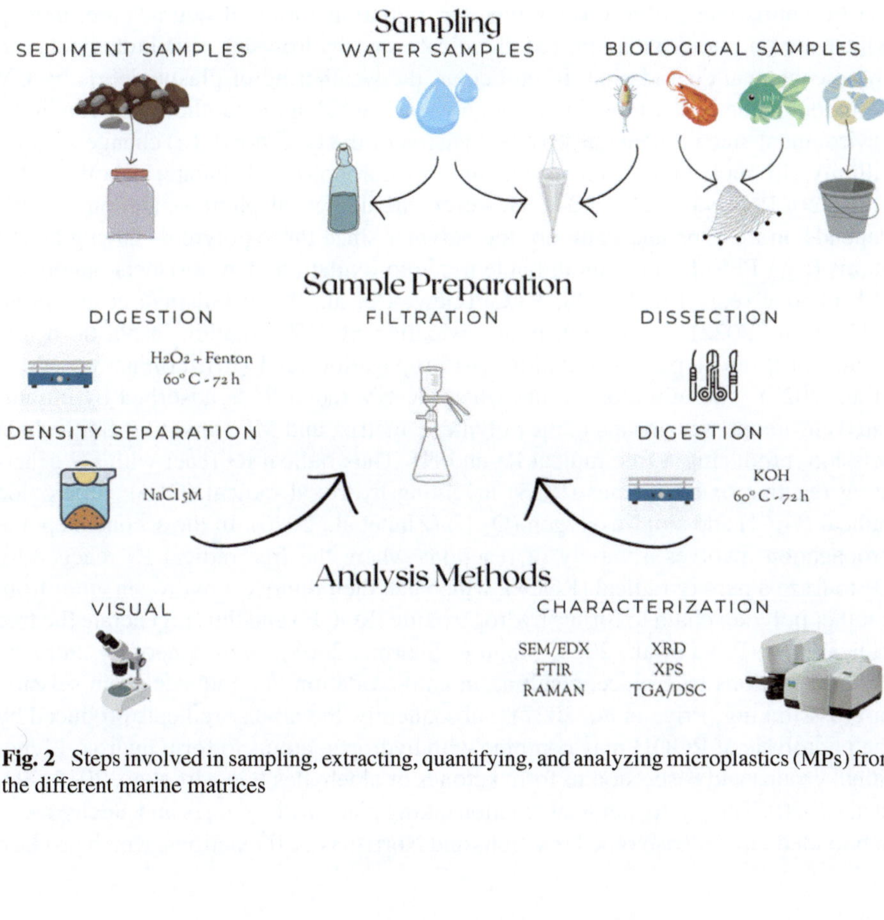

Fig. 2 Steps involved in sampling, extracting, quantifying, and analyzing microplastics (MPs) from the different marine matrices

5 Degradation of MPs in the Marine Environment

It was previously mentioned that MPs exhibit a broad polymeric composition, and may also contain a range of chemical additives in their matrix to confer physical and chemical properties according to the function of commercial plastic products (Forero-López et al., 2022a). In Table 1, we present a summary of current studies investigating the degradation processes of MPs under various aging conditions, considering mechanical, chemical, and biological perspectives. Upon entering the aquatic environment, plastic debris undergoes alterations in its physical and chemical attributes. These changes encompass modifications in crystallinity, wettability, chemical composition, surface charge, molecular weight reduction, color alteration, and the loss of mechanical integrity, including Young's modulus, among other factors (Rocha-Santos et al., 2022). These physical and chemical changes occur when plastic debris is exposed to marine environmental variables (e.g., salinity, temperature, pH, and dissolved organic matter (DOM)) for a certain time (Forero-López et al., 2022a; Liu et al, 2020a, b; Tang et al., 2020) (Fig. 3). Other important environmental parameters that affect plastic properties are solar radiation and humidity (aqueous medium). Plastic debris can suffer various types or initiation forms of degradation, among which are (i) photooxidation, (ii) chemical (e.g., hydrolysis), (iii) biological, and (iv) mechanical degradation. In particular, the weathering of plastic debris by UV radiation is one of the most important non-biological aging methods of MPs in the environment since it generates (i) the incision of the C–C bond, (ii) change of crystallinity, (iii) molecular weight reduction, (iv) morphological change, and others (Du et al., 2021; Priya et al., 2022). However, the degree of photo-oxidation of MPs depends on the type and nature of the polymer since those polymers having hetero atoms (e.g., PET, PU) are more liable to photo-oxidation than polymers having C–C backbone (e.g., PE, PP, PS, PVC) (Gewert et al., 2015; Gijsman et al., 1999; Priya et al., 2022). In general, photo-oxidation by UV radiation in plastic debris occurs in three steps: (i) initiation, (ii) propagation, and (iii) termination (Priya et al., 2022). The initiation occurs when the UV radiation is adsorbed by unsaturated chromophores groups in the polymeric matrix, and MPs tend to suffer a chain scission, producing a free radical R• and H•. Thus radical R• react with O_2 generating reactive oxide species (ROS) involving hydroxyl radical (OH•), superoxide radical (O_2^{2-}) and singlet oxygen (O_2^{-}•) (Zhu et al., 2020). In the second step, the propagation involves a variety of reactions where the free radical R• reacts with O_2 to form a peroxy radical (ROO•), which can then remove a hydrogen atom from another polymer chain to form a hydroperoxide (ROOH) and thus regenerate the free radical (R•) (Priya et al., 2022; Singh & Sharma, 2008). Simultaneously, intricate radical reactions take place, resulting in auto-oxidation that induces chain scission or cross-linking (Priya et al., 2022). Subsequently, the alkoxy radicals produced by the photolysis of ROOH may combine with hydrogen atoms to form hydroxyl functional groups and β-scission to form ketones or aldehydes (Gewert et al., 2015; Zhu et al., 2020). The photochemical reaction taking place with ketones and aldehydes, is subdivided into Norrish type I reactions and Norrish type II reactions, which produce an end carboxyl group (R-CO-CH_3) (Gardette et al., 2013; Priya et al., 2022). Finally, the termination of photooxidation occurs when the radicals combine to form inert products (Priya et al., 2022). However, the degradation of some polymers, such as Polyethylene (PE), depending on the presence of catalyst residues and other impurities, will often act as receptors for free radicals (Lewin & Pearce, 1998), whereas the PVC is the most sensitive one to photodegradation due to the presence of saturated chemical bonds (Priya et al., 2022).

The initiation of the MPs photodegradation pathway can occur preferentially on the surface of plastic particles at micrometer depth (around 100 μm) (Ter Halle et al., 2017), where the surface oxidized layer that contains functional groups (ketone, carboxylic acid, and ester functional groups) is eroded, and a layer of non-oxidized plastic particle is exposed, continuing with the process of oxidation of a new surface exposed (Priya et al., 2022; Ter Halle et al., 2017). In this way, infrared spectroscopy, such as micro Fourier transform interferometer (μ-FTIR) and X-ray photoelectron spectroscopy (XPS), are techniques commonly employed to study and characterize the chemical composition or chemical changes on the surface of plastic waste (Pizarro-Ortega et al., 2022).

Moreover, during the weathering of plastic debris in the aquatic environment, it can leach other emerging pollutants (e.g., plasticizers, antioxidants, UV stabilizers) and act as a carrier of others (e.g., heavy metals, organic pollutants) present in the water column (Costigan et al., 2022; Do et al., 2022; Liu et al., 2022a, b). The interaction between MPs and organic and inorganic pollutants present in the water column can occur through multiple sorption processes, such as adsorption through mechanisms of (i) cation exchange, (ii) electrostatic interactions, and (iii) complexation with functional groups (such as OH, amino, and carboxyl) (Torres et al., 2021; Martín et al., 2022). However, depending on the concentration of organic pollutants in the environment, both processes can take place: (i) the adsorption predominance at low concentrations, and/or (ii) the absorption at high concentrations of pollutants (Fu et al., 2021; Hartmann et al., 2017). Although, according to Fred-Ahmadu et al. (2020), it is impossible to distinguish between adsorption and absorption because the two processes usually occur simultaneously. Both hydrophobic interactions and π-π interactions of the aromatic rings composing the polymer chain are the most common mechanisms by which MPs adsorb organic pollutants present in the aquatic environment, providing available sites, such as hydrogen bonds and halogen bonds (Fu et al., 2021), for the settling of contamination on the surface (Forero-López et al., 2022a; Li et al., 2020). However, the sorption behavior of MPs is influenced by environmental factors such as pH, DOM, salinity, contaminant concentration, and ion coexistence. Therefore, the adsorption characteristics of contaminants onto MPs are intertwined with their physical and chemical attributes, level of weathering, contaminant concentration, and prevailing environmental conditions within the water column (Liu et al., 2022a, b). On the other hand, the plastisphere can enhance the accumulation of pollutants such as heavy metals in MPs covered by the biofilm (Liu et al., 2021; Forero-López et al., 2022b), as well as retard the weathering process in the environment (Amaral-Zettler et al., 2020; Liu et al., 2020a, b). Nevertheless, the development of this ecological niche can be controlled by combined factors of

Table 1 Current studies on the degradation mechanisms of microplastics (MPs) under various aging conditions

Experiment	Degradation process	MPs type	Aging condition	Reference
9 years of exposure to the marine environment and 28 days of UV radiation chamber in the laboratory	Photooxidation	HDPE	Natural and Artificial	Ghanadi and Padhye (2024)
In-situ bioremediation using indigenous microbial consortium (8 bacteria)	Microbiological	PET, PS	Natural	Auta et al. (2022)
16 h autoclaving at 140 °C in an oven with Fenton solution ($FeSO_4$-$7H_2O$, 4 mM) and H_2O_2	Hydrothermal	HDPE	Artificial	Hu et al. (2022)
Accelerated aging process by irradiation and mechanical stirring	Photooxidation	PE, PP, PS (marine debris)	Natural/ Artificial	Sorasan et al. (2022)
Plastic bags were collected from beaches and characterized by SEM and ATR-FTIR	Physicochemical	LDPE oxo-degradable plastic bags	Natural	Tziourrou et al. (2021)
2 weeks in water under visible light irradiation of zinc oxide nanorods (ZnO NRs)	Photooxidation	PP beads	Artificial	Uheida et al. (2021)
10 weeks of pilot-scale system simulating dry and marine conditions	Physicochemical	LDPE, HDPE, PS, PP, PET	Artificial	Fernández-González et al. (2021)
Electro-Fenton technology based on TiO_2/graphite (TiO_2/C) cathode in water	Chemical	PVC	Artificial	Miao et al. (2020)
8 weeks in stirred batch reactors in the dark with artificial seawater	Mechanical	PE beads	Artificial	Da Costa et al. (2018)

(continued)

Table 1 (continued)

Experiment	Degradation process	MPs type	Aging condition	Reference
6 months on laboratory-simulated onshore and nearshore conditions	Physicochemical	HDPE films	Artificial	Kalogerakis et al. (2017)
12 months of benthic conditions at 10 m depth	Mechanical and biological	PP, PE, Ny	Natural	Welden and Cowie (2017)

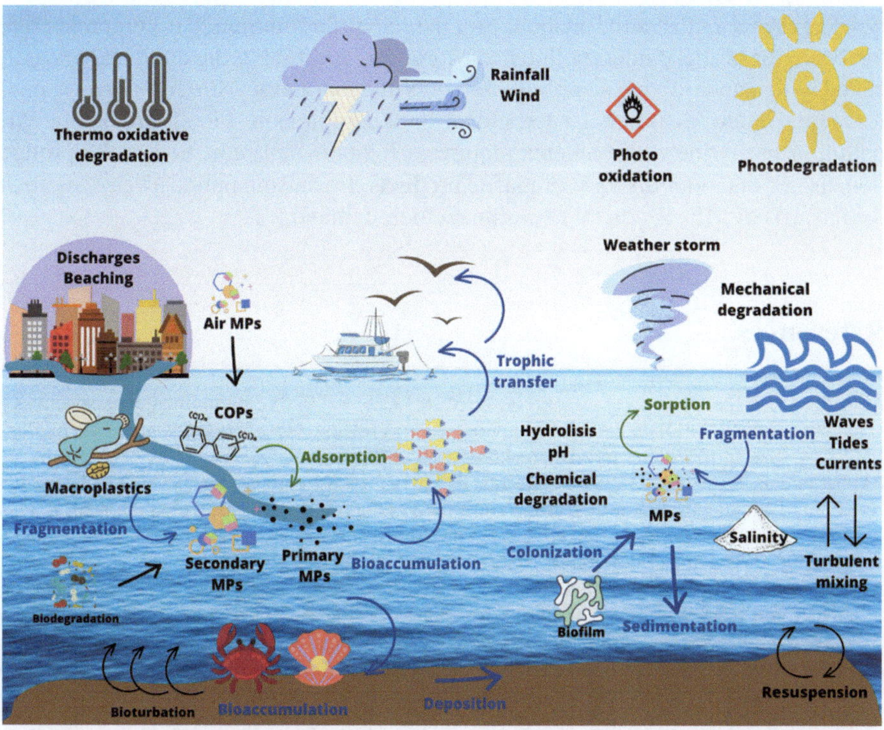

Fig. 3 Summary of the entry pathways (land-based and sea-based sources) and transformation patterns (physical, chemical and biological factors) of microplastics (MPs) once they enter the marine environment

plastic debris substrata such as surface characteristics (e.g., hydrophobicity), type, and chemical composition (Yu et al., 2023a, b).

6 Conclusions

In this chapter, we have provided an overview and discussion of the factors relevant to the study of MPs concerning the marine environment. We have highlighted the current advancements in understanding MPs sources, transportation mechanisms, and analytical techniques. It is evident that MPs are extensively distributed in marine environments, instigating a cascade of adverse effects on organisms and disrupting the balance of various processes, including chemical, biological, and geochemical phenomena in both water columns and sediments. The ability of MPs to adsorb, leach, and transport traditional and emerging pollutants, as well as pathogens, while exhibiting diverse physicochemical properties (size, shape, crystallinity, polymer type), presents a substantial challenge for the scientific community in comprehending the behavior of these micropollutants. Moreover, addressing the complexities associated with standardizing sampling protocols and extracting MPs from diverse environmental matrices is crucial for global data comparison. Finally, reducing MPs pollution in marine environments requires a fundamental reduction in the production, usage, and poor disposal of plastic products. Increasing public awareness about the hazards of MPs is crucial to eliminate their demand.

References

Ahmad, M., Li, J. L., Wang, P. D., Hozzein, W. N., & Li, W. J. (2020). Environmental perspectives of microplastic pollution in the aquatic environment: A review. *Marine Life Science & Technology, 2*, 414–430. https://doi.org/10.1007/s42995-020-00056-w

Akhbarizadeh, R., Moore, F., & Keshavarzi, B. (2019). Investigating microplastics bioaccumulation and biomagnification in seafood from the Persian Gulf: A threat to human health? *Food Additives & Contaminants, Part a, Chemistry, Analysis, Control, Exposure & Risk Assessment, 36*, 1696–1708. https://doi.org/10.1080/19440049.2019.1649473

Alberghini, L., Truant, A., Santonicola, S., Colavita, G., & Giaccone, V. (2022). Microplastics in fish and fishery products and risks for human health: A review. *International Journal of Environmental Research and Public Health, 20*, 789. https://doi.org/10.3390/ijerph20010789

Allen, S., Allen, D., Phoenix, V. R., Le-Roux, G., Durántez-Jiménez, P., Simonneau, A., Binet, S., & Galop, D. (2019). Atmospheric transport and deposition of microplastics in a remote mountain catchment. *Nature Geoscience, 12*, 339–344. https://doi.org/10.1038/s41561-019-0335-5

Amaral-Zettler, L. A., Zettler, E. R., & Mincer, T. J. (2020). Ecology of the plastisphere. *Nature Reviews. Microbiology, 18*, 139–151. https://doi.org/10.1038/s41579-019-0308-0

Amesho, K. T. T., Chinglenthoiba, C., Samsudin, M. S. A. B., Lani, M. N., Pandey, A., Desa, M. N. M., & Suresh, V. (2023). Microplastics in the environment: An urgent need for coordinated waste management policies and strategies. *Journal of Environmental Management, 344*, 118713. https://doi.org/10.1016/j.jenvman.2023.118713

Andrady, A. L. (2011). Microplastics in the marine environment. *Marine Pollution Bulletin, 62*, 1596–1605. https://doi.org/10.1016/j.marpolbul.2011.05.030

Aragaw, T. A., De-la-Torre, G. E., & Teshager, A. A. (2022). Personal protective equipment (PPE) pollution driven by the COVID-19 pandemic along the shoreline of Lake Tana, Bahir Dar, Ethiopia. *The Science of the Total Environment, 820*, 153261. https://doi.org/10.1016/j.scitotenv.2022.153261

Athapaththu, A. M. A. I. K., Thushari, G. G. N., Dias, P. C. B., Abeygunawardena, A. P., Egodauyana, K. P. U. T., Liyanage, N. P. P., Pitawala, H. M. J. C., & Senevirathna, J. D. M. (2020). Plastics in surface water of southern coastal belt of Sri Lanka (Northern Indian Ocean): Distribution and characterization by FTIR. *Marine Pollution Bulletin, 161*(Pt A), 111750. https://doi.org/10.1016/j.marpolbul.2020.111750

Auta, H. S., Abioye, O. P., Aransiola, S. A., Bala, J. D., Chukwuemeka, V. I., Hassan, A., Aziz, A., & Fauziah, S. H. (2022). Enhanced microbial degradation of PET and PS microplastics under natural conditions in mangrove environments. *Journal of Environmental Management, 304*, 114273. https://doi.org/10.1016/j.jenvman.2021.114273

Bai, C. L., Liu, L. Y., Hu, Y. B., Zeng, E. Y., & Guo, Y. (2022). Microplastics: A review of analytical methods, occurrence and characteristics in food, and potential toxicities to biota. *The Science of the Total Environment, 806*(Pt 1), 150263. https://doi.org/10.1016/j.scitotenv.2021.150263

Barboza, L. G. A., Vieira, L. R., Branco, V., Figueiredo, N., Carvalho, F., Carvalho, C., & Guilhermino, L. (2018). Microplastics cause neurotoxicity, oxidative damage and energy-related changes and interact with the bioaccumulation of mercury in the European seabass, *Dicentrarchus labrax* (Linnaeus, 1758). *Aquatic Toxicology, 195*, 49–57. https://doi.org/10.1016/j.aquatox.2017.12.008

Bertoldi, C., Lara, L. Z., Mizushima, F. A. L., Martins, F. C. G., Battisti, M. A., Hinrichs, R., & Fernandes, A. N. (2021). First evidence of microplastic contamination in the freshwater of Lake Guaíba, Porto Alegre, Brazil. *The Science of the Total Environment, 759*, 143503. https://doi.org/10.1016/j.scitotenv.2020.143503

Bosker, T., Behrens, P., & Vijver, M. G. (2017). Determining global distribution of microplastics by combining citizen science and in-depth case studies. *Integrated Environmental Assessment and Management, 13*(3), 536–541. https://doi.org/10.1002/ieam.1908

Botterell, Z. L. R., Bergmann, M., Hildebrandt, N., Krumpen, T., Steinke, M., Thompson, R. C., & Lindeque, P. K. (2022). Microplastic ingestion in zooplankton from the Fram Strait in the Arctic. *Science of the Total Environment, 831*, 154886. https://doi.org/10.1016/j.scitotenv.2022.154886

Boucher, J., & Friot, D. (2017). *Primary Microplastics in the Oceans: A Global Evaluation of Sources* (Vol. 10). IUCN.

Cole, M., Lindeque, P., Halsband, C., & Galloway, T. S. (2011). Microplastics as contaminants in the marine environment: A review. *Marine Pollution Bulletin, 62*, 2588–2597. https://doi.org/10.1016/j.marpolbul.2011.09.025

Colen, C. V., Moereels, L., Vanhove, B., Vrielinck, H., & Moens, T. (2021). The biological plastic pump: Evidence from a local case study using blue mussel and infaunal benthic communities. *Environmental Pollution, 274*, 115825. https://doi.org/10.1016/j.envpol.2020.115825

Colombo, C. V., Fernández-Severini, M. D., Forero-López, A. D., Ardusso, M. G., Rimondino, G. N., Malanca, F. E., & Buzzi, N. S. (2023). Microplastics in commercial seafood: *Pleoticus muelleri* as a case study in an estuarine environment highly affected by human pressure (Southwestern Atlantic). *Environmental Research, 216*, 114738. https://doi.org/10.1016/j.envres.2022.114738

Cordova, M. R., & Hernawan, U. E. (2018). Microplastics in Sumba Waters, East Nusa Tenggara. *IOP Conference Series. Earth and Environmental Science, 162*, 012023. https://doi.org/10.1088/1755-1315/162/1/012023

Corinaldesi, C., Canensi, S., Dell'Anno, A., Tangherlini, M., Di Capua, I., Varrella, S., Willis, T. J., Cerrano, C., & Danovaro, R. (2021). Multiple impacts of microplastics can threaten marine habitat-forming species. *Communications Biology, 4*, 431. https://doi.org/10.1038/s42003-021-01961-1

Costigan, E., Collins, A., Hatinoglu, M. D., Bhagat, K., MacRae, J., Perreault, F., & Apul, O. (2022). Adsorption of organic pollutants by microplastics: Overview of a dissonant literature. *Journal of Hazardous Materials Advances, 6*, 100091. https://doi.org/10.1016/j.hazadv.2022.100091

Da Costa, J. P., Nunes, A. R., Santos, P. S. M., Girão, A. V., Duarte, A. C., & Rocha-Santos, T. (2018). Degradation of polyethylene microplastics in seawater: Insights into the environmental degradation of polymers. *Journal of Environmental Science and Health. Part a, Toxic/hazardous*

Substances & Environmental Engineering, 53(9), 866–875. https://doi.org/10.1080/10934529.2018.1455381

Dai, Z., Zhang, H., Zhou, Q., Tian, Y., Chen, T., Tu, C., Fu, C., & Luo, Y. (2018). Occurrence of microplastics in the water column and sediment in an inland sea affected by intensive anthropogenic activities. *Environmental Pollution (barking, Essex: 1987), 242*(Pt B), 1557–1565. https://doi.org/10.1016/j.envpol.2018.07.131

Daniel, D. B., Ashraf, P. M., Thomas, S. N., & Thomson, K. T. (2021). Microplastics in the edible tissues of shellfishes sold for human consumption. *Chemosphere, 264*, 128554. https://doi.org/10.1016/j.chemosphere.2020.128554

De-la-Torre, G. E., Pizarro-Ortega, C. I., Dioses-Salinas, D. C., Castro-Loayza, J., Smith Sanchez, J., Meza-Chuquizuta, C., Espinoza-Morriberón, D., Rakib, M. R. J., Ben-Haddad, M., & Dobaradaran, S. (2022). Are we underestimating floating microplastic pollution? A quantitative analysis of two sampling methodologies. *Marine Pollution Bulletin, 178*, 113592. https://doi.org/10.1016/j.marpolbul.2022.113592

Dellisanti, W., Leung, M. M. L., Lam, K. W. K., Wang, Y., Hu, M., Lo, H. S., & Fang, J. K. H. (2023). A short review on the recent method development for extraction and identification of microplastics in mussels and fish, two major groups of seafood. *Marine Pollution Bulletin, 186*, 114221. https://doi.org/10.1016/j.marpolbul.2022.114221

Deng, L., Li, G., Peng, S., Wu, J., & Che, Y. (2022). Microplastics in personal care products: Exploring public intention of usage by extending the theory of planned behaviour. *Science of the Total Environment, 848*, 157782. https://doi.org/10.1016/j.scitotenv.2022.157782

Ding, J., Jiang, F., Li, J., Wang, Z., Sun, C., Wang, Z., Fu, L., Ding, N. X., & He, C. (2019). Microplastics in the coral reef systems from Xisha Islands of South China Sea. *Environmental Science & Technology, 53*, 8036–8046. https://doi.org/10.1021/acs.est.9b01452

Do, A. T. N., Ha, Y. J., & Kwon, J. H. (2022). Leaching of microplastic-associated additives in aquatic environments: A critical review. *Environmental Pollution, 305*, 119258. https://doi.org/10.1016/j.envpol.2022.119258

dos-Santos-Coentro, J. T. G. (2022). *Waterbirds as sentinels of microplastic and plastic additive pollution in coastal systems in southern Portugal*. Universidade de Coimbra, Faculdade de Ciências e Tecnologia, Departamento de ciências da vida.

Du, H., Xie, Y., & Wang, J. (2022). Environmental impacts of microplastics on fishery products: An overview. *Gondwana Research: International Geoscience Journal, 108*, 213–220. https://doi.org/10.1016/j.gr.2021.08.013

Du, H., Xie, Y., & Wang, J. (2021). Microplastic degradation methods and corresponding degradation mechanism: Research status and future perspectives. *Journal of Hazardous Materials, 418*, 126377. https://doi.org/10.1016/j.jhazmat.2021.126377

Efimova, I., Bagaeva, M., Bagaev, A., Kileso, A., & Chubarenko, I. P. (2018). Secondary microplastics generation in the sea swash zone with coarse bottom sediments: Laboratory experiments. *Frontiers in Marine Science, 5*, 313. https://doi.org/10.3389/fmars.2018.00313

Erni-Cassola, G., Gibson, M. I., Thompson, R. C., & Christie-Oleza, J. A. (2017). Lost, but found with Nile red: A novel method for detecting and quantifying small microplastics (1 mm to 20 μm) in environmental samples. *Environmental Science & Technology, 51*, 13641–13648. https://doi.org/10.1021/acs.est.7b04512

Gall, S. C., & Thompson, R. C. (2015). The impact of debris on marine life. *Marine Pollution Bulletin, 92*, 170–179. https://doi.org/10.1016/j.marpolbul.2014.12.041

Fackelmann, G., & Sommer, S. (2019). Microplastics and the gut microbiome: How chronically exposed species may suffer from gut dysbiosis. *Marine Pollution Bulletin, 143*, 193–203. https://doi.org/10.1016/j.marpolbul.2019.04.030

Fernández-González, V., Andrade-Garda, J. M., López-Mahía, P., & Muniategui-Lorenzo, S. (2021). Impact of weathering on the chemical identification of microplastics from usual packaging polymers in the marine environment. *Analytica Chimica Acta, 1142*, 179–188. https://doi.org/10.1016/j.aca.2020.11.002

Forero-López, A. D., Rimondino, G. N., Truchet, D. M., Colombo, C. V., Buzzi, N. S., Malanca, F. E., Spetter, C. V., & Fernández-Severini, M. D. (2021). Occurrence, distribution, and characterization of suspended microplastics in a highly impacted estuarine wetland in Argentina. *The Science of the Total Environment, 785*, 147141. https://doi.org/10.1016/j.scitotenv.2021.147141

Forero-López, A. D., Fabiani, M., Lassalle, V. L., Spetter, C. V., & Fernández-Severini, M. D. (2022a). Critical review of the characteristics, interactions, and toxicity of micro/nanomaterials pollutants in aquatic environments. *Marine Pollution Bulletin, 174*, 113276. https://doi.org/10.1016/j.marpolbul.2021.113276

Forero-López, A. D., Brugnoni, L. I., Abasto, B., Rimondino, G. N., Lassalle, V. L., Ardusso, M. G., Nazzarro, M. S., Martinez, A. M., Spetter, C. V., Biancalana, F., & Fernández-Severini, M. D. (2022b). Plastisphere on microplastics: In situ assays in an estuarine environment. *Journal of Hazardous Materials, 440*, 129737. https://doi.org/10.1016/j.jhazmat.2022.129737

Frank, Y. A., Vorobiev, D. S., Vorobiev, E. D., Samarinova, A. A., Antsiferov, D. V., & Strezov, V. (2023). Ability of benthic oligochaetes to bury microplastics in aquatic bottom sediments. *The Science of the Total Environment, 857*, 159687. https://doi.org/10.1016/j.scitotenv.2022.159687

Fred-Ahmadu, O. H., Bhagwat, G., Oluyoye, I., Benson, N. U., Ayejuyo, O. O., & Palanisami, T. (2020). Interaction of chemical contaminants with microplastics: Principles and perspectives. *The Science of the Total Environment, 706*, 135978. https://doi.org/10.1016/j.scitotenv.2019.135978

Fu, L., Li, J., Wang, G., Luan, Y., & Dai, W. (2021). Adsorption behavior of organic pollutants on microplastics. *Ecotoxicology and Environmental Safety, 217*, 112207. https://doi.org/10.1016/j.ecoenv.2021.112207

Gao, C. M., Cao, Z., Yan, C. B., & Zhu, G. P. (2021). Microplastic characteristics and distribution of gastrointestinal contents of striped Antarctic fish in the waters around the South Shetland Islands, Antarctica. *Journal of Fisheries of China* 1–9.

Gao, S., Yan, K., Liang, B., Shu, R., Wang, N., & Zhang, S. (2022). The different ways microplastics from the water column and sediment accumulate in fish in Haizhou Bay. *The Science of the Total Environment, 854*, 158575. https://doi.org/10.1016/j.scitotenv.2022.158575

Garcés-Ordóñez, O., Saldarriaga-Vélez, J. F., Espinosa-Díaz, L. F., Canals, M., Sánchez-Vidal, A., & Thiel, M. (2022). A systematic review on microplastic pollution in water, sediments, and organisms from 50 coastal lagoons across the globe. *Environmental Pollution, 315*, 120366. https://doi.org/10.1016/j.envpol.2022.120366

Gardette, M., Perthue, A., Gardette, J. L., Janecska, T., Földes, E., Pukánszky, B., & Therias, S. (2013). Photo- and thermal-oxidation of polyethylene: Comparison of mechanisms and influence of unsaturation content. *Polymer Degradation and Stability, 98*, 2383–2390. https://doi.org/10.1016/j.polymdegradstab.2013.07.017

Gewert, B., Plassmanna, M. M., & MacLeod, M. (2015). Pathways for degradation of plastic polymers floating in the marine environment. *Environmental Science: Processes & Impacts, 17*, 1513–1521. https://doi.org/10.1039/C5EM00207A

Geyer, R., Jambeck, J. R., & Lavender-Law, K. (2017). Production, use, and fate of all plastics ever made. *Science Advances, 3*, 7. https://doi.org/10.1126/sciadv.1700782

Ghanadi, M., & Padhye, L. P. (2024). Revealing the long-term impact of photodegradation and fragmentation on HDPE in the marine environment: Origins of microplastics and dissolved organics. *Journal of Hazardous Materials, 465*, 133509. https://doi.org/10.1016/j.jhazmat.2024.133509

Gijsman, P., Meijers, G., & Vitarelli, G. (1999). Comparison of the UV-degradation chemistry of polypropylene, polyethylene, polyamide 6 and polybutylene terephthalate. *Polymer Degradation and Stability, 65*, 433–441. https://doi.org/10.1016/s0141-3910(99)00033-6

Gola, D., Tyagi, P. K., Arya, A., Chauhan, N., Agarwal, M., Singh, S. K., & Gola, S. (2021). The impact of microplastics on marine environment: A review. *Environmental Nanotechnology, Monitoring & Management, 16*, 100552. https://doi.org/10.1016/j.enmm.2021.100552

Goswami, P., Vinithkumar, N. V., & Dharani, G. (2020). First evidence of microplastics bioaccumulation by marine organisms in the Port Blair Bay, Andaman Islands. *Marine Pollution Bulletin, 155*, 111163. https://doi.org/10.1016/j.marpolbul.2020.111163

Hale, R. C., Seeley, M. E., La-Guardia, M. J., Mai, L., & Zeng, L. Y. (2020). A global perspective on microplastics. *Journal of Geophysical Research: Oceans, 125*, 1. https://doi.org/10.1029/2018JC014719

Haave, M., Lorenz, C., Primpke, S., & Gerdts, G. (2019). Different stories told by small and large microplastics in sediment - first report of microplastic concentrations in an urban recipient in Norway. *Marine Pollution Bulletin, 141*, 501–513. https://doi.org/10.1016/j.marpolbul.2019.02.015

Hahladakis, J. N., Velis, C. A., Weber, R., Iacovidou, E., & Purnell, P. (2018). An overview of chemical additives present in plastics: Migration, release, fate and environmental impact during their use, disposal and recycling. *Journal of Hazardous Materials, 344*, 179–199. https://doi.org/10.1016/j.jhazmat.2017.10.014

Harlacher, J. M., Emmons, C., Hanson, B., Olsen, D., Matkin, C., Phan, S., Luscombe, C., da-Silva, D., Park, L., & Parsons, K. M. (2023). Evidence of microplastics in marine top predators: Microplastic particles isolated from the feces of fish-eating killer whales. *SSRN Electronic Journal*. https://doi.org/10.2139/ssrn.4364184.

Hartmann, N. B., Rist, S., Bodin, J., Jensen, L. H. S., Schmidt, S. N., Mayer, P., Meibom, A., & Baun, A. (2017). Microplastics as vectors for environmental contaminants: Exploring sorption, desorption, and transfer to biota: Microplastics as contaminant vectors: Exploring the processes. *Integrated Environmental Assessment and Management, 13*, 488–493. https://doi.org/10.1002/ieam.1904

Hiemstra, A. F., Rambonnet, L., Gravendeel, B., & Schilthuizen, M. (2021). The effects of COVID-19 litter on animal life. *Animal Biology, 71*, 215–231. https://doi.org/10.1163/15707563-bja10052

Hoffmann, L., Eggers, S. L., Allhusen, E., Katlein, C., & Peeken, I. (2020). Interactions between the ice algae *Fragillariopsis cylindrus* and microplastics in sea ice. *Environment International, 139*, 105697. https://doi.org/10.1016/j.envint.2020.105697

Hu, K., Zhou, P., Yang, Y., Hall, T., Nie, G., Yao, Y., Duan, X., & Wang, S. (2022). Degradation of microplastics by a thermal Fenton reaction. *ACS ES&T Engineering, 2*(1), 110–120. https://doi.org/10.1021/acsestengg.1c00323

Hu, M., & Palić, D. (2020). Micro- and nano-plastics activation of oxidative and inflammatory adverse outcome pathways. *Redox Biology, 37*, 101620. https://doi.org/10.1016/j.redox.2020.101620

Issac, M. N., & Kandasubramanian, B. (2021). Effect of microplastics in water and aquatic systems. *Environmental Science and Pollution Research, 28*, 19544–19562. https://doi.org/10.1007/s11356-021-13184-2

Ita-Nagy, D., Vázquez-Rowe, I., & Kahhat, R. (2022). Developing a methodology to quantify mismanaged plastic waste entering the ocean in coastal countries. *Journal of Industrial Ecology, 26*, 2108–2122. https://doi.org/10.1111/jiec.13349

Jalón-Rojas, I., Romero-Ramírez, A., Fauquembergue, K., Rossignol, L., Cachot, J., Sous, D., & Morin, B. (2022). Effects of biofilms and particle physical properties on the rising and settling velocities of microplastic fibers and sheets. *Environmental Science & Technology, 56*, 8114–8123. https://doi.org/10.1021/acs.est.2c01302

Jinhui, S., Sudong, X., Yan, N., Xia, P., Jiahao, Q., & Yongjian, X. (2019). Effects of microplastics and attached heavy metals on growth, immunity, and heavy metal accumulation in the yellow seahorse, *Hippocampus Kuda* Bleeker. *Marine Pollution Bulletin, 149*, 110510. https://doi.org/10.1016/j.marpolbul.2019.110510

Jovanovic, B. (2017). Ingestion of microplastics by fish and its potential consequences from a physical perspective. *Integrated Environmental Assessment and Management, 13*, 510–515. https://doi.org/10.1002/ieam.1913

Kalogerakis, N., Karkanorachaki, K., Kalogerakis, G. C., Triantafyllidi, E. I., Gotsis, A. D., Partsinevelos, P., Fava, F. (2017). Microplastics generation: Onset of fragmentation of polyethylene films in marine environment mesocosms. *Frontiers in Marine Science, 4*. https://doi.org/10.3389/fmars.2017.00084

Karlsson, T. M., Kärrman, A., Rotander, A., & Hassellöv, M. (2020). Comparison between manta trawl and in situ pump filtration methods, and guidance for visual identification of microplastics in surface waters. *Environmental Science and Pollution Research International, 27*, 5559–5571. https://doi.org/10.1007/s11356-019-07274-5

Kim, M. J., Kim, J. A., Song, J. A., Kho, K. H., & Choi, C. Y. (2023). Synthetic microfiber exposure negatively affects reproductive parameters in male medaka (*Oryzias latipes*). *General and Comparative Endocrinology, 334*, 114216. https://doi.org/10.1016/j.ygcen.2023.114216

Kramm, J., Steinhoff, S., Werschmöller, S., Völker, B., & Völker, C. (2022). Explaining risk perception of microplastics: Results from a representative survey in Germany. *Global Environmental Change, 73*, 102485. https://doi.org/10.1016/j.gloenvcha.2022.102485

Kukulka, T., Proskurowski, G., Morét-Ferguson, S., Meyer, D. W., & Law, K. L. (2012). The effect of wind mixing on the vertical distribution of buoyant plastic debris. *Journal of Geophysical Research: Oceans, 39*, 7. https://doi.org/10.1029/2012GL051116

Kunz, A., Schneider, F., Anthony, N., & Lin, H. T. (2023). Microplastics in rivers along an urban-rural gradient in an urban agglomeration: Correlation with land use, potential sources and pathways. *Environmental Pollution, 321*, 121096. https://doi.org/10.1016/j.envpol.2023.121096

Kurtela, A., & Antolović, N. (2019). The Problem of plastic waste and microplastics in the seas and oceans: Impact on marine organisms. *Croatian Journal of Fisheries, 77*, 51–56. https://doi.org/10.2478/cjf-2019-0005

Laursen, S. N., Fruergaard, M., & Andersen, T. J. (2022). Rapid flocculation and settling of positively buoyant microplastic and fine-grained sediment in natural seawater. *Marine Pollution Bulletin, 178*, 113619. https://doi.org/10.1016/j.marpolbul.2022.113619

Lee, H., Kim, S., Sin, A., Kim, G., Khan, S., Nadagouda, M. N., Sahle-Demessie, E., & Han, C. (2023). Pretreatment methods for monitoring microplastics in soil and freshwater sediment samples: A comprehensive review. *The Science of the Total Environment, 871*, 161718. https://doi.org/10.1016/j.scitotenv.2023.161718

Lestari, P., Trihadiningrum, Y., & Warmadewanthi, I. D. A. A. (2023). Investigation of microplastic ingestion in commercial fish from Surabaya river, Indonesia. *Environmental Pollution (barking, Essex: 1987), 331*, 121807. https://doi.org/10.1016/j.envpol.2023.121807

Lewin, M., & Pearce, E. M. (Eds.). (1998). *Handbook of Fiber Chemistry (Second edition, Revised and Expanded)* (2nd ed.). CRC Press.

Li, D., Ji, J., Yuan, Y., & Wang, D. (2020). Toxicity comparison of nanopolystyrene with three metal oxide nanoparticles in nematode *Caenorhabditis elegans*. *Chemosphere, 245*, 125625. https://doi.org/10.1016/j.chemosphere.2019.125625

Li, J., Shan, E., Zhao, J., Teng, J., & Wang, Q. (2023a). The factors influencing the vertical transport of microplastics in marine environments: A review. *Science of the Total Environment, 870*, 161893. https://doi.org/10.1016/j.scitotenv.2023.161893

Li, X., Bao, L., Wei, Y., Zhao, W., Wang, F., Liu, X., Su, H., & Zhang, R. (2023b). Occurrence, bioaccumulation, and risk assessment of microplastics in the aquatic environment: A review. *Water, 15*(9), 1768. https://doi.org/10.3390/w15091768

Lima, C. D. M., Melo-Júnior, M., Schwamborn, S. H. L., Kessler, F., Oliveira, L. A., Ferreira, B. P., Mugrabe, G., Frias, J., & Neumann-Leitão, S. (2023). Zooplankton exposure to microplastic contamination in a estuarine plume-influenced region, in Northeast Brazil. *Environmental Pollution, 322*, 121072. https://doi.org/10.1016/j.envpol.2023.121072

Lindeque, P. K., Cole, M., Coppock, R. L., Lewis, C. N., Miller, R. Z., Watts, A. J. R., Wilson-McNeal, A., Wright, S. L., & Galloway, T. S. (2020). Are we underestimating microplastic abundance in the marine environment? A comparison of microplastic capture with nets of different mesh-size. *Environmental Pollution (barking, Essex: 1987), 265*(Pt A), 114721. https://doi.org/10.1016/j.envpol.2020.114721

Liu, K., Zhang, F., Song, Z., Zong, C., Wei, N., & Li, D. (2019). A novel method enabling the accurate quantification of microplastics in the water column of the deep ocean. *Marine Pollution Bulletin, 146*, 462–465. https://doi.org/10.1016/j.marpolbul.2019.07.008

Liu, P., Zhan, X., Wu, X., Li, J., Wang, H., & Gao, S. (2020a). Effect of weathering on environmental behavior of microplastics: Properties, sorption and potential risks. *Chemosphere, 242*, 125193. https://doi.org/10.1016/j.chemosphere.2019.125193

Liu, K., Courtene-Jones, W., Wang, X., Song, Z., Wei, N., & Li, D. (2020b). Elucidating the vertical transport of microplastics in the water column: A review of sampling methodologies and distributions. *Water Research, 186*, 116403. https://doi.org/10.1016/j.watres.2020.116403

Liu, Z., Adyel, T. M., Miao, L., You, G., Liu, S., & Hou, J. (2021). Biofilm influenced metal accumulation onto plastic debris in different freshwaters. *Environmental Pollution (barking, Essex: 1987), 285*, 117646. https://doi.org/10.1016/j.envpol.2021.117646

Liu, Y., Zhang, K., Xu, S., Yan, M., Tao, D., Chen, L., Wei, L., Wu, C., Liu, G., & Lam, P. K. S. (2022a). Heavy metals in the "plastisphere" of marine microplastics: Adsorption mechanisms and composite risk. *Gondwana Research, 108*, 171–180. https://doi.org/10.1016/j.gr.2021.06.017

Liu, S., Huang, J., Zhang, W., Shi, L., Yi, K., Yu, H., Zhang, C., Li, S., & Li, J. (2022b). Microplastics as a vehicle of heavy metals in aquatic environments: A review of adsorption factors, mechanisms, and biological effects. *Journal of Environmental Management, 302*, 113995. https://doi.org/10.1016/j.jenvman.2021.113995

Liu, W., Chen, X., Liang, T., Mu, T., Ding, Y., Liu, Y., & Liu, X. (2023). Varying abundance of microplastics in tissues associates with different foraging strategies of coastal shorebirds in the Yellow Sea. *The Science of the Total Environment, 866*, 161417. https://doi.org/10.1016/j.scitotenv.2023.161417

López-Martínez, S., Giménez-Luque, E., Molina-Pardo, J. L., Manzano-Medina, S., Arribas-Arias, H., Gavara, R., Morales-Caselles, C., & L Rivas, M. (2023). Macro and microplastic ingestion by marine megafauna. *SSRN Electronic Journal*. https://doi.org/10.2139/ssrn.4335655.

Lorenz, C., Roscher, L., Meyer, M. S., Hildebrandt, L., Prume, J., Löder, M. G. J., Primpke, S., & Gerdts, G. (2019). Spatial distribution of microplastics in sediments and surface waters of the southern North Sea. *Environmental Pollution, 252*, 1719–1729. https://doi.org/10.1016/j.envpol.2019.06.093

Lundebye, A. K., Lusher, A. L., & Bank, M. S. (2022). Marine microplastics and seafood: Implications for food security (Chap. 5). *Environmental Contamination Remediation and Management*, 131–154. https://doi.org/10.1007/978-3-030-78627-4_5.

Lusher, A. L., Burke, A., O'Connor, I., & Officer, R. (2014). Microplastic pollution in the Northeast Atlantic Ocean: Validated and opportunistic sampling. *Marine Pollution Bulletin, 88*, 325–333. https://doi.org/10.1016/j.marpolbul.2014.08.023

Ma, Y., Huang, A., Cao, S., Sun, F., Wang, L., Guo, H., & Ji, R. (2016). Effects of nanoplastics and microplastics on toxicity, bioaccumulation, and environmental fate of phenanthrene in fresh water. *Environmental Pollution, 219*, 166–173. https://doi.org/10.1016/j.envpol.2016.10.061

Maghsodian, Z., Sanati, A. M., Tahmasebi, S., Shahriari, M. H., & Ramavandi, B. (2022). Study of microplastics pollution in sediments and organisms in mangrove forests: A review. *Environmental Research, 208*, 112725. https://doi.org/10.1016/j.envres.2022.112725

Mammo, F. K., Amoah, I. D., Gani, K. M., Pillay, L., Ratha, S. K., Bux, F., & Kumari, S. (2020). Microplastics in the environment: Interactions with microbes and chemical contaminants. *Science of the Total Environment, 743*, 140518. https://doi.org/10.1016/j.scitotenv.2020.140518

Mao, X., Xu, Y., Cheng, Z., Yang, Y., Guan, Z., Jiang, L., & Tang, K. (2022). The impact of microplastic pollution on the ecological environment: a review. *Frontiers in Biosciences-Landmark, 27*, 46. https://doi.org/10.31083/j.fbl2702046.

Martín, J., Santos, J. L., Aparicio, I., & Alonso, E. (2022). Microplastics and associated emerging contaminants in the environment: Analysis, sorption mechanisms and effects of co-exposure.

Trends in Environmental Analytical Chemistry, 35, e00170. https://doi.org/10.1016/j.teac.2022.e00170

Mattsson, K., Ekvall, M. T., Hansson, L., Linse, S., Malmendal, A., & Cedervall, T. (2015). Altered behavior, physiology, and metabolism in fish exposed to polystyrene nanoparticles. *Environmental Science & Technology, 49*, 553–561. https://doi.org/10.1021/es5053655

Mattsson, K., Ekstrand, E., Granberg, M., Hassellöv, M., & Magnusson, K. (2022). Comparison of pre-treatment methods and heavy density liquids to optimize microplastic extraction from natural marine sediments. *Scientific Reports, 12*, 15459. https://doi.org/10.1038/s41598-022-19623-5

Meijer, L. J. J., Emmerik, T. V., van Der Ent, R., Schmidt, C., & Leberton, L. (2021). More than 1000 rivers account for 80% of global riverine plastic emissions into the ocean. *Science Advances, 7*, 18. https://doi.org/10.1126/sciadv.aaz5803

Murphy, F., Ewins, C., Carbonnier, F., & Quinn, B. (2016). Wastewater treatment works (WwTW) as a source of microplastics in the aquatic environment. *Environmental Science & Technology, 50*(11), 5800–5808. https://doi.org/10.1021/acs.est.5b05416

Miao, F., Liu, Y., Gao, M., Yu, X., Xiao, P., Wang, M., Wang, S., & Wang, X. (2020). Degradation of polyvinyl chloride microplastics via an electro-Fenton-like system with a TiO_2/graphite cathode. *Journal of Hazardous Materials, 399*, 123023. https://doi.org/10.1016/j.jhazmat.2020.123023

Navarro, A., Luzardo, O. P., Gómez, M., Acosta-Dacal, A., Martínez, I., Felipe-de-la-Rosa, J., Macías-Montes, A., Suárez-Pérez, A., & Herrera, A. (2023). Microplastics ingestion and chemical pollutants in seabirds of Gran Canaria (Canary Islands, Spain). *Marine Pollution Bulletin, 186*, 114434. https://doi.org/10.1016/j.marpolbul.2022.114434

Nelms, S. E., Galloway, T. S., Godley, B. J., Jarvis, D. S., & Lindeque, P. K. (2018). Investigating microplastic trophic transfer in marine top predators. *Environmental Pollution, 238*, 999–1007. https://doi.org/10.1016/j.envpol.2018.02.016

Ng, K. L., Suk, K. F., Cheung, K. W., Shek, R. H. T., Chan, S. M. N., Tam, N. F. Y., Cheung, S. G., Fang, J. K. H., & Lo, H. S. (2022). Macroalgal morphology mediates microplastic accumulation on thallus and in sediments. *Science of the Total Environment, 825*, 153987. https://doi.org/10.1016/j.scitotenv.2022.153987

Niu, L., Li, Y., Li, Y., Hu, Q., Wang, C., Hu, J., Zhang, W., Wang, L., Zhang, C., & Zhang, H. (2021). New insights into the vertical distribution and microbial degradation of microplastics in urban river sediments. *Water Research, 188*, 116449. https://doi.org/10.1016/j.watres.2020.116449

O'Donovan, S., Mestre, N. C., Abel, S., Fonseca, T. G., Carteny, C. C., Cormier, B., Keiter, S. H., & Bebianno, M. J. (2018). Ecotoxicological effects of chemical contaminants adsorbed to microplastics in the clam *Scrobicularia plana. Frontiers in Marine Science, 5*, 143. https://doi.org/10.3389/fmars.2018.00143

Ogunola, S. O., Reis-Santos, P., Wootton, N., & Gillanders, B. M. (2022). Microplastics in decapod crustaceans sourced from Australian seafood markets. *Marine Pollution Bulletin, 179*, 113706. https://doi.org/10.1016/j.marpolbul.2022.113706

Okeke, E. S., Ezeorba, T. P. C., Chen, Y., Mao, G., Feng, W., & Wu, X. (2022). Ecotoxicological and health implications of microplastic-associated biofilms: A recent review and prospect for turning the hazards into benefits. *Environmental Science and Pollution Research, 29*, 70611–70634. https://doi.org/10.1007/s11356-022-22612-w

Okoffo, E. D., O'Brien, S., O'Brien, J. W., Tscharke, B. J., & Thomas, K. V. (2019). Wastewater treatment plants as a source of plastics in the environment: A review of occurrence, methods for identification, quantification and fate. *Environmental Science: Water Research & Technology, 5*, 1908–1931. https://doi.org/10.1039/C9EW00428A

Oliveira, A. M., Silva, A. L. P., Soares, A. M. V. M., Barceló, D., Duart, A. C., & Rocha-Santos, T. (2023). Current knowledge on the presence, biodegradation, and toxicity of discarded face masks in the environment. *Journal of Environmental Chemical Engineering, 11*, 109308. https://doi.org/10.1016/j.jece.2023.109308

Oceans Asia. (2020). COVID-19 Facemasks & marine plastic pollution. Retrieved March 27, 2023, from https://oceansasia.org/covid-19-facemasks/.

Pan, Z., Liu, Q., Jiang, R., Li, W., Sun, X., Lin, H., Jiang, S., & Huang, H. (2021). Microplastic pollution and ecological risk assessment in an estuarine environment: The Dongshan Bay of China. *Chemosphere, 262*, 127876. https://doi.org/10.1016/j.chemosphere.2020.127876

Peller, J., Nevers, M. B., Byappanahalli, M., Nelson, C., Babu, B. G., Evans, M. A., Kostelnik, E., Keller, M., Johnston, J., & Shidler, S. (2021). Sequestration of microfibers and other microplastics by green algae, Cladophora, in the US Great Lakes. *Environmental Pollution, 276*, 116695. https://doi.org/10.1016/j.envpol.2021.116695

Phuong, N. N., Fauvelle, V., Grenz, C., Ourgaud, M., Schmidt, N., Strady, E., & Sempéré, R. (2021). Highlights from a review of microplastics in marine sediments. *The Science of the Total Environment, 777*, 146225. https://doi.org/10.1016/j.scitotenv.2021.146225

Pizarro-Ortega, C. I., Dioses-Salinas, D. C., Fernández Severini, M. D., Forero López, A. D., Rimondino, G. N., Benson, N. U., Dobaradaran, S., & De-la-Torre, G. E. (2022). Degradation of plastics associated with the COVID-19 pandemic. *Marine Pollution Bulletin, 176*, 113474. https://doi.org/10.1016/j.marpolbul.2022.113474

Priya, K. L., Renjith, K. L., Joseph, C. J., Indu, M. S., Srinivas, R., & Haddout, S. (2022). Fate, transport and degradation pathway of microplastics in aquatic environment—A critical review. *Regional Studies in Marine Science, 56*, 102647. https://doi.org/10.1016/j.rsma.2022.102647

Ribeiro, V. V., Póvoa, A. A., De-la-Torre, G. E., & Castro, I. B. (2022). Indexing anthropogenic litter as a contamination gradient from rivers to beaches in southeast Brazil. *Journal of Coastal Research, 38*, 1172–1180. https://doi.org/10.2112/JCOASTRES-D-22A-00005.1

Rocha-Santos, T., Costa, M. F., & Mouneyrac, C. (Eds.). (2022). *Handbook of Microplastics in the Environment*. Springer International Publishing.

Sadia, M., Mahmood, A., Ibrahim, M., Irshad, M. K., Quddusi, A. H. A., Bokhari, A., Mubashir, M., Chuah, L. F., & Show, P. L. (2022). Microplastics pollution from wastewater treatment plants: A critical review on challenges, detection, sustainable removal techniques and circular economy. *Environmental Technology & Innovation, 28*, 102946. https://doi.org/10.1016/j.eti.2022.102946

Sagawa, N., Kawaai, K., & Hinata, H. (2018). Abundance and size of microplastics in a coastal sea: Comparison among bottom sediment, beach sediment, and surface water. *Marine Pollution Bulletin, 133*, 532–542. https://doi.org/10.1016/j.marpolbul.2018.05.036

Shamskhany, A., & Karimpour, S. (2022). Entrainment and vertical mixing of aquatic microplastics in turbulent flow: The coupled role of particle size and density. *Marine Pollution Bulletin, 184*, 114160. https://doi.org/10.1016/j.marpolbul.2022.114160

Sharifinia, M., Bahmanbeigloo, Z. A., Keshavarzifard, M., Khanjani, M. H., & Lyons, B. P. (2020). Microplastic pollution as a grand challenge in marine research: A closer look at their adverse impacts on the immune and reproductive systems. *Ecotoxicology and Environmental Safety, 204*, 111109. https://doi.org/10.1016/j.ecoenv.2020.111109

Shim, W. J., & Thomposon, R. C. (2015). Microplastics in the Ocean. *Archives of Environmental Contamination and Toxicology, 69*, 265–268. https://doi.org/10.1007/s00244-015-0216-x

Silva, A. L. P., Prata, J. C., Mouneyrac, C., Barcelò, D., Duarte, A. C., & Rocha-Santos, T. (2021). Risks of Covid-19 face masks to wildlife: Present and future research needs. *Science of the Total Environment, 792*, 148505. https://doi.org/10.1016/j.scitotenv.2021.148505

Singh, B., & Sharma, N. (2008). Mechanistic implications of plastic degradation. *Polymer Degradation and Stability, 93*, 561–584. https://doi.org/10.1016/j.polymdegradstab.2007.11.008

Sorasan, C., Edo, C., González-Pleiter, M., Fernández-Piñas, F., Leganés, F., Rodríguez, A., & Rosal, R. (2022). Ageing and fragmentation of marine microplastics. *The Science of the Total Environment, 827*, 154438. https://doi.org/10.1016/j.scitotenv.2022.154438

Sun, J., Peng, Z., Zhu, Z. R., Fu, W., Da, X., & Ni, B. J. (2022). The atmospheric microplastics deposition contributes to microplastic pollution in urban waters. *Water Research, 225*, 119116. https://doi.org/10.1016/j.watres.2022.119116

Sun, Z. L., Sun, L. Q., & Wang, B. (2023). Microplastic pollution in the yellow river basin: Current status and control strategy. *Water, Air & Soil Pollution, 234*, 30. https://doi.org/10.1007/s11270-022-06038-1

Talvitie, J., Mikola, A., Koistinen, A., & Setälä, O. (2017). Solutions to microplastic pollution—Removal of microplastics from wastewater effluent with advanced wastewater treatment technologies. *Water Research, 123*, 401–407. https://doi.org/10.1016/j.watres.2017.07.005

Tang, S., Lin, L., Wang, X., Feng, A., & Yu, A. (2020). Pb(II) uptake onto nylon microplastics: Interaction mechanism and adsorption performance. *Journal of Hazardous Materials, 386*, 121960. https://doi.org/10.1016/j.jhazmat.2019.121960

Tang, K. H. D. (2021). Interactions of microplastics with persistent organic pollutants and the ecotoxicological effects: A review. *Tropical Aquatic and Soil Pollution, 1*, 24–34. https://doi.org/10.53623/tasp.v1i1.11.

Tang, Y., Liu, Y., Chen, Y., Zhang, W., Zhao, J., He, S., Yang, C., Zhang, T., Tang, C., Zhang, C., & Yang, Z. (2021). A review: Research progress on microplastic pollutants in aquatic environments. *Science of the Total Environment, 766*, 142572. https://doi.org/10.1016/j.scitotenv.2020.142572

Tang, L., Feng, J. C., Li, C., Liang, J., Zhang, S., & Yang, Z. (2023). Global occurrence, drivers, and environmental risks of microplastics in marine environments. *Journal of Environmental Management, 329*, 116961. https://doi.org/10.1016/j.jenvman.2022.116961

Ter-Halle, A., Ladirat, L., Gendre, X., Goudouneche, D., Pusineri, C., Routaboul, C., Tenailleau, C., Duployer, B., & Emile, P. (2016). Understanding the fragmentation pattern of marine plastic debris. *Environmental Science & Technology, 50*, 5668–5675. https://doi.org/10.1021/acs.est.6b00594

Ter Halle, A., Ladirat, L., Martignac, M., Mingotaud, A. F., Boyron, O., & Perez, E. (2017). To what extent are microplastics from the open ocean weathered? *Environmental Pollution (barking, Essex: 1987), 227*, 167–174. https://doi.org/10.1016/j.envpol.2017.04.051

Thushari, G. G. N., & Senevirathna, J. D. M. (2020). Plastic pollution in the marine environment. *Heliyon, 6*, 8. https://doi.org/10.1016/j.heliyon.2020.e04709

Torres, F. G., Dioses-Salinas, D. C., Pizarro-Ortega, C. I., & De-la-Torre, G. E. (2021). Sorption of chemical contaminants on degradable and non-degradable microplastics: Recent progress and research trends. *The Science of the Total Environment, 757*, 143875. https://doi.org/10.1016/j.scitotenv.2020.143875

Tošić, T. N., Vruggink, M., & Vesman, A. (2020). Microplastics quantification in surface waters of the Barents, Kara and White Seas. *Marine Pollution Bulletin, 161*(Pt A), 111745. https://doi.org/10.1016/j.marpolbul.2020.111745

Truchet, D. M., Ardusso, M. G., Forero-López, A. D., Rimondino, G. N., Buzzi, N. S., Malanca, F., Spetter, C. V., & Fernández-Severini, M. D. (2022). Tracking synthetic microdebris contamination in a highly urbanized estuary through crabs as sentinel species: An ecological trait-based approach. *Science of the Total Environment, 837*, 155631. https://doi.org/10.1016/j.scitotenv.2022.155631

Tsang, Y. Y., Mak, C. W., Liebich, C., Lam, S. W., Sze, E. T. P., & Chan, K. M. (2017). Microplastic pollution in the marine waters and sediments of Hong Kong. *Marine Pollution Bulletin, 115*, 20–28. https://doi.org/10.1016/j.marpolbul.2016.11.003

Turner, A. (2021). Paint particles in the marine environment: An overlooked component of microplastics. *Water Research X, 12*, 100110. https://doi.org/10.1016/j.wroa.2021.100110

UNEP. (2014). The United Nations environment programme annual report.

Tziourrou, P., Kordella, S., Ardali, Y., Papatheodorou, G., & Karapanagioti, H. K. (2021). Microplastics formation based on degradation characteristics of beached plastic bags. *Marine Pollution Bulletin, 169*, 112470. https://doi.org/10.1016/j.marpolbul.2021.112470

Uheida, A., Mejía, H. G., Abdel-Rehim, M., Hamd, W., & Dutta, J. (2021). Visible light photocatalytic degradation of polypropylene microplastics in a continuous water flow system. *Journal of Hazardous Materials, 406*, 124299. https://doi.org/10.1016/j.jhazmat.2020.124299

Vázquez-Rowe, I., Ita-Nagy, D., & Kahhat, R. (2021). Microplastics in fisheries and aquaculture: Implications to food sustainability and safety. *Current Opinion in Green and Sustainable Chemistry, 29*, 100464. https://doi.org/10.1016/j.cogsc.2021.100464

Vethaak, A. D. & Legler, J. (2021). Microplastics and human health. *Science, 371*, 672–674. https://www.science.org/doi/abs/10.1126/science.abe5041.

Vieira, K. S., Neto, J. A. B., Crapez, M. A. C., Gaylarde, C., Da-Silva-Pierri, B., Saldaña-Serrano, B., Bainy, A. C. D., Nogueira, D. J., & Fonseca, E. M. (2021). Occurrence of microplastics and heavy metals accumulation in native oysters *Crassostrea gasar* in the Paranaguá estuarine system, Brazil. *Marine Pollution Bulletin, 166*, 112225. https://doi.org/10.1016/j.marpolbul.2021.112225

Villagran, D. M., Truchet, D. M., Buzzi, N. S., Forero-Lopez, A. D., & Fernández-Severini, M. D. (2020). A baseline study of microplastics in the burrowing crab (*Neohelice granulata*) from a temperate southwestern Atlantic estuary. *Marine Pollution Bulletin, 150*, 110686. https://doi.org/10.1016/j.marpolbul.2019.110686

Vivekanand, A. C., Mohapatra, S., & Tyagi, V. K. (2021). Microplastics in aquatic environments: Challenges and perspectives. *Chemosphere, 228*, 131151. https://doi.org/10.1016/j.chemosphere.2021.131151

Walkinshaw, C., Lindeque, P. K., Thompson, R., Tolhurst, T., & Cole, M. (2020). Microplastics and seafood: Lower trophic organisms at highest risk of contamination. *Ecotoxicology and Environmental Safety, 190*, 110066. https://doi.org/10.1016/j.ecoenv.2019.110066

Wang, T., Li, B., Zou, X., Wang, Y., Li, Y., Xu, Y., Mao, L., Zhang, C., & Yu, W. (2019). Emission of primary microplastics in mainland China: Invisible but not negligible. *Water Research, 162*, 214–224. https://doi.org/10.1016/j.watres.2019.06.042

Wang, T., Wang, L., Chen, Q., Kalogerakis, N., Ji, R., & Ma, Y. (2020a). Interactions between microplastics and organic pollutants: Effects on toxicity, bioaccumulation, degradation, and transport. *Science of the Total Environment, 748*, 142427. https://doi.org/10.1016/j.scitotenv.2020.142427

Wang, X., Li, C., Liu, K., Zhu, L., Song, Z., & Li, D. (2020b). Atmospheric microplastic over the South China Sea and East Indian Ocean: Abundance, distribution and source. *Journal of Hazardous Materials, 389*, 121846. https://doi.org/10.1016/j.jhazmat.2019.121846

Wang, X., Bolan, N., Tsang, D. C. W., Sarkar, B., Bradney, L., & Li, Y. (2021). A review of microplastics aggregation in aquatic environments: Influence factors, analytical methods, and environmental implications. *Journal of Hazardous Materials, 402*, 123496. https://doi.org/10.1016/j.jhazmat.2020.123496

Weideman, E. A., Perold, V., Donnarumma, V., Suaria, G., & Ryan, P. G. (2023). Proximity to coast and major rivers influence the density of floating microplastics and other litter in east African coastal waters. *Marine Pollution Bulletin, 188*, 114644. https://doi.org/10.1016/j.marpolbul.2023.114644

Welden, N. A., & Cowie, P. R. (2017). Degradation of common polymer ropes in a sublittoral marine environment. *Marine Pollution Bulletin, 118*(1–2), 248–253. https://doi.org/10.1016/j.marpolbul.2017.02.072

Wichmann, D., Delandmeter, P., & Van-Sebille, E. (2019). Influence of near-surface currents on the global dispersal of marine microplastic. *Journal of Geophysical Research: Oceans, 124*, 8. https://doi.org/10.1029/2019JC015328

Wright, S. L., Ulke, J., Font, A., Chan, K. L. A., & Kelly, F. J. (2020). Atmospheric microplastic deposition in an urban environment and an evaluation of transport. *Environment International, 136*, 105411. https://doi.org/10.1016/j.envint.2019.105411

Wu, H., Hou, L., & Wang, X. (2023). A review of microplastic pollution in aquaculture: Sources, effects, removal strategies and prospects. *Ecotoxicology and Environmental Safety, 252*, 114567. https://doi.org/10.1016/j.ecoenv.2023.114567

Xie, Y. S., Zhang, Q. K., Zhang, L. S., Zhong, S., Liu, J. Y., Hou, X. Y., Wang, H., Feng, J. P., Liang, Z. H., & Chen, W. Y. (2020). Impact of a sewage treatment plant on the accumulation of microplastics in freshwater organisms in the Lijiang River of the Guilin urban section. *Environmental Science, 41*, 4999–5007. https://doi.org/10.13227/j.hjkx.202003075.

Yu, H., Shao, J., Jia, H., Gang, D., Ma, B., & Hu, C. (2023). Characteristics and influencing factors of microplastics in snow in the inner Mongolia plateau, China. *Engineering (Beijing, China)*. https://doi.org/10.1016/j.eng.2023.02.007.

Yu, Y., Miao, L., Adyel, T. M., Waldschläger, K., Wu, J., & Hou, J. (2023b). Aquatic plastisphere: Interactions between plastics and biofilms. *Environmental Pollution (barking, Essex: 1987), 322*, 121196. https://doi.org/10.1016/j.envpol.2023.121196

Zettler, E. R., Mincer, T. J., & Amaral-Zettler, L. A. (2013). Life in the "plastisphere": microbial communities on plastic marine debris. *Environmental Science & Technology, 47*, 7137–7146. https://doi.org/10.1021/es401288x

Zhang, X., Li, S., Liu, Y., Yu, K., Zhang, H., Yu, H., & Jiang, J. (2012). Neglected microplastics pollution in the nearshore surface waters derived from coastal fishery activities in Weihai, China. *Science of the Total Environment, 768*, 144484. https://doi.org/10.1016/j.scitotenv.2020.144484

Zhou, C., Bi, R., Su, C., Liu, W., & Wang, T. (2022). The emerging issue of microplastics in the marine environment: A bibliometric analysis from 2004 to 2020. *Marine Pollution Bulletin, 179*, 113712. https://doi.org/10.1016/j.marpolbul.2022.113712

Zhu, K., Jia, H., Sun, Y., Dai, Y., Zhang, C., Guo, X., Wang, T., & Zhu, L. (2020). Long-term phototransformation of microplastics under simulated sunlight irradiation in aquatic environments: Roles of reactive oxygen species. *Water Research, 173*, 115564. https://doi.org/10.1016/j.watres.2020.115564

Chapter 8
Microplastics and Food Safety

Gabriel Enrique De-la-Torre, Diana Carolina Dioses-Salinas, Melisa D. Fernández Severini, and Ana D. Forero López

Abstract The widespread distribution of microplastics (MPs) in the environment poses an evident threat to the well-being of aquatic and terrestrial organisms, many of which are farmed or caught for human consumption. Under this context, MP ingestion seems inevitable. The nature of MPs and their interaction with external contaminants, including organic and inorganic chemicals and pathogens, could derive implications for food security. The present chapter addresses the factors involved in MPs and food security by covering their interaction with chemical contaminants and pathogens, as well as recent studies on the concentration of MPs in seafood and crops and their respective sources. Current studies are still limited primarily to seafood (mostly, fish and mollusks), while other important sources of food, such as vegetables, livestock, and poultry products remain poorly investigated. Furthermore, MP surveys in food products focus on the identification of suspected particles (confirming polymer composition), while overlooking their association with other contaminants, including those that may leach out from food-contact plastic materials. The latter is of particular interest to food safety due to their toxicity. The current literature is still far from being sufficient to estimate realistic MP and MP-related contaminant consumption in humans at a large scale and their potential chronic health effects.

Keywords Plastic debris · Synthetic polymer · Food security · Nanoplastics · Human consumption

G. E. De-la-Torre (✉) · D. C. Dioses-Salinas
Grupo de Investigación de Biodiversidad, Medio Ambiente y Sociedad, Universidad San Ignacio de Loyola, Lima, Perú
e-mail: gabriel.delatorre@usil.pe

M. D. Fernández Severini · A. D. Forero López
Instituto Argentino de Oceanografía (IADO), CONICET/UNS, CCT-Bahía Blanca, Camino La Carrindanga, km 7.5, Edificio E1, Bahía Blanca, B8000FWB Buenos Aires, Argentina

A. K. Mishra et al. (eds.), *Microplastics*, https://doi.org/10.1007/978-981-97-6461-7_8

1 Introduction

The widespread distribution of microplastics (MPs) in virtually any environmental compartment has raised environmental health concerns (Ajith et al., 2020; De-la-Torre et al. 2023a; Valdiviezo Gonzales et al., 2024). However, as MPs make their way along the food web, species meant for human consumption will inevitably be contaminated (Colombo et al., 2023; Lehel & Murphy, 2021). Thus, the food safety implications of MPs are a rather important focus that needs to be addressed (Rubio-Armendáriz et al., 2022; Al Mamun et al. 2023). Food safety is defined as the conditions and measures along the whole food supply chain (from production to preparation and storage) to ensure that food is served safely and prevent the development of food-borne illnesses (Yemane & Tamene, 2022). Failing to comply with food safety standards is related to the development of diseases and malnutrition (Nordhagen et al., 2022). Given that MPs carry various chemical components in their polymeric matrix used as additives and serve as vectors for a wide range of chemical pollutants (Deng et al., 2021; Torres et al., 2021), their incorporation into foods and drinking water could compromise food safety (Barboza et al., 2018; De-la-Torre, 2020). In the present chapter, the hazardous and pathogenic implications of MPs are summarized, as well as the recent studies on seafood and vegetables or fruit contamination and their respective sources.

2 Associated Contaminants and Pathogens

2.1 Chemical Contaminants

It is widely known that various chemical pollutants, such as polycyclic aromatic hydrocarbons (PAHs) (Conesa, 2022), heavy metals (Besson et al., 2020), pharmaceuticals (Santos et al., 2021), and perfluoroalkyl substances (PFASs) (Llorca et al., 2018), among others, are able to adsorb on the surface of MPs. The adsorption of chemical contaminants occurs through one or multiple mechanisms, such as hydrophobic, electrostatic, and π–π interactions, hydrogen bonding, Van de Waals forces, and pore filling (Torres et al., 2021). On the other hand, the adsorption is governed by several factors involving the characteristics of the MP (e.g., polymer composition, crystallinity, surface charge, porosity), the environment (e.g., pH, temperature, salinity, organic matter), and the sorbate (chemical properties) (Agboola & Benson, 2021). Due to the diversity of MPs and chemical contaminants in the environment, as well as the multiple factors that govern this phenomenon, the adsorption of chemical contaminants is a rather complex process. However, the occurrence of pollutant-loaded MPs cannot be denied. Field research has demonstrated that MPs are found to contain polychlorinated biphenyls (PCBs), polybrominated diphenyl ethers (PBDEs), PAHs (Lo et al., 2019; Yeo et al., 2020), pesticides (Camacho et al., 2019), and heavy metals (Forero-López et al., 2022). Considering that MPs are easily uptaken by most organisms, they can presumably increase the bioavailability of external contaminants. However, Sørensen et al. (2020) demonstrated that, unlike free-dissolved PAHs, PAHs adsorbed on MPs are not significantly bioavailable to marine copepods. Hence, the absorption of chemical contaminants is not always translated into higher bioavailability. Furthermore, the model carried out by Koelmans et al. (2016) indicates that bioaccumulation of hydrophobic organic chemicals is significantly more pronounced from natural prey consumption than contaminated MPs. Nevertheless, the ability to adsorb chemical contaminants remains a matter of concern, particularly for those that may bioaccumulate and compromise food safety. For instance, PFASs are compounds categorized as carcinogenic compounds and possible endocrine disruptors (Steenland & Winquist, 2021). While the partition behavior of perfluorooctanesulfonate (PFOS) and perfluorooctanesulfonamide (FOSA) on MPs has been investigated (Wang et al., 2015), the review by Torres and De-la-Torre (2022) on the presence of PFASs in consumable species and food products indicates that multiple studies agree that PFASs concentrations are linked to the proximity of factories and fluorochemical industries. Under this context, the contribution of MPs as PFAS carriers and their influence on the resulting concentration in edible species is unknown and probably negligible.

There is extensive scientific evidence indicating that MPs are packed with chemical contaminants. However, the path from MP absorption to reaching foods readily sold for consumption is diffuse and rather complex to investigate (Fig. 1). There is still insufficient information to state that these associated contaminants are taking part in food safety implications linked to the presence of MPs. However, research on this matter needs to better develop these contamination pathways.

2.2 Pathogens

MPs are suitable substrates for the colonization of biofilm-forming microbial communities, including potentially pathogenic species (Zettler et al., 2013). Such structure is referred to as the "plastisphere." The variables that govern the microbial structure of the plastisphere are related to the type of polymer, seasonality, and environmental factors (Dudek et al., 2020; Oberbeckmann et al., 2018). Furthermore, the stages of the plastisphere formation can be summarized into (1) fragmentation of plastic debris (MP formation), (2) surface conditioning by the deposition of biomacromolecules, (3) microbial biomass transport to MPs, (4) quorum sensing, and (5) formation of the plastisphere (Kaur et al., 2022).

Field and laboratory studies have investigated the bacterial colonization of different types of MPs. For instance, a 14-day microcosm experiment to investigate the bacterial colonization of low-density polyethylene (LDPE) MPs showed a rapid successional formation of bacterial assemblages mainly dominated by *Arcobacter* and *Colwellia* species. On the other hand, Tong et al. (2022) reported that PE MPs and *Helicobacter pylori* act synergistically causing rapid bacterial colonization of gastric mucosal epithelial cells and entry into mice tissues, raising concerns for

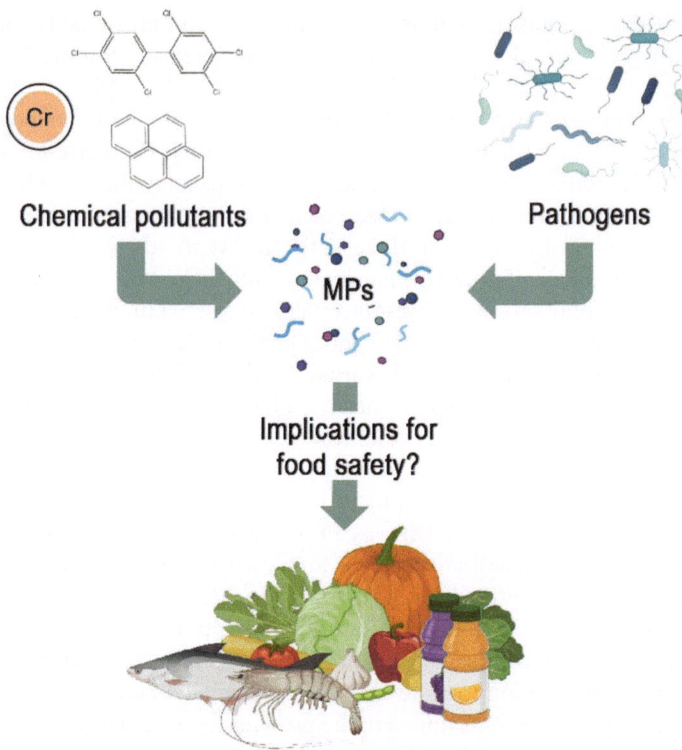

Fig. 1 Schematic representation of the possible pathways by which chemical pollutants and pathogens make their way into foods through MPs adsorption or colonization

human health. The pathogenic bacteria *Aeromonas salmonicida*, known for causing disease in some fish species (Kaattari & Piganelli, 1996), was previously identified on floating MPs from the North Adriatic Sea (Viršek et al., 2017); thus, raising concerns about the possibility of MP spreading diseases among consumable species. Other pathogenic bacteria, such as *Vibrio* spp. and *Escherichia coli*, and invasive algal species have been isolated from floating MPs elsewhere (Casabianca et al., 2019; Kaur et al., 2022; Kirstein et al., 2016; Muniz Silva et al., 2019; Zettler et al., 2013).

Like in the case of chemical contaminants, the bridge between a group of colonized MPs and contaminated food is unknown. Furthermore, multiple questions remain unanswered. For instance, it is not clear whether MP–microbe complexes are generally more harmful than microbes or MPs alone. Yang et al. (2023) suggest that MP–microbe complexes are more easily eliminated by phagocytes than microbes alone. The reasoning behind this is based on the effective immune response mechanisms despite exhibiting an unconventional entry route. All-in-all, the adhesion of chemical pollutants and pathogens on MPs seems like a fair topic to include in the discussion of MPs and food safety. Previous reviews have also addressed these topics (Barboza et al., 2018; De-la-Torre, 2020). Nevertheless, a realistic interpretation should be carefully considered, as well as recognizing the current knowledge gaps.

3 MP Contamination in Foods

3.1 Seafood

Seafood is the most investigated food in terms of MP contamination, probably because the oceans are recognized as the biggest MP (and plastic debris in general) sink (García Rellán et al., 2023). Indeed, consumable marine species, such as fish and mollusks, are important sources of proteins and micronutrients and are part of the traditional human diet worldwide (Gephart, 2019). However, MP contamination has been investigated in edible species extracted from the wild, farmed, or directly purchased for consumption (Fernández Severini et al., 2020; Colombo et al., 2023; De-la-Torre et al. 2023b). The latter is of particular interest because these samples have gone through every step of the food supply chain and may exhibit MP concentrations that are directly consumed (however, preparation and cooking may influence MP content). Another important factor is eating behavior. For instance, most fish tend to accumulate MPs along their gastrointestinal tracts (GITs) or gills depending on their feeding behavior (Zazouli et al., 2022). However, fish is mostly eviscerated, and the GITs are discarded before preparation. Thus, examining fish fillets may be the most suitable way to investigate MP's contamination with a food safety focus. The presence of MPs in fish fillets may be related to atmospheric deposition or cross-contamination during handling instead of direct intake by the fish in the ocean. Significantly lower MP levels have been reported in fish muscle than in the GIT or gills, regardless of the species or sampling point (Hasan et al., 2023). Thus, suggesting a relatively low MP consumption from most fishes. However, in some cases, the whole body of the fish is eaten entirely, such as canned sardines. Akhbarizadeh et al. (2020) reported an overall mean concentration of 25.6 MPs per can of tuna (*Thunnus tonggol, Thunnus albacares*) and mackerel (*Scombermorus commerson*) sold at Iranian markets. Table 1 summarizes MP studies involving seafood directly sold at fisheries, markets, or commercial producers.

Another important and widely consumed type of seafood is bivalves. Unlike most fishes, the soft tissues of bivalves are generally eaten entirely, regardless of the cooking method. This indicates a more pronounced influence of MP uptake and bioaccumulation on the consumed proportion. In other cases, bivalves are eaten raw or seasoned without an advanced cooking method (De-la-Torre et al. 2022a). The number of MPs per gram of consumable bivalve tissue is highly variable across studies and species (Table 1). One important factor influencing MP contamination may be food processing. For instance, Li et al. (2018) reported a relatively high MP concentration in pre-cooked mussels (1.4 MPs per gram) than in those supplied live

Table 1 Summary of the studies investigating MP contamination in seafood obtained from markets or commercial producers

Country of purchase	Species	Organ analyzed	Mean concentration	Dominant type	Polymers identified	References
Fish						
Iran	*Thunnus tonggol* *Thunnus albacares* *Scombermorus commerson*	Whole (canned)	25.6 MPs/can	Fiber	PET, PS, PP, PS-PP, PS-PET, Nylon, PVC, LDPE	Akhbarizadeh et al. (2020)
Malaysia	*Chelon subviridis* *Johnius belangerii* *Rastrelliger kanagurta* *Stolephorus waitei*	Dried flesh	–	Fragment	PP, PE, PS, PET, Nylon	Karami et al. (2017)
China	*Pseudosciaena Polyactis* *Pleuronichthys cornutus* *Sebastes schlegeli* *Liza haematocheila*	Digestive tract	0.09–0.58 MPs/g	Fiber	Rayon, PET, CPE, PE-PP, PTFE, PS	Ding et al. (2019)
Malaysia Australia	Sardines Sprats	Whole (canned)	–	Fragment	PP, PET, PE, PVC	Karami et al. (2018)

(continued)

Table 1 (continued)

Country of purchase	Species	Organ analyzed	Mean concentration	Dominant type	Polymers identified	References
UK	*Melanogrammus aeglefinus* *Dicentrarchus labrax* *Pleuronectes platessa* *Scromber scombrus*	Flesh	0.18–0.77 MPs/g flesh	Fibers	Alkyd resin, PE, PET, Cellophane, PVAE, PET, PE-PP, PAN, PA	Akoueson et al. (2020)
Bivalves						
UK	*Zygochlamys patagonica* *Pecten maximus*	Soft tissues	0.17–0.32 MPs/g tissues	Fibers	Alkyd resin, PE, PET, Cellophane, PVAE, PET, PE-PP, PAN, PA	Akoueson et al. (2020)
China	*Mytilus galloprovincialis* *Ruditapes philippinarum* *Mactra veneriformis*	Soft tissues	0.16–0.74 MP/g tissues	Fiber	Rayon, PET, CPE, PE-PP, PTFE, PS	Ding et al. (2019)
Peru	*Aulacomya atra*	Soft tissues	0.56 MPs/g tissues	Fiber	Polyester, PP, PE, PS	De-la-Torre et al. (2022a)
Korea	*Crassostrea gigas* *Mytilus edulis* *Tapes philippinarum* *Patinopecten yessoensis*	Soft tissues	0.15 MPs/g tissues	Fragment	PE, Polyester, Acrylic, PP, PS, PET, Nylon, PVA, PVA, among others	Cho et al. (2019)

(continued)

Table 1 (continued)

Country of purchase	Species	Organ analyzed	Mean concentration	Dominant type	Polymers identified	References
China	*Mytilus galloprovincialis*	Soft tissues	3.2–7.1 MPs/g tissues	Fiber	Cellophane, PP, PTFE	Ding et al. (2018)
	Chlamys farreri		5.2–19.4 MPs/g tissues			
UK	*Mytilus edulis*	Soft tissues (pre-cooked)	1.4 MPs/g tissues	Fiber	PP, Polyester, Rayon, Acrylic, PE, Propylene glycol ricinoleate, Nitrile rubber	Li et al. (2018)
		Soft tissues (live)	0.9 MPs/g tissues			
Peru	*Anadara tuberculosa*	Soft tissues	1.6 MPs/g tissues	Fiber	PE, Polyester, PVC, PS, Pam, PP, Enzacryl polyacetal, Phenol resin, Nylon, PAA, PAN	Aguirre-Sanchez et al. (2024)
China	*Scapharca subcrenata* *Tegillarca granosa* *Mytilus galloprovincialis* *Patinopecten yessoensis* *Alectryonella plicatula* *Sinonovacula constricta* *Ruditapes philippinarum* *Meretrix lusoria* *Cyclina sinensis*	Soft tissues	2.1–10.5 MPs/g tissues	Fiber	PE, PET, PA	Li et al. (2015)

(continued)

Table 1 (continued)

Country of purchase	Species	Organ analyzed	Mean concentration	Dominant type	Polymers identified	References
China	*Mytilus galloprovincialis* *Perna viridis* *Ruditapes philippinarum* *Crassostrea gigas* *Sinonovacula constricta* *Scapharca subcrenata* *Meretrix lusoria* *Busycon canaliculatu*	Soft tissues	0.8–4.4 MPs/g tissues	Fiber	Rayon, CPE, PVC, PET, PVDF, PTFE, among others	Ding et al. (2020)
Crustaceans						
Peru	*Ucides occidentalis*	Gills and stomach	1.94 MPs/g tissues	Fiber	PE, Polyester, PVC, PS, Pam, PP, Enzacryl polyacetal, Phenol resin, Nylon, PAA, PAN	Aguirre-Sanchez et al. (2024)

(0.9 MPs per gram) obtained from UK markets. As such, processing techniques and hygiene standards along the entire food supply chain may play an important role in the resulting MP concentration.

3.2 *Vegetables and Fruits*

More recently, it was demonstrated that the smallest MPs (a few microns in size) and nanoplastics (<1 μm; NPs) could be uptaken by plant species through the roots and possibly translocate and bioaccumulate in various organs (Sun et al., 2021; Wang et al. 2022a, 2022b). The factors influencing phytotoxicity are related to the characteristics of the MPs/NPs (e.g., concentration, size, polymer, hydrophobicity, and shape) and the plant (e.g., growth stage, growing media, species, and environmental conditions) (Roy et al., 2023). Phytotoxicity studies have shown that polylactic acid (PLA) MPs reduced *Lolium perenne* (perennial ryegrass) germination and shoot height, while high-density PE MPs reduced the biomass of the endogenic rosy-tipped earthworm (*Aporrectodea rosea*) in the soil (Boots et al., 2019). Similarly, PE MPs inhibit photosynthetic parameters, chlorophyll content, and lettuce growth (*Lactuca sativa*) (Gao et al., 2019). Nevertheless, the long-term accumulation of MPs/NPs in farmlands and agricultural soils requires further investigation to determine whether crop yields could be compromised and, thus, affect food security.

A different concern is the possibility of ingesting MPs/NPs accumulated in vegetables and fruits as a result of translocation. However, precisely analyzing NPs in organic matrices remains a significant analytical challenge (de Bruin et al., 2022). Oliveri Conti et al. (2020) applied scanning electron microscopy (SEM) coupled with an X energy dispersion detector (EDX) to quantify MPs <10 μm in a variety of fruits and vegetables, resulting in concentrations ranging from 50,550 to 189,550 MPs per gram of edible content. The details of the quantification method are described in Ferrante et al. (2020). Instead of microscopic methods, Dessì et al. (2021) used a thermal degradation-based method, namely pressurized liquid extraction combined with double-shot pyrolysis gas chromatography–mass spectrometry (GC/MS), to quantify MPs in market rice. The method has been developed to quantify MPs from complex matrices, such as biosolids (Okoffo et al. 2020a, 2020b), but can be applied to other matrices like foods. Their results indicated that instant pre-cooked rice had a significantly higher MP concentration (283 μg/g) than uncooked packages (67 μg/g), consisting of PE, PP, and PET (Dessì et al., 2021). MP consumption by the Australian population was estimated at around 1 g per person per year from rice consumption alone. The results obtained from studies applying unconventional methods to quantify the smallest MPs/NPs fractions in foods are rather promising. However, this type of research is still lacking for consumable products. Prospectively, laboratories are expected to continue developing reliable methods to quantify the unaccounted number of consumed MPs from a wide range of products and foods.

4 Sources of Contamination

Food contamination has been reported multiple times in the literature. One crucial point to understand is the key sources of MPs that lead to their human consumption (An et al., 2020). MPs are emitted from multiple sources and transported long distances via natural forces (e.g., wind and water streams), ultimately reaching remote areas and rural lands (Lwanga et al., 2022). However, several sources of contamination are more pronounced in different environments.

Understanding the sources, circulation, and fate of MPs in the world's oceans is rather complex (Yang et al., 2021). However, the majority of MPs found in almost any marine environmental matrix (e.g., surface water, bottom sediments, water column) and species display a fibrous shape (Abelouah et al., 2022; Ben-Haddad et al., 2022; De-la-Torre et al. 2022b), synthetic textiles are regarded as a major source of MPs. Poorly treated sewage and wastewater treatment plants (WWTPs) are the main contributors of microfibers to the marine environment. MPs are released during laundry, estimated in the tens to hundreds of thousands per wash depending on the type of clothing, and time, among several factors (De Falco et al., 2019; Napper & Thompson, 2016; Wang et al., 2023). In WWTPs, microfibers are removed at different stages, although tertiary WWTPs are the most effective ones, reaching removal efficiencies of >90% (Iyare et al., 2020; Tang & Hadibarata, 2021). Nevertheless, a significant number of MPs are still constantly released into the oceans, particularly in metropolitan areas and cities. Another important driver is the fragmentation of larger plastic debris, which is emitted through a wide range of sources. Recent estimations suggest that more than 1,000 rivers account for 80% of global plastic emissions, ranging from 0.8 million to 2.7 million tons per year (Meijer et al., 2021). Plastic debris is then subject to environmental weathering and fragmentation, leading to the release of secondary MPs (Andrady et al., 2022). Additionally, sea-based sources (e.g., incorrect disposal of fishing and aquaculture gear) may account for an important source of plastic litter and, subsequently, MPs in certain areas (Daniel et al., 2020).

The sources of contamination in farmlands are categorized into (1) agricultural practices (soil amendments and irrigation), and (2) atmospheric deposition (Dioses-Salinas et al., 2020). Agricultural soils are treated with several soil amendments, such as compost, sewage sludge, chemical fertilizers, and guano, among others, to incorporate nutrients that promote plant growth and development, and increase crop yield (Clements & Bihn, 2019). For instance, sewage sludge is widely used as an alternative to chemical fertilizers due to its nutrient and organic matter content (Lamastra et al., 2018). However, there is some debate on the possible risks of continuously applying sewage sludge on agricultural land due to the presence of several contaminants, including MPs (Ekane et al., 2021). For instance, MP content in sewage sludge from a WWTP in Ireland ranged from 4,196 to 15,385 MPs per kg of dried sludge depending on the type of treatment process (Mahon et al., 2017). Investigations on sewage sludge are vast and continuously being developed due to their relevancy as potential sources of MPs worldwide (Franco et al., 2023; Nayon Hossain et al.,

2023; Okoffo et al., 2023; Wei et al., 2022), as well as the contribution of WWTP technology and sources (Hajji et al., 2023). MPs in other types of soil amendments, like compost, have also been reported (Braun et al., 2021; van Schothorst et al., 2021). Mulching has also been recognized as a possible source of MPs in agricultural soils (Sa'adu and Farsang 2023), although their use is not as widespread as solid amendments. The use of an irrigation method and water source depends on several geographic, atmospheric, and economic factors. Some commonly used methods are surface, underground, sprinkler, and drip irrigation. Zhou et al. (2023) showed that fields irrigated by surface irrigation with water from the Huangshui River, China, presented MP concentrations almost ten times higher than non-irrigated fields. This is because rivers are exceptional carriers and transports of plastic debris downstream from the upper river basin. Similar studies have also reported a significant contribution of MPs coming from river water on irrigated farmlands (Kundu et al., 2022; Yu et al., 2021) highly altered by seasonality and atmospheric events, like typhoons (Jiang et al., 2023). Surface irrigation may also alter the vertical distribution of MPs in soil, where small fibers may have higher vertical mobility (Liu et al., 2023). The second major source of contamination is through atmospheric deposition. Most farmlands are far from urbanized areas and lack a latent source of contamination. However, MPs are able to follow air mass trajectories for tens of kilometers or more and be deposited (or precipitated) in pristine or remote areas (Allen et al., 2019; Huang et al., 2021). Meteorological studies suggest that precipitation is the major driver of atmospheric deposition of MPs (Roblin et al., 2020), reaching remote areas, including farmlands. The contamination of farmlands may subsequently expose livestock and poultry animals that feed in open fields, accompanied by common animal feed (Bahrani et al., 2024). However, the occurrence of MPs in livestock in yet poorly investigated (Sheriff et al., 2023) (Fig. 2).

5 Conclusions

The factors involved in the discussion of MPs and food security are multiple. In recent years, significant efforts have been made to shed light on the details of such discussion. However, there are still many loose ends that need to be addressed. In the present chapter, the variables involved in MP research that concern food security were summarized and discussed. MPs are presented as pathogen and contaminant reservoirs due to the absorption of various contaminants, as well as ubiquitous contaminants across environments and species, including those meant for human consumption. Nevertheless, estimations and current studies are still insufficient to land on a solid position regarding food security. Further research is needed to obtain realistic MP intake estimations that account for all types of meals, particularly those that are poorly investigated, such as livestock and poultry products (e.g., meat, eggs, and milk). Furthermore, studies should aim to determine the pathogens and chemical contaminants that harbor MPs present in foods readily sold for consumption and the migration of chemicals in food-contact plastic materials.

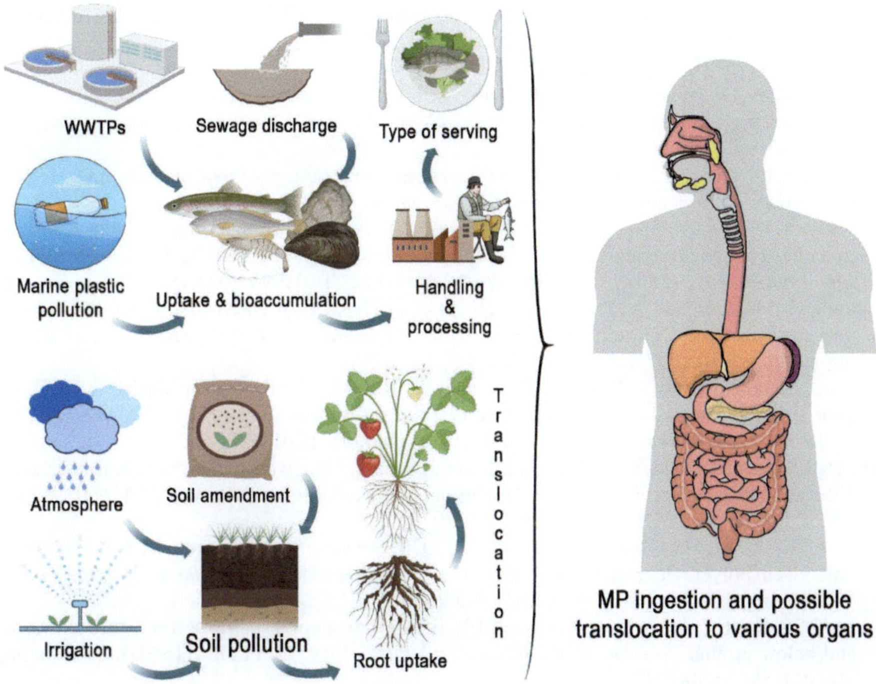

Fig. 2 Summary of the main sources of contamination leading to the ingestion of contaminated food

References

Abelouah, M. R., Ben-Haddad, M., Hajji, S., et al. (2022). Floating microplastics pollution in the Central Atlantic Ocean of Morocco: Insights into the occurrence, characterization, and fate. *Marine Pollution Bulletin, 182*, 113969. https://doi.org/10.1016/j.marpolbul.2022.113969

Agboola, O. D., & Benson, N. U. (2021). Physisorption and chemisorption mechanisms influencing micro (nano) plastics-organic chemical contaminants interactions: A review. *Frontiers in Environmental Science, 9*, 167. https://doi.org/10.3389/FENVS.2021.678574/BIBTEX

Aguirre-Sanchez, A., Purca, S., Cole, M., et al. (2024). Prevalence of microplastics in Peruvian mangrove sediments and edible mangrove species. *Marine Pollution Bulletin, 200*, 116075. https://doi.org/10.1016/j.marpolbul.2024.116075

Ajith, N., Arumugam, S., Parthasarathy, S., et al. (2020). Global distribution of microplastics and its impact on marine environment—A review. *Environmental Science and Pollution Research, 27*, 25970–25986. https://doi.org/10.1007/s11356-020-09015-5

Akhbarizadeh, R., Dobaradaran, S., Nabipour, I., et al. (2020). Abundance, composition, and potential intake of microplastics in canned fish. *Marine Pollution Bulletin, 160*, 111633. https://doi.org/10.1016/J.MARPOLBUL.2020.111633

Akoueson, F., Sheldon, L. M., Danopoulos, E., et al. (2020). A preliminary analysis of microplastics in edible versus non-edible tissues from seafood samples. *Environmental Pollution, 263*, 114452. https://doi.org/10.1016/j.envpol.2020.114452

Al Mamun, A., Prasetya, T. A. E., Dewi, I. R., & Ahmad, M. (2023). Microplastics in human food chains: Food becoming a threat to health safety. *Science of the Total Environment, 858*, 159834. https://doi.org/10.1016/j.scitotenv.2022.159834

Allen, S., Allen, D., Phoenix, V. R., et al. (2019). Atmospheric transport and deposition of microplastics in a remote mountain catchment. *Nature Geoscience, 12*, 339–344. https://doi.org/10.1038/s41561-019-0335-5

An, L., Liu, Q., Deng, Y., et al. (2020). Sources of microplastic in the environment. In D. He & Y. Luo (Eds.), *Microplastics in Terrestrial Environments* (pp. 143–159). Springer.

Andrady, A. L., Barnes, P. W., Bornman, J. F., et al. (2022). Oxidation and fragmentation of plastics in a changing environment; from UV-radiation to biological degradation. *Science of the Total Environment, 851*, 158022. https://doi.org/10.1016/J.SCITOTENV.2022.158022

Bahrani, F., Mohammadi, A., Dobaradaran, S., et al. (2024). Occurrence of microplastics in edible tissues of livestock (cow and sheep). *Environmental Science and Pollution Research, 31*, 22145. https://doi.org/10.1007/s11356-024-32424-9

Barboza, L. G. A., Dick Vethaak, A., Lavorante, B. R. B. O., et al. (2018). Marine microplastic debris: An emerging issue for food security, food safety and human health. *Marine Pollution Bulletin, 133*, 336–348. https://doi.org/10.1016/j.marpolbul.2018.05.047

Ben-Haddad, M., Abelouah, M. R., Hajji, S., et al. (2022). The wedge clam Donax trunculus L., 1758 as a bioindicator of microplastic pollution. *Marine Pollution Bulletin, 178*, 113607. https://doi.org/10.1016/J.MARPOLBUL.2022.113607

Besson, M., Jacob, H., Oberhaensli, F., et al. (2020). Preferential adsorption of Cd, Cs and Zn onto virgin polyethylene microplastic versus sediment particles. *Marine Pollution Bulletin, 156*, 111223. https://doi.org/10.1016/j.marpolbul.2020.111223

Boots, B., Russell, C. W., & Green, D. S. (2019). Effects of microplastics in soil ecosystems: Above and below ground. *Environmental Science and Technology, 53*, 11496–11506. https://doi.org/10.1021/acs.est.9b03304

Braun, M., Mail, M., Heyse, R., & Amelung, W. (2021). Plastic in compost: Prevalence and potential input into agricultural and horticultural soils. *Science of the Total Environment, 760*, 143335. https://doi.org/10.1016/j.scitotenv.2020.143335

Camacho, M., Herrera, A., Gómez, M., et al. (2019). Organic pollutants in marine plastic debris from Canary Islands beaches. *Science of the Total Environment, 662*, 22–31. https://doi.org/10.1016/j.scitotenv.2018.12.422

Casabianca, S., Capellacci, S., Giacobbe, M. G., et al. (2019). Plastic-associated harmful microalgal assemblages in marine environment. *Environmental Pollution, 244*, 617–626. https://doi.org/10.1016/j.envpol.2018.09.110

Cho, Y., Shim, W. J., Jang, M., et al. (2019). Abundance and characteristics of microplastics in market bivalves from South Korea. *Environmental Pollution, 245*, 1107–1116. https://doi.org/10.1016/j.envpol.2018.11.091

Clements, D. P., & Bihn, E. A. (2019). The impact of food safety training on the adoption of good agricultural practices on farms (Chap. 16). In D. Biswas & S. A. Micallef (Eds.), *Safety and Practice for Organic Food* (pp. 321–344). Elsevier B.V.

Colombo, C. V., Fernández-Severini, M. D., Forero-López, A. D., et al. (2023). Microplastics in commercial seafood: Pleoticus muelleri as a case study in an estuarine environment highly affected by human pressure (Southwestern Atlantic). *Environmental Research, 216*, 114738. https://doi.org/10.1016/j.envres.2022.114738

Conesa, J. A. (2022). Adsorption of PAHs and PCDD/Fs in microplastics: A review. *Microplastics, 1*, 346–358. https://doi.org/10.3390/microplastics1030026

Daniel, D. B., Thomas, S. N., & Thomson, K. T. (2020). Assessment of fishing-related plastic debris along the beaches in Kerala Coast, India. *Marine Pollution Bulletin, 150*, 110696. https://doi.org/10.1016/j.marpolbul.2019.110696

De-la-Torre, G. E. (2020). Microplastics: An emerging threat to food security and human health. *Journal of Food Science and Technology, 57*, 1601–1608. https://doi.org/10.1007/s13197-019-04138-1

De-la-Torre, G. E., Dioses-Salinas, D. C., Pizarro-Ortega, C. I., et al. (2023). Plastic and paint debris in marine protected areas of Peru. *Science of the Total Environment, 901*, 165788. https://doi.org/10.1016/j.scitotenv.2023.165788

De-la-Torre, G. E., Dobaradaran, S., Urizar Garfias Reyes, D. F., & Dioses-Salinas, D. C. (2023b). Micro- and nanoplastics in foods. In: *Reference Module in Food Science*. Elsevier B.V.

De-la-Torre, G. E., Laura, R. P., Mendoza-Castilla, L. M. (2022a). Abundance and characteristics of microplastics in market bivalve Aulacomya Atra (Mytilidae: Bivalvia). *Acta Biológica Colombiana, 27*. https://doi.org/10.15446/abc.v27n2.88832.

De-la-Torre, G. E., Pizarro-Ortega, C. I., Dioses-Salinas, D. C., et al. (2022). Are we underestimating floating microplastic pollution? A quantitative analysis of two sampling methodologies. *Marine Pollution Bulletin, 178*, 113592. https://doi.org/10.1016/J.MARPOLBUL.2022.113592

de Bruin, C. R., de Rijke, E., van Wezel, A. P., & Astefanei, A. (2022). Methodologies to characterize, identify and quantify nano- and sub-micron sized plastics in relevant media for human exposure: A critical review. *Environmental Science Advances, 1*, 238. https://doi.org/10.1039/d1va00024a

De Falco, F., Di Pace, E., Cocca, M., & Avella, M. (2019). The contribution of washing processes of synthetic clothes to microplastic pollution. *Scientific Reports, 9*, 6633. https://doi.org/10.1038/s41598-019-43023-x

Deng, H., Su, L., Zheng, Y., et al. (2021). Crack patterns of environmental plastic fragments. *Environmental Science and Technology, 56*, 6399–6414. https://doi.org/10.1021/acs.est.1c08100

Dessì, C., Okoffo, E. D., O'Brien, J. W., et al. (2021). Plastics contamination of store-bought rice. *Journal of Hazardous Materials, 416*, 125778. https://doi.org/10.1016/j.jhazmat.2021.125778

Ding, J., Li, J., Sun, C., et al. (2019). Detection of microplastics in local marine organisms using a multi-technology system. *Analytical Methods, 11*, 78–87. https://doi.org/10.1039/c8ay01974f

Ding, J., Li, J., Sun, C., et al. (2020). An examination of the occurrence and potential risks of microplastics across various shellfish. *Science of the Total Environment, 739*, 139887. https://doi.org/10.1016/j.scitotenv.2020.139887

Ding, J. F., Li, J. X., Sun, C. J., et al. (2018). Separation and identification of microplastics in digestive system of bivalves. *Chinese Journal of Analytical Chemistry, 46*, 690–697. https://doi.org/10.1016/S1872-2040(18)61086-2

Dioses-Salinas, D. C., Pizarro-Ortega, C. I., & De-la-Torre, G. E. (2020). A methodological approach of the current literature on microplastic contamination in terrestrial environments: Current knowledge and baseline considerations. *Science of the Total Environment, 730*, 139164. https://doi.org/10.1016/j.scitotenv.2020.139164

Dudek, K. L., Cruz, B. N., Polidoro, B., & Neuer, S. (2020). Microbial colonization of microplastics in the Caribbean Sea. *Limnology and Oceanography Letters, 5*, 5–17. https://doi.org/10.1002/lol2.10141

Ekane, N., Barquet, K., & Rosemarin, A. (2021). Resources and risks: Perceptions on the application of sewage sludge on agricultural land in Sweden, a case study. *Frontiers in Sustainable Food Systems, 5*, 647780. https://doi.org/10.3389/fsufs.2021.647780

Fernández Severini, M. D., Buzzi, N. S., Forero López, A. D., et al. (2020). Chemical composition and abundance of microplastics in the muscle of commercial shrimp Pleoticus muelleri at an impacted coastal environment (Southwestern Atlantic). *Marine Pollution Bulletin, 161*, 111700. https://doi.org/10.1016/j.marpolbul.2020.111700

Ferrante, M., Oliveri Conti, G., & Zuccarello, P. (2020). Patent method for the extraction and determination of micro- and nano- plastics in organic and inorganic matrix samples: An application on vegetals. *MethodsX, 7*, 100989. https://doi.org/10.1016/J.MEX.2020.100989

Forero-López, A. D., Brugnoni, L. I., Abasto, B., et al. (2022). Plastisphere on microplastics: In situ assays in an estuarine environment. *Journal of Hazardous Materials, 440*, 129737. https://doi.org/10.1016/j.jhazmat.2022.129737

Franco, A. A., Martín-García, A. P., Egea-Corbacho, A., et al. (2023). Assessment and accumulation of microplastics in sewage sludge at wastewater treatment plants located in Cádiz, Spain. *Environmental Pollution, 317*, 120689. https://doi.org/10.1016/j.envpol.2022.120689

Gao, M., Liu, Y., & Song, Z. (2019). Effects of polyethylene microplastic on the phytotoxicity of di-n-butyl phthalate in lettuce (Lactuca sativa L. var. ramosa Hort). *Chemosphere, 237*, 124482. https://doi.org/10.1016/j.chemosphere.2019.124482

García Rellán, A., Vázquez Ares, D., Vázquez Brea, C., et al. (2023). Sources, sinks and transformations of plastics in our oceans: Review, management strategies and modelling. *Science of the Total Environment, 854*, 158745. https://doi.org/10.1016/j.scitotenv.2022.158745

Gephart, J. A. (2019). Global seafood trade. In P. Ferranti, E. M. Berry, & J. R. Anderson (Eds.), *Encyclopedia of Food Security and Sustainability* (pp. 93–97). Elsevier.

Hajji, S., Ben-Haddad, M., Abelouah, M. R., et al. (2023). Occurrence, characteristics, and removal of microplastics in wastewater treatment plants located on the Moroccan Atlantic: The case of Agadir metropolis. *Science of the Total Environment, 862*, 160815. https://doi.org/10.1016/J.SCITOTENV.2022.160815

Hasan, J., Dristy, E. Y., Anjumanara, et al. (2023). Dried fish more prone to microplastics contamination over fresh fish—Higher potential of trophic transfer to human body. Ecotoxicology *Environmental Safety, 250*, 114510. https://doi.org/10.1016/j.ecoenv.2023.114510.

Huang, Y., He, T., Yan, M., et al. (2021). Atmospheric transport and deposition of microplastics in a subtropical urban environment. *Journal of Hazardous Materials, 416*, 126168. https://doi.org/10.1016/j.jhazmat.2021.126168

Iyare, P. U., Ouki, S. K., & Bond, T. (2020). Microplastics removal in wastewater treatment plants: A critical review. *Environmental Science Water Research & Technology, 6*, 2664–2675. https://doi.org/10.1039/D0EW00397B

Jiang, J.-J., Hanun, J. N., Chen, K.-Y., et al. (2023). Current levels and composition profiles of microplastics in irrigation water. *Environmental Pollution, 318*, 120858. https://doi.org/10.1016/j.envpol.2022.120858

Kaattari, S. L., & Piganelli, J. D. (1996). The specific immune system: Humoral defense. *Fish Physiology, 15*, 207–254. https://doi.org/10.1016/S1546-5098(08)60275-7

Karami, A., Golieskardi, A., Choo, C. K., et al. (2018). Microplastic and mesoplastic contamination in canned sardines and sprats. *Science of the Total Environment, 612*, 1380–1386. https://doi.org/10.1016/j.scitotenv.2017.09.005

Karami, A., Golieskardi, A., Ho, Y. B., et al. (2017). Microplastics in eviscerated flesh and excised organs of dried fish. *Scientific Reports, 7*, 5473. https://doi.org/10.1038/s41598-017-05828-6

Kaur, K., Reddy, S., Barathe, P., et al. (2022). Microplastic-associated pathogens and antimicrobial resistance in environment. *Chemosphere, 291*, 133005. https://doi.org/10.1016/j.chemosphere.2021.133005

Kirstein, I. V., Kirmizi, S., Wichels, A., et al. (2016). Dangerous hitchhikers? Evidence for potentially pathogenic Vibrio spp. on microplastic particles. *Marine Environmental Research, 120*, 1–8. https://doi.org/10.1016/j.marenvres.2016.07.004

Koelmans, A. A., Bakir, A., Burton, A., & Janssen, C. R. (2016). Microplastic as a vector for chemicals in the aquatic environment: critical review and model-supported reinterpretation of empirical studies. *Environmental Science & Technology, 50*, 3315–3326. https://doi.org/10.1021/acs.est.5b06069

Kundu, M. N., Komakech, H. C., & Lugomela, G. (2022). Analysis of macro- and microplastics in riverine, riverbanks, and irrigated farms in Arusha, Tanzania. *Archives of Environmental Contamination and Toxicology, 82*, 142–157. https://doi.org/10.1007/s00244-021-00897-1

Lamastra, L., Suciu, N. A., & Trevisan, M. (2018). Sewage sludge for sustainable agriculture: contaminants' contents and potential use as fertilizer. *Chemical and Biological Technologies in Agriculture, 5*, 10. https://doi.org/10.1186/s40538-018-0122-3

Lehel, J., & Murphy, S. (2021). Microplastics in the food chain: Food safety and environmental aspects. In P. de Voogt (Ed.), *Reviews of Environmental Contamination and Toxicology* (Vol. 259, pp. 1–49). Cham: Springer.

Li, J., Green, C., Reynolds, A., et al. (2018). Microplastics in mussels sampled from coastal waters and supermarkets in the United Kingdom. *Environmental Pollution, 241*, 35–44. https://doi.org/10.1016/j.envpol.2018.05.038

Li, J., Yang, D., Li, L., et al. (2015). Microplastics in commercial bivalves from China. *Environmental Pollution, 207*, 190–195. https://doi.org/10.1016/j.envpol.2015.09.018

Liu, Y., Liu, Y., Li, Y., et al. (2023). Effects of irrigation on the fate of microplastics in typical agricultural soil and freshwater environments in the upper irrigation area of the Yellow River. *Journal of Hazardous Materials, 447*, 130766. https://doi.org/10.1016/j.jhazmat.2023.130766

Llorca, M., Schirinzi, G., Martínez, M., et al. (2018). Adsorption of perfluoroalkyl substances on microplastics under environmental conditions. *Environmental Pollution, 235*, 680–691. https://doi.org/10.1016/j.envpol.2017.12.075

Lo, H. S., Wong, C. Y., Tam, N. F. Y., & Cheung, S. G. (2019). Spatial distribution and source identification of hydrophobic organic compounds (HOCs) on sedimentary microplastic in Hong Kong. *Chemosphere, 219*, 418–426. https://doi.org/10.1016/j.chemosphere.2018.12.032

Lwanga, E. H., Beriot, N., Corradini, F., et al. (2022). Review of microplastic sources, transport pathways and correlations with other soil stressors: A journey from agricultural sites into the environment. *Chemical and Biological Technology in Agriculture, 9*, 20. https://doi.org/10.1186/S40538-021-00278-9

Mahon, A. M., O'Connell, B., Healy, M. G., et al. (2017). Microplastics in sewage sludge: Effects of treatment. *Environmental Science and Technology, 51*, 810–818. https://doi.org/10.1021/acs.est.6b04048

Meijer, L. J. J., van Emmerik, T., van der Ent, R., et al. (2021). More than 1000 rivers account for 80% of global riverine plastic emissions into the ocean. *Science Advances, 7*, eaaz5803. https://doi.org/10.1126/sciadv.aaz5803

Muniz Silva, M., Carvalho Maldonado, G., Oliveira Castro, R., et al. (2019). Dispersal of potentially pathogenic bacteria by plastic debris in Guanabara Bay, RJ, Brazil. *Marine Pollution Bulletin, 141*, 561–568. https://doi.org/10.1016/j.marpolbul.2019.02.064

Napper, I. E., & Thompson, R. C. (2016). Release of synthetic microplastic plastic fibres from domestic washing machines: Effects of fabric type and washing conditions. *Marine Pollution Bulletin, 112*, 39–45. https://doi.org/10.1016/j.marpolbul.2016.09.025

Nayon Hossain, M., Rahman, M. M., Afrin, S., et al. (2023). Identification and quantification of microplastics in agricultural farmland soil and textile sludge in Bangladesh. *Science of the Total Environment, 858*, 160118. https://doi.org/10.1016/j.scitotenv.2022.160118

Nordhagen, S., Lambertini, E., DeWaal, C. S., et al. (2022). Integrating nutrition and food safety in food systems policy and programming. *Global Food Security, 32*, 100593. https://doi.org/10.1016/j.gfs.2021.100593

Oberbeckmann, S., Kreikemeyer, B., & Labrenz, M. (2018). Environmental factors support the formation of specific bacterial assemblages on microplastics. *Frontiers in Microbiology, 8*, 2709. https://doi.org/10.3389/fmicb.2017.02709

Okoffo, E. D., Rauert, C., & Thomas, K. V. (2023). Mass quantification of microplastic at wastewater treatment plants by pyrolysis-gas chromatography–mass spectrometry. *Science of the Total Environment, 856*, 159251. https://doi.org/10.1016/j.scitotenv.2022.159251

Okoffo, E. D., Ribeiro, F., O'Brien, J. W., et al. (2020). Identification and quantification of selected plastics in biosolids by pressurized liquid extraction combined with double-shot pyrolysis gas chromatography–mass spectrometry. *Science of the Total Environment, 715*, 136924. https://doi.org/10.1016/J.SCITOTENV.2020.136924

Okoffo, E. D., Tscharke, B. J., O'Brien, J. W., et al. (2020). Release of plastics to Australian Land from biosolids end-use. *Environmental Science Technology, 54*, 15132–15141. https://doi.org/10.1021/acs.est.0c05867

Oliveri Conti, G., Ferrante, M., Banni, M., et al. (2020). Micro- and nano-plastics in edible fruit and vegetables. The first diet risks assessment for the general population. *Environmental Research, 187*, 109677. https://doi.org/10.1016/J.ENVRES.2020.109677

Roblin, B., Ryan, M., Vreugdenhil, A., & Aherne, J. (2020). Ambient atmospheric deposition of anthropogenic microfibers and microplastics on the Western Periphery of Europe (Ireland). *Environmental Science and Technology, 54*, 11100–11108. https://doi.org/10.1021/acs.est.0c04000

Roy, T., Dey, T. K., & Jamal, M. (2023). Microplastic/nanoplastic toxicity in plants: An imminent concern. *Environmental Monitoring and Assessment, 195*, 27. https://doi.org/10.1007/s10661-022-10654-z

Rubio-Armendáriz, C., Alejandro-Vega, S., Paz-Montelongo, S., et al. (2022). Microplastics as emerging food contaminants: A challenge for food safety. *International Journal of Environmental Research and Public Health, 19*, 1174. https://doi.org/10.3390/ijerph19031174

Sa'adu, I., & Farsang, A. (2023). Plastic contamination in agricultural soils: A review. *Environmental Sciences Europe, 35*, 13. https://doi.org/10.1186/s12302-023-00720-9

Santos, L. H. M. L. M., Rodríguez-Mozaz, S., & Barceló, D. (2021). Microplastics as vectors of pharmaceuticals in aquatic organisms—An overview of their environmental implications. *Case Stud in Chemical and Environmental Engineering, 3*, 100079. https://doi.org/10.1016/j.cscee.2021.100079

Sheriff, I., Yusoff, M. S., Manan, T. S. B. A., & Koroma, M. (2023). Microplastics in manure: Sources, analytical methods, toxicodynamic, and toxicokinetic endpoints in livestock and poultry. *Environmental Advances, 12*, 100372. https://doi.org/10.1016/j.envadv.2023.100372

Sørensen, L., Rogers, E., Altin, D., et al. (2020). Sorption of PAHs to microplastic and their bioavailability and toxicity to marine copepods under co-exposure conditions. *Environmental Pollution, 258*, 113844. https://doi.org/10.1016/j.envpol.2019.113844

Steenland, K., & Winquist, A. (2021). PFAS and cancer, a scoping review of the epidemiologic evidence. *Environmental Research, 194*, 110690. https://doi.org/10.1016/j.envres.2020.110690

Sun, H., Lei, C., Xu, J., & Li, R. (2021). Foliar uptake and leaf-to-root translocation of nanoplastics with different coating charge in maize plants. *Journal of Hazardous Materials, 416*, 125854. https://doi.org/10.1016/J.JHAZMAT.2021.125854

Tang, K. H. D., & Hadibarata, T. (2021). Microplastics removal through water treatment plants: Its feasibility, efficiency, future prospects and enhancement by proper waste management. *Environmental Challenges, 5*, 100264. https://doi.org/10.1016/j.envc.2021.100264

Tong, X., Li, B., Li, J., et al. (2022). Polyethylene microplastics cooperate with Helicobacter pylori to promote gastric injury and inflammation in mice. *Chemosphere, 288*, 132579. https://doi.org/10.1016/j.chemosphere.2021.132579

Torres, F. G., & De-la-Torre, G. E. (2022). Per- and polyfluoroalkyl substances (PFASs) in consumable species and food products. *Journal of Food Science and Technology, 60*(9), 2319–2336. https://doi.org/10.1007/s13197-022-05545-7

Torres, F. G., Dioses-Salinas, D. C., Pizarro-Ortega, C. I., & De-la-Torre, G. E. (2021). Sorption of chemical contaminants on degradable and non-degradable microplastics: Recent progress and research trends. *Science of the Total Environment, 757*, 143875. https://doi.org/10.1016/j.scitotenv.2020.143875

Valdiviezo Gonzales, L., Ortiz Ojeda, P., Espinoza Morriberón, D., et al. (2024). Influence of the geographic location and house characteristics on the concentration of microplastics in indoor dust. *Science of the Total Environment, 917*, 170353. https://doi.org/10.1016/j.scitotenv.2024.170353

van Schothorst, B., Beriot, N., Huerta Lwanga, E., & Geissen, V. (2021). Sources of light density microplastic related to two agricultural practices: The use of compost and plastic mulch. *Environments, 8*, 36. https://doi.org/10.3390/environments8040036

Viršek, M. K., Lovšin, M. N., Koren, Š, et al. (2017). Microplastics as a vector for the transport of the bacterial fish pathogen species Aeromonas salmonicida. *Marine Pollution Bulletin, 125*, 301–309. https://doi.org/10.1016/j.marpolbul.2017.08.024

Wang, C., Chen, W., Zhao, H., et al. (2023). Microplastic fiber release by laundry: A comparative study of hand-washing and machine-washing. *ACS Environmental Science and Technology Water, 1*, 147–155. https://doi.org/10.1021/acsestwater.2c00462

Wang, F., Feng, X., Liu, Y., et al. (2022). Micro(nano)plastics and terrestrial plants: Up-to-date knowledge on uptake, translocation, and phytotoxicity. *Resources, Conservation and Recycling, 185*, 106503. https://doi.org/10.1016/J.RESCONREC.2022.106503

Wang, F., Shih, K. M., & Li, X. Y. (2015). The partition behavior of perfluorooctanesulfonate (PFOS) and perfluorooctanesulfonamide (FOSA) on microplastics. *Chemosphere, 119*, 841–847. https://doi.org/10.1016/j.chemosphere.2014.08.047

Wang, W., Yuan, W., Xu, E. G., et al. (2022). Uptake, translocation, and biological impacts of micro(nano)plastics in terrestrial plants: Progress and prospects. *Environmental Research, 203*, 111867. https://doi.org/10.1016/j.envres.2021.111867

Wei, W., Shi, X., Zhang, Y.-T., et al. (2022). Microplastics in sewage sludge of wastewater treatment. In B.-J. Ni, Q. Xu, & W. Wei (Eds.), *Microplastics in Urban Water Management* (pp. 147–173). Verlag: Wiley.

Yang, H., Chen, G., & Wang, J. (2021). Microplastics in the marine environment: Sources, fates, impacts and microbial degradation. *Toxics, 9*, 41. https://doi.org/10.3390/toxics9020041

Yang, W., Li, Y., & Boraschi, D. (2023). Association between microorganisms and microplastics: How does it change the host-pathogen interaction and subsequent immune response? *International Journal of Molecular Science, 24*, 4065. https://doi.org/10.3390/ijms24044065

Yemane, B., & Tamene, A. (2022). Understanding domestic food safety: An investigation into self-reported food safety practice and associated factors in Southern Ethiopian households. *Environ Health Insights, 16*, 11786302221103880. https://doi.org/10.1177/11786302221110388

Yeo, B. G., Takada, H., Yamashita, R., et al. (2020). PCBs and PBDEs in microplastic particles and zooplankton in open water in the Pacific Ocean and around the coast of Japan. *Marine Pollution Bulletin, 151*, 110806. https://doi.org/10.1016/j.marpolbul.2019.110806

Yu, L., Zhang, J., Liu, Y., et al. (2021). Distribution characteristics of microplastics in agricultural soils from the largest vegetable production base in China. *Science of the Total Environment, 756*, 143860. https://doi.org/10.1016/j.scitotenv.2020.143860

Zazouli, M., Nejati, H., Hashempour, Y., et al. (2022). Occurrence of microplastics (MPs) in the gastrointestinal tract of fishes: A global systematic review and meta-analysis and meta-regression. *Science of the Total Environment, 815*, 152743. https://doi.org/10.1016/j.scitotenv.2021.152743

Zettler, E. R., Mincer, T. J., & Amaral-Zettler, L. A. (2013). Life in the "plastisphere": Microbial communities on plastic marine debris. *Environmental Science and Technology, 47*, 7137–7146. https://doi.org/10.1021/es401288x

Zhou, Y., Jing, J., Yu, R., et al. (2023). Microplastics in plateau agricultural areas: Spatial changes reveal their source and distribution characteristics. *Environmental Pollution, 319*, 121006. https://doi.org/10.1016/j.envpol.2023.121006

Zhu, [illegible]

Zhou, [illegible]

Zhou, [illegible]

Zhu, [illegible]

Chapter 9
Exploring the Core Aspects of Microplastic Pollution: Origin, Distribution, Imprint, Detection, and Policy Solutions

Shikha Jyoti Borah, Abhijeet Kumar Gupta, Vinod Kumar, Akanksha Gupta, Ravi Kant, and Ravinder Kumar

Abstract In the midst of escalating global concern surrounding microplastic pollution, this chapter takes the reader on illuminating the intricate core aspects of microplastic. It delves into fundamental components of microplastic pollution, including their sources, profound ecological imprint, detection techniques, and the dynamic regulatory measures implemented throughout the globe. It emphasizes the numerous repercussions of microplastic contamination on aquatic and terrestrial ecosystems. More importantly, the life-threatening impact of microplastic on human health potentially through ingestion, inhalation, and dermal contact has been highlighted. Additionally, it provides insights into the visionary policies and regulations of entities like the European Union, the progressive strategies of the United Nations Environment Programme, and the forward-thinking approach of developing nations, like the Indian Government, propelling global responses to combat microplastic pollution. This chapter offers a valuable overview of the fundamental key points associated with microplastic pollution and a roadmap of the solutions adopted by policymakers toward a sustainable future.

Keywords Microplastic · Ecological imprint · Policy · Regulations

---

S. J. Borah · A. K. Gupta · V. Kumar (✉)
Sustainable Energy & Environmental Nanotechnology (SEEN) Group, Special Centre for Nano Sciences, Jawaharlal Nehru University, Delhi 110067, India
e-mail: kumarv@mail.jnu.ac.in

A. Gupta
Department of Science and Technology, Delhi 110016, India

R. Kant
Department of Chemistry, Zakir Hussain Delhi College, Delhi 110002, India

R. Kumar (✉)
Department of Chemistry, Gurukula Kangri (Deemed to be University), Haridwar 249404, Uttarakhand, India
e-mail: ravinder.kumar@gkv.ac.in

A. K. Mishra et al. (eds.), *Microplastics*, https://doi.org/10.1007/978-981-97-6461-7_9

Abbreviations

MPs	Microplastics
SUPs	Single-use plastics
WWTPs	Wastewater treatment plants
SEM	Scanning electron microscopy
TEM	Transmission electron microscopy
FTIR	Fourier-transform infrared spectroscopy
EPR	Extended producer responsibility
EU	European Union
UNEP	United Nations Environment Program
ECHA	European Chemical Agency
GPWM	Global Partnership on Waste Management
GESAMP	Joint Group of Experts on Scientific Aspects of Marine Environmental Protection
SDG	Sustainable Development Goals
UN	United Nations
UNEA	United Nations Environment Assembly

1 Introduction

Microplastic pollution has been recognized as a burgeoning concern, raising alarm among scientists, environmentalists, and the general community due to the exponentially growing usage of plastic products in various sectors of the world. Typically, microplastics (MPs) refer to diminutive particles having a size less than 5 mm (Hartmann et al., 2019). These small particles have the potential to pervade various environmental matrices, encompassing aquatic and terrestrial systems, as well as atmospheric environments. This can exert both direct or unintended effects on the ecosystem and human health, necessitating urgent and effective mitigation strategies.

Numerous recent scientific investigations have emphasized the pronounced risk of freshwater and marine animals ingesting MPs, resulting in physical complications and mortality. Various marine species have exhibited deleterious effects on the reproductive health (Sharifinia et al., 2020), growth and development processes, and other biological functions as a consequence of MP ingestion (Galloway, 2015). In freshwater environments, MPs can cause reduced growth in photoautotrophs, higher mortality for certain invertebrates, and genetic alteration in amphibians (Castro-Castellon et al., 2022). Furthermore, the infiltration of MPs into the food chain has emerged as a significant concern for human health, as bioaccumulation of MPs in human tissues and organs has been observed, which can yield long-term health ramifications (Chen et al., 2023; Prata et al., 2020; Zhang et al., 2022a). In response to these escalating consequences of MPs, several international organizations and governments across different countries have proactively adopted comprehensive policies

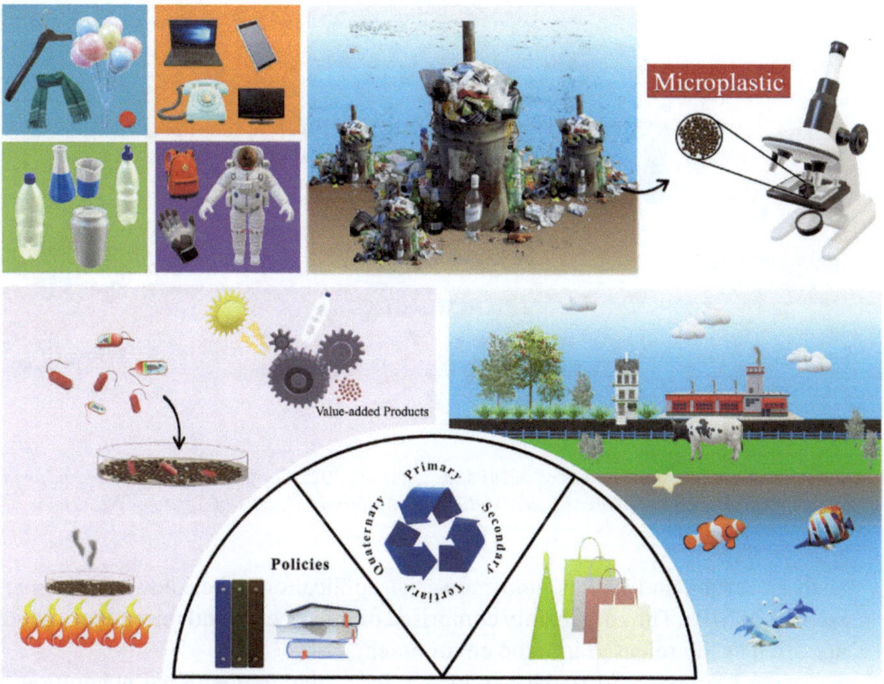

Fig. 1 A concise overview of the escalating microplastic pollution in the ecosystem and its core aspects

and initiatives aimed at mitigating single-use plastics (SUPs) and the amelioration of waste management practices. Figure 1 summarizes the core aspects of microplastic pollution. It demonstrates growing plastic waste contributes to the generation of MPs, producing significant detrimental consequences on aquatic and terrestrial ecosystems. Thus, the mitigation of MPs through the adoption of strict regulatory measures and the development of sustainable methods for MP mitigation from the environment is of utmost importance.

1.1 Sources

MP pollution is a rapidly advancing environmental threat, demanding a comprehensive understanding of the sources governing MP production. Gaining insights into these sources is vital to disentangle their complex source–pathway–sink interrelationships. The most prevalent origins of MP can be classified into primary and secondary sources (Borah et al., 2023) (Fig. 2).

a. Primary Sources: Sources, which intentionally manufacture MPs, deliberately produce MPs for various purposes such as in cosmetics, personal care products

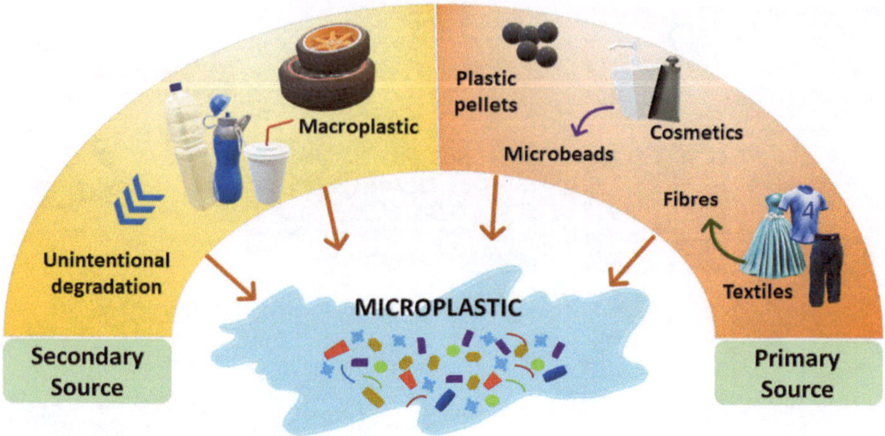

Fig. 2 Primary and secondary sources of MPs. Borah et al., 2023 *Copyright © 2023, The Author(s), under exclusive licence to Springer Science Business Media, LLC, part of Springer Nature* ()

such as scrubs and facials, and industrial applications, are known as primary sources of MPs. They are mainly comprised of microbeads and microfibers which are often easily released into the environment.

b. Secondary Sources: Secondary sources refer to the unintentional generation of MPs through the breakdown of bulk plastic materials. Plastic debris can undergo fragmentation due to natural physical, chemical, or biological stressors, operating independently of artificial external factors, ultimately transforming bulk plastic debris into microplastic.

1.2 Global Distribution

The distribution of MPs across the globe through different routes and the factors involved in its transportation are important aspects of MP pollution. Recent research reveals that MP is on its way to be ubiquitously present in nearly all global ecosystems, encompassing ocean, glaciers, freshwater systems, soil, and atmosphere, signifying their extensive distribution throughout the planet. Currently, wastewater treatment plants (WWTPs), plastic microbeads, synthetic textile production, and many other MP sources have been put under the scrutiny of researchers and environmentalists (Tadsuwan & Babel, 2022).

Untreated or inadequately treated wastewater from WWTPs and industrial facilities can harbor MPs from various sources. MPs can be released into their neighboring environment, including water bodies such as rivers and streams. MPs can be discharged into water bodies or infiltrate terrestrial soil. In marine environments, ocean currents play a significant role in MP transportation, resulting in their global distribution. Similarly, within freshwater systems, microplastics can be transported by the flow of rivers and streams, consequently accumulating in lakes and reservoirs (Lu et al., 2023). Interestingly, recent studies have provided evidence that MPs can also be present in tap water, drinking water, and bottled water (Muhib et al., 2023; Sol et al., 2023; Tse et al., 2022).

Furthermore, MPs can leak into the air through various routes. Urban and industrial locations are major areas where high atmospheric MP content is present; while it has been hypothesized that remote areas may also be equivalently contaminated over the next few years. Subsequently, these atmospheric MPs can remain suspended or deposit over a variety of surfaces (atmospheric deposition), and may further undergo short/long-range transport through influential factors such as wind speed and direction (Roblin et al., 2020). Finally, deposited atmospheric MPs can create toxic repercussions. These pollutants are capable of depositing in a number of habitats, with humans being particularly susceptible to inhalation of suspended atmospheric MPs. Table 1 is a compilation of the MPs detected at various locations, depicting their global distribution.

In recent years, the deposition of MPs in remote and seemingly pristine areas, such as mountainous regions (Parolini et al., 2021; Talukdar et al., 2023; Yang et al., 2022), arctic regions (Hamilton et al., 2021; Primpke et al., 2022), Antarctic regions (Rota et al., 2022), and glaciers (Dong et al., 2021; Zhang et al., 2022b) has grabbed a surprising pivotal point for researchers. Despite the absence of significant plastic consumption and human interventions in these regions, studies have revealed unexpected and noteworthy levels of MPs. Recent research indicates that atmospheric deposition is a prominent route for detecting MPs in isolated areas (Evangeliou et al., 2020). Consequently, there is a crucial need to assess the occurrence of MPs in distant regions and understand the various trajectories, including wind and water transport.

2 Microplastic Imprint

2.1 Aquatic and Terrestrial System

Various organisms may ingest MPs under the assumption that they are food sources. Ingestion of MPs has been reported to cause physical injuries, blockages in the digestive tract, and hinder nutrient absorption, potentially causing malnutrition and mortality. Besides, MPs can infiltrate the soil, thereby altering soil properties, and causing harm to microorganisms. Some of the major impacts of MPs on both terrestrial and aquatic systems have been highlighted.

a. Bioaccumulation and Biomagnification

MPs can penetrate the food chain and cause bioaccumulation, where they gradually accumulate in the tissues of living organisms. Additionally, biomagnification phenomena may also take place in which MP concentration increases as they ascend

Table 1 Studies carried out across the globe for understanding the widespread deposition of MPs at different locations

MP	Size	Shape	Deposition	Location	References
PP, RY, PES, PS	50–2210 μm	Fibers, fragments, granules	0.013–0.063 N/m^3	South China Sea	Ding et al., (2021)
PET, PAN, PP, PA	50 μm–5 mm	Fibers, fragments, films and microbeads	51–178 particles/m^2/d	Rooftop, China	Huang et al., (2021)
PE and PP	51–67 μm	Fibers	$1.1 \times 10^3 \pm 0.06 \times 10^3$ to $3.5 \times 10^3 \pm 0.3 \times 10^3$ particles/m^2/day	Urban waters, China	Sun et al., (2022)
PP, PET, PE, PS, PVC	≤30 μm	Fibers and fragments	178 (±79) MP/m^2/day	Central Pyrenees	Allen et al., (2021)
MP	<100 μm	Amorphous fragments	10^5 items m^{-2} d^{-1}	Dining/drinking venues, China	Fang et al., (2022)
PE, PET, PES, HDPE, LDPE	50–1910 μm	Fibers and fragments	0.39 ± 0.39 to 4.91 ± 2.48 MPs/L	Aosta Valley, Western Italian Alps	Parolini et al., (2021)
PES and PP	5–5000 μm	Fibers, films, and fragments	0 m^{-2}d^{-1} to 30 m^{-2}d^{-1}	Coastal zone, Southern Baltic area	Szewc et al., (2021)
PP, PE, PVC	300–5000 μm	Fibers, films and fragments	71–917 items m^{-2} d^{-1}	South of Vietnam	Truong et al., (2021)
PS, PE, PES and Polybutadiene	500–1000 μm	Foam, Fragment and Fibers	3–40 particles m^{-2}d^{-1}	Jakarta, Indonesia	Purwiyanto et al., (2022)
PS and PE	≤50 μm	Fragment, films, fibers	365 particles m^{-2}d^{-1}	Saharan-, North African	Allen et al., (2019)
PET, PA, and PVF	<1000 μm	Fibers	3261.22 ± 2847.99 particles ·m^{-2}d^{-1}	Shanghai, China	Jia et al., 2022)
PET	–	Fibers	0.0046 to 0.064 items/m^3	North-western Pacific Ocean	Ding et al., (2022)
PP, PS, PVC, PE and PET	<11 μm	–	10–367 N m^{-2} day^{-1}	Weser River catchment	Kernchen et al., (2022)
PET	50–500 μm	Fragments and fibers	353.83 N m^{-2} d^{-1}	Lanzhou, China	Liu et al., (2022)

PP: Polyethylene; PAN: Polyacrylonitrile; RY: Rayon; PES: Polyester; PS: Polystyrene; PET: Polyethylene Terephthalate; PP: Polypropylene; PVF: Polyvinyl Fluoride; PA: Polyamide; PVC: Polyvinyl Chloride the food chain. This suggests that apex predators in the food chain may face increased vulnerability (Borah et al., 2023) (Fig. 3). Bioaccumulation of MPs can lead to oxidative stress, inflammation, and damage to cells and tissues.

b. Reproduction and Growth

The ability to reproduce may also be hindered as a consequence of MP exposure. Recent studies reveal hindered larval development and decreased hatching successes. This can have detrimental effects on the dynamics of marine species under long-term exposure to MPs. Ingested MPs can cause a lowered ability to absorb nutrients, resulting in lower reproduction rates, and reduced growth rates. Additionally, the accumulation of MPs in tissues can directly affect the health and growth of offspring, leading to reduced survival rates.

c. Chemical Contamination

Plastic products often contain chemical additives, which are used to enhance their various physical and chemical properties. For instance, bisphenol A has been largely used in various plastic products. The release of plastic additives into the environment alongside MPs can lead to enhanced toxicity. Additionally, MPs may have the ability to interact with pesticides, heavy metals, nano-waste, and other hazardous contaminants, resulting in potentially harmful repercussions. Organisms at the lower end of the food chain ingest MPs, and the associated toxins and chemicals can be transferred up the food chain to higher level predators.

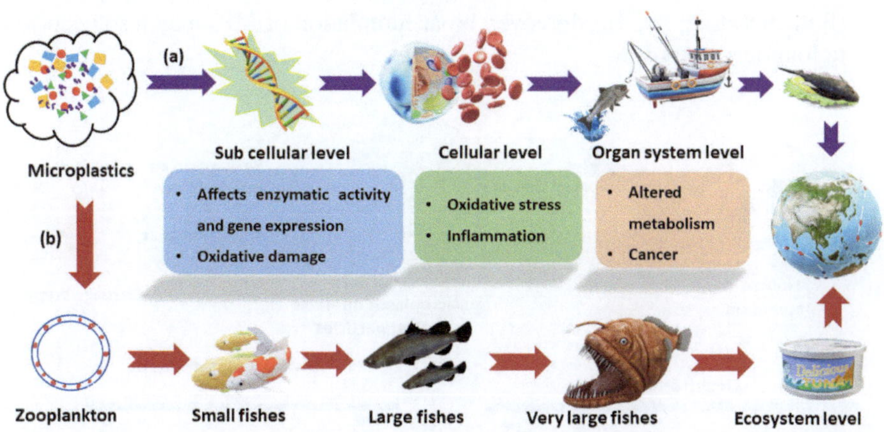

Fig. 3 The flow of MPs from lower level to higher level organisms in the food chain. Borah et al., 2023 *Copyright © 2023, The Author(s), under exclusive licence to Springer Science Business Media, LLC, part of Springer Nature* ()

2.2 Human Health

Microplastics present an emerging hazard to human health due to their trivial size and ubiquitous dispersion in different environments. Figure 4 demonstrates the myriad pathways that facilitate human exposure to MPs. Through multiple exposure routes, including consumption, inhalation, and dermal absorption, these pollutants can infiltrate the human physiological system. As research continues, comprehending, assessing, and mitigating the risks of MPs becomes imperative for ensuring public safety and environmental well-being.

Humans can be affected by MPs through various pathways:

a. Ingestion: MPs can contaminate seafood, bottled water, and the soil. This indicates that consuming MP-contaminated seafood and drinking water, or eating crops grown in microplastic-contaminated soil can directly or indirectly expose human lifestyle to MPs (Alberghini et al., 2022; Patil et al., 2022). Once ingested, MPs may lead to bioaccumulation in the gastrointestinal tract, causing inflammation and physical damage.
b. Inhalation: Airborne MPs can be inhaled, leading to respiratory exposure. Recent reports have concluded a high concentration of fibrous MPs (Vianello et al., 2019; Yuan et al., 2022). Additionally, young children have been reported to be at higher risk of MP inhalation.
c. Health and Gut: Although research has not achieved advanced stages, there are primitive studies through mouse models and human cell lines, which have reported adverse health effects such as inflammation and oxidative stress (Bahadur et al., 2023). Moreover, bioaccumulation of MPs may also occur due to long-term exposure.

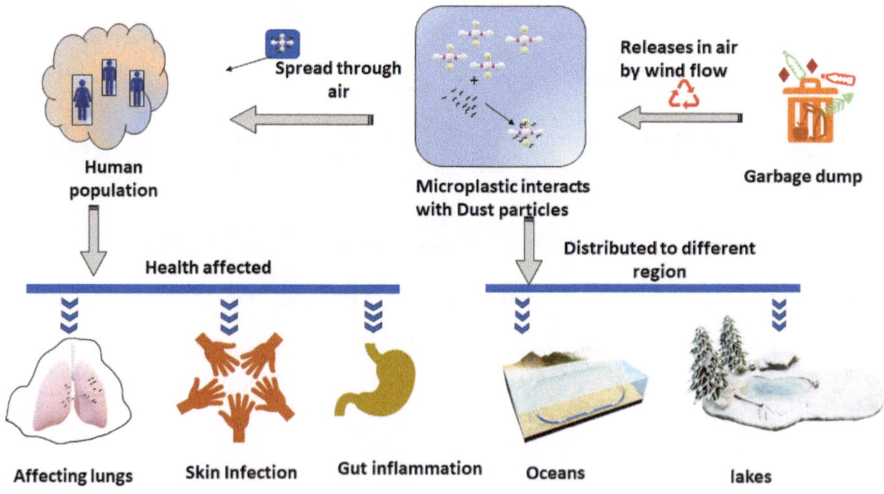

Fig. 4 Interaction of MPs with the human health and the environment

d. Chemical Contamination: MPs may function as carriers for the transportation of hazardous plastic additives, heavy metals, organic contaminants, airborne viruses such as COVID-19, and other environmental pollutants into the human body (Amato-Lourenço et al., 2022; Pandey et al., 2021). These chemicals and co-contaminants can potentially trigger a wide range of detrimental consequences on human health, including cancer, cardiovascular, and respiratory illness (Osman et al., 2023).

3 Detection and Analysis

Numerous advanced techniques are incorporated for the detection and analysis of MPs to comprehensively elucidate their spatial dispersal, prevalence, and possible repercussions. The key techniques involved in the MPs research and investigations have been summarized.

a. Optical microscopy: Optical microscopy enables visual identification and characterization of MPs to determine their shape, size, and morphology (Huang et al., 2023). This provides essential information on individual particles and often acts as the initial step in MP detection and analysis.
b. Electron Microscopy: Scanning electron microscopy (SEM) and transmission electron microscopy (TEM) enable high-magnification images for even very small particles (Huang et al., 2023). SEM, in particular, confirms the morphology, texture, and shape of MPs through high-resolution images. Additionally, elemental composition and components of chemical additives or other co-contaminants can be analyzed through energy-dispersive X-ray spectroscopy (EDS).
c. Infrared Spectroscopy: Fourier-transform infrared spectroscopy (FTIR) serves as a vital method in characterizing the chemical composition of MPs, enabling the differentiation of distinct plastic polymers, and ensuring accurate identification (Huang et al., 2023). Furthermore, integrating microscopy with FTIR, in the form of μ-FTIR, offers a direct analytical approach for MPs, providing comprehensive insights into their chemical attributes. Thus, this integration enhances precision and in-depth analysis.
d. Raman Spectroscopy: This technique is used to analyze the chemical structure of different polymers through molecular vibrations (Huang et al., 2023). Similar to μ-FTIR, micro-Raman Spectroscopy combines spectroscopy with Raman, enabling in-situ characterization with enhanced sensitivity.
e. Pyrolysis Gas Chromatography-Mass spectrometry: In this technique, pyrolysis is employed to thermally degrade MP, resulting in the release of gases. The analysis of these liberated gases allows a comprehensive assessment of the chemical composition of MP (Huang et al., 2023).

4 Policy and Regulations

An important global trend in the last ten years has been the adoption of environmental legal authorities to reduce the impact of plastic waste. Current available science and technology may not be able to completely eradicate MPs generated from the enormous litter accumulated due to the never-ending usage of plastics. However, it is truly possible to reduce the usage and accumulation of plastic waste through the adoption of strict policies. The recent consensus among global experts, leaders, and environmentalists underscores the necessity for the development of robust plastic policies. Effective measures should not only encompass recycling requirements but also entail restrictions on plastic manufacturing and extended producer responsibility (EPR). Given the possible outcomes of all forms of plastic waste, effective strategies must be devised to holistically address the challenges posed by plastic waste management (Seay & Ternes, 2022).

Nearly half a decade ago, the United Nations (UN) highlighted, in its study "Legal Limits on Single-Use Plastics and Microplastics: A Global Review of National Laws and Regulations" published on December 5, 2018, that out of the 192 countries examined, 127 have passed laws governing plastic bags, most often limiting their use in free retail distribution. It also reported that out of the 192 countries, 27 countries had imposed a strict ban on SUPs, an effort that reduces the leakage of SUPs, such as plastic water bottles and polythene carry bags into different ecosystems. In the same, it was reported that 63 nations had introduced extended producer responsibility measures, while 29 had imposed various taxes on SUPs (UNEP, 2018a). Some of the recent policies, amendments, and rules imposed by the international organizations, European Union (EU) and United Nations Environment Program (UNEP), have been briefly discussed. This is also followed by a brief discussion on some of the actions and policies adopted by the Indian government to eliminate plastics.

4.1 The European Union

When it comes to sustainability and circularity, the EU has been setting the pace. The restriction procedure was initiated by REACH (acronym of the EU chemical regulation), when the European Commission urged the European Chemical Agency (ECHA) to draft a restriction dossier on MPs in November 2017 (European Commission, 2017; European Union, 2007). European nations have placed an efficient plastic waste management system (Jambeck et al., 2015). During the period of 2012–2014, these efforts enabled significant improvement in plastic recycling, raising the recycling rate from 34.7 to 39.5% (Lam et al., 2018). Additionally, the European Commission simultaneously unveiled a new project to lower the unintended MP discharge into the environment along with the European Green Deal (European Commission, 2015). It seeks to:

- Prevent the unintended release of MPs, implement labeling, standardization, certification, and regulatory measures, as well as strategies to increase the capture of MPs at all pertinent points in a product's lifecycle.
- Develop and standardize measurement techniques for accidentally released MPs, particularly from tires and textiles, and offer standardized details on the concentrations of MPs in seawater.
- Address the information gaps about the danger and prevalence of MPs in the air, food, and water supplies.

More recently, the Circular Economy Action Plan was adopted by the EU on March 11, 2020. It covers textiles, electronics, construction, buildings, vehicles, packaging, batteries, plastics, food, water, and nutrients. This approach primarily targets MPs, especially those that are inadvertently emitted from tires and fabrics. One of the major objectives of this plan is to conduct a risk assessment associated with the generation and release of MPs into the environment, as well as explore alternatives such as bio-based plastics (Seay & Ternes, 2022).

4.2 United Nations Environment Programme

In 2010, the Global Partnership on Waste Management (GPWM) under the United Nations Environment Programme (UNEP) established a platform for collaboration with international and non-governmental entities for effective waste management. This initiative is considered a viable approach to address plastic pollution in the marine environment through coordinated efforts and alternative strategies. To combat marine plastic pollution, UNEP has played a significant part in synchronizing initiatives within the UN. Collaborating with the Joint Group of Experts on Scientific Aspects of Marine Environmental Protection (GESAMP) and various UN agencies, UNEP actively combats plastic pollution (Kershaw et al., 2019). As a part of its efforts to achieve Sustainable Development Goals (SDG) Indicator 14.1.1b, UNEP actively works for the development of sub-indicators to report marine plastics. GESAMP has proposed potential sub-indicators, which include regulating and reporting beach litter, water column plastics, MPs, floating and seabed plastic, and ingestion of plastics (Walker, 2021). Another effort to reduce pollution and create a cleaner environment is Sustainable Materials Management (UNEP, 2018b). Moreover, to limit the amount of plastic in the environment, a secondary market for recycled materials can also be established (Thushari & Senevirathna, 2020).

In a recent report, titled "For people and planet," the UNEP has developed a new strategy (Medium Term Strategy 2022–2025) which is designed to address critical global challenges, including climate change, loss in biodiversity, and pollution of various forms, during the period from 2022 to 2025 (UNEP, 2022a). In its 5th session, the United Nations Environment Assembly (UNEA) recognized that plastic pollution includes MPs, and thereby adopted the resolution "End Plastic Pollution" on March 2nd,2022 (UNEP, 2022b). It affirmed the necessity of creating globally strong cooperative and coordinated governance for eliminating the long-term plastic life cycle and avoiding further plastic-derived detriment towards all ecosystems.

4.3 Indian Government

India has demonstrated a strong determination to combat the environmental pollution stemming from SUP waste. As a proactive participant, at the 4th UNEA, India spearheaded the development of a resolution concerning the growing SUP pollution. This highlights India's recognition of the urgent necessity for the international community to prioritize and address the multifaceted challenges posed by SUP waste. A recent amendment (Plastic Waste Management (Amendment) Rules, 2021) was drafted in August 2021 and published in view of the growing MP contamination (MOEFCC, 2021). These rules include:

(I) SUPs such as PS and expanded polystyrene, as well as products like plastic sticks of earbuds, balloon sticks, flags, cups and plates, cutlery like forks, spoons, straws and knives, trays, wrapping or packing films around candy boxes, invitation cards, and cigarette packets, as well as plastic or PVC banners smaller than 100 microns, are prohibited from July 1st,2022.

(II) To reduce the littering of lightweight plastic carry bags, the thickness of plastic carry bags has been adjusted from 50 to 75 μ with effect from September 30, 2021. It was further adjusted to 100 and 20 μ with effect from December 31, 2022. As a result of the increased thickness, such plastic carriers can be reused.

In alignment with the 2016 Plastic Waste Management Rules, specific provisions have been placed to manage plastic packaging waste in an environmentally sustainable manner. While certain SUP items are subject to phase-out measures, plastic packaging waste that falls outside this scope must be collected and handled in an ecologically responsible manner. The management of such plastic wastes has been included under the EPR of the manufacturer, trader, and product owner. Furthermore, the Swachh Bharat Mission is also working towards creating improvements in the waste management infrastructure. The Indian Government also emphasizes on the importance of education and awareness among children and the youth who essentially shape the future of our planet. The India Plastic Challenge—Hackathon 2021 was a targeted event designed for Higher Educational Institutions and startups that have received recognition under the Startup India Initiative. Its goal is to emphasize and promote innovations to develop alternatives to SUPs and determine adoptable measures and solutions to plastic waste management.

For efficiently abolishing SUPs and the successful execution of Plastic Waste Management Rules, 2016, various Indian states and UTs have been instructed to form a Special Task Force. Various actions have been adopted for improving the application of various Plastic Waste Management Rules. For instance, Kerala is one of the Indian states, which has embarked on its mission to develop a plastic-free environment (Joshi & Ahmed, 2016). As a part of the project "Zero-waste

Kovalam," a "Plastic-free" campaign has been launched at the Kovalam beach by Kerala tourism. People are encouraged to use plastic alternatives like cloth and paper bags. Additionally, plants for recycling plastic wastes to oil have also been set up.

However, despite the global recognition and emphasis on plastic waste management and the adoption of improved recycling and mitigation approaches, recent scientific studies reveal that a substantial portion of environmental plastic is still derived by MPs shed from plastics. Capturing these MPs is neither envisaged nor achievable under current strategies (NASEM, 2022). Thus, mitigating MPs shed into our ecosystems needs to be further regulated by imposing more specific legislation regarding the manufacture and usage of primary and secondary sources of MPs.

5 Key Recap Points

This chapter serves as a stepping stone in advancing our knowledge and awareness of MPs, providing a solid foundation for future research aimed at mitigating MPs. The following are the key points highlighted in this exploration of the intricate world of MPs:

- The widespread distribution of MPs can be attributed to both intentional and unintentional generation of MPs. Transportation of MPs is influenced by various factors, such as river water flow, ocean currents, soil infiltration, and wind speed and direction.
- MPs can leave a significant imprint on diverse aquatic and terrestrial ecosystems, posing risks to organisms across all levels of the food chain.
- Human health faces potential risks due to inhalation, ingestion, and chemical contamination of MPs, demanding in-depth investigations to develop a comprehensive understanding.
- Highly precise and accurate analytical techniques are essential for microplastic detection, often requiring a combination of multiple methods to enhance research efforts.
- Effective mitigation of MPs must begin with sound plastic waste management. Furthermore, strengthening existing policies and regulations associated with plastic pollution, with a primary focus on MP pollution, is imperative to emphasize its urgency.

6 Conclusion

This chapter provides insights into the fundamental facets of microplastic pollution, emphasizing its global significance and multifaceted influence on both the environment and human health. In-depth analysis of sources, detection methodologies, and robust regulatory frameworks play a crucial role in achieving effective MP mitigation. A holistic approach, involving collaborative efforts between policymakers, scientists, industries, and the public is essential for achieving fruitful advancements. Adopting sustainable practices, adopting the 7Rs (Rethink, Refuse, Reduce, Reuse, Repair, Recycle, and Rot), and promoting innovations for the removal of MPs from different systems are vital steps in combatting MP pollution.

Acknowledgements Authors SJB and AKG are grateful to Jawaharlal Nehru University for financial assistance.

References

Alberghini, L., Truant, A., Santonicola, S., Colavita, G., & Giaccone, V. (2022). Microplastics in fish and fishery products and risks for human health: A review. *International Journal of Environmental Research and Public Health, 20*, 789.

Allen, S., Allen, D., Phoenix, V. R., Le Roux, G., Durántez Jiménez, P., Simonneau, A., Binet, S., & Galop, D. (2019). Atmospheric transport and deposition of microplastics in a remote mountain catchment. *Nature Geoscience, 12*, 339–344. https://doi.org/10.1038/s41561-019-0335-5

Allen, D., Allen, S., Le Roux, G., Simonneau, A., Galop, D., & Phoenix, V. R. (2021). Temporal archive of atmospheric microplastic deposition presented in ombrotrophic peat. *Environmental Science & Technology Letters, 8*, 954–960. https://doi.org/10.1021/acs.estlett.1c00697

Amato-Lourenço, L. F., de Souza Xavier Costa, N., Dantas, K. C., dos Santos Galvão, L., Moralles, F. N., Lombardi, S. C. F. S., Júnior, A. M., Lindoso, J. A. L., Ando, R. A., Lima, F. G., Carvalho-Oliveira, R., Mauad, T. (2022). Airborne microplastics and SARS-CoV-2 in total suspended particles in the area surrounding the largest medical centre in Latin America. *Environmental Pollution, 292*, 118299. https://doi.org/10.1016/j.envpol.2021.118299.

Bahadur, K. C. P., Maharjan, A., Acharya, M., Lee, D., Kusma, S., Gautam, R., Kwon, J.-T., Kim, C., Kim, K., & Kim, H. (2023). Polytetrafluorethylene microplastic particles mediated oxidative stress, inflammation, and intracellular signaling pathway alteration in human derived cell lines. *Science of the Total Environment* 165295.

Borah, S. J., Gupta, A. K., Gupta, A., Bhawna, K. S., Sharma, R., Kumar, R., Kumar, P., Dubey, K. K., Kaushik, S., Mishra, A. K., & Kumar, V. (2023). Grasping the supremacy of microplastic in the environment to understand its implications and eradication: A review. *Journal of Materials Science, 58*, 1–30. https://doi.org/10.1007/s10853-023-08806-8

Castro-Castellon, A. T., Horton, A. A., Hughes, J. M. R., Rampley, C., Jeffers, E. S., Bussi, G., & Whitehead, P. (2022). Ecotoxicity of microplastics to freshwater biota: Considering exposure and hazard across trophic levels. *Science of the Total Environment, 816*, 151638.

Chen, S., Yang, J.-L., Zhang, Y.-S., Wang, H.-Y., Lin, X.-Y., Xue, R.-Y., Li, M.-Y., Li, S.-W., Juhasz, A. L., Ma, L. Q., Zhou, D.-M., & Li, H.-B. (2023). Microplastics affect arsenic bioavailability by altering gut microbiota and metabolites in a mouse model. *Environmental Pollution, 324*, 121376. https://doi.org/10.1016/j.envpol.2023.121376

European Commission. (2015). Microplastics. Retrieved August 22, 2022 from https://echa.europa.eu/documents/10162/17233/echa_annex_xv_restriction_proposals_en.pdf/ed07424a-328d-88e0-b7c6-412251426582?t=1576513180293.

European Commission. (2017). Request to the European Chemical Agency to prepare restriction proposals conforming to the requirements of Annex XV to REACH. Ref. ARES (2017)5463573-09/11/2017

Ding, Y., Zou, X., Wang, C., Feng, Z., Wang, Y., Fan, Q., & Chen, H. (2021). The abundance and characteristics of atmospheric microplastic deposition in the northwestern South China Sea in the fall. *Atmospheric Environment, 253*, 118389. https://doi.org/10.1016/j.atmosenv.2021.118389

Ding, J., Sun, C., He, C., Zheng, L., Dai, D., & Li, F. (2022). Atmospheric microplastics in the Northwestern Pacific Ocean: Distribution, source, and deposition. *Science of the Total Environment, 829*, 154337. https://doi.org/10.1016/j.scitotenv.2022.154337

Dong, H., Wang, L., Wang, X., Xu, L., Chen, M., Gong, P., & Wang, C. (2021). Microplastics in a remote Lake Basin of the Tibetan plateau: Impacts of atmospheric transport and glacial melting. *Environmental Science and Technology, 55*, 12951–12960.

Evangeliou, N., Grythe, H., Klimont, Z., Heyes, C., Eckhardt, S., Lopez-Aparicio, S., & Stohl, A. (2020). Atmospheric transport is a major pathway of microplastics to remote regions. *Nature Communications, 11*, 3381. https://doi.org/10.1038/s41467-020-17201-9

Fang, M., Liao, Z., Ji, X., Zhu, X., Wang, Z., Lu, C., Shi, C., Chen, Z., Ge, L., Zhang, M., Dahlgren, R. A., & Shang, X. (2022). Microplastic ingestion from atmospheric deposition during dining/drinking activities. *Journal of Hazardous Materials, 432*, 128674. https://doi.org/10.1016/j.jhazmat.2022.128674

Galloway, T. S. (2015). Micro-and nano-plastics and human health. *Marine anthropogenic litter* (pp. 343–366). Springer.

Hamilton, B. M., Bourdages, M. P. T., Geoffroy, C., Vermaire, J. C., Mallory, M. L., Rochman, C. M., & Provencher, J. F. (2021). Microplastics around an Arctic seabird colony: Particle community composition varies across environmental matrices. *Science of the Total Environment, 773*, 145536.

Hartmann, N. B., Hüffer, T., Thompson, R. C., Hassellöv, M., Verschoor, A., Daugaard, A. E., Rist, S., Karlsson, T., Brennholt, N., Cole, M., Herrling, M. P., Hess, M. C., Ivleva, N. P., Lusher, A. L., & Wagner, M. (2019). Are we speaking the same language? Recommendations for a definition and categorization framework for plastic debris. *Environmental Science and Technology, 53*, 1039–1047. https://doi.org/10.1021/acs.est.8b05297

Huang, Z., Hu, B., & Wang, H. (2023). Analytical methods for microplastics in the environment: A review. *Environmental Chemistry Letters, 21*, 383–401. https://doi.org/10.1007/s10311-022-01525-7

Huang, Y., He, T., Yan, M., Yang, L., Gong, H., Wang, W., Qing, X., & Wang, J. (2021). Atmospheric transport and deposition of microplastics in a subtropical urban environment. *Journal of Hazardous Material, 416*, 126168. https://doi.org/10.1016/j.jhazmat.2021.126168.

Jambeck, J. R., Geyer, R., Wilcox, C., Siegler, T. R., Perryman, M., Andrady, A., Narayan, R., & Law, K. L. (2015). Plastic waste inputs from land into the ocean. *Science, 347*, 768–771.

Jia, Q., Duan, Y., Han, X., Sun, X., Munyaneza, J., Ma, J., & Xiu, G. (2022). Atmospheric deposition of microplastics in the megalopolis (Shanghai) during rainy season: Characteristics, influence factors, and source. *Science of the Total Environment, 847*, 157609. https://doi.org/10.1016/j.scitotenv.2022.157609

Joshi, R., & Ahmed, S. (2016). Status and challenges of municipal solid waste management in India: A review. *Cogent Environmental Science, 2*, 1139434.

Kernchen, S., Löder, M. G. J., Fischer, F., Fischer, D., Moses, S. R., Georgi, C., Nölscher, A. C., Held, A., & Laforsch, C. (2022). Airborne microplastic concentrations and deposition across the Weser River catchment. *Science of the Total Environment, 818*, 151812. https://doi.org/10.1016/j.scitotenv.2021.151812.

Kershaw, P. J., Turra, A., Galgani, F., et al. (2019). Guidelines for the monitoring and assessment of plastic litter and microplastics in the ocean.

Lam, C.-S., Ramanathan, S., Carbery, M., Gray, K., Vanka, K. S., Maurin, C., Bush, R., & Palanisami, T. (2018). A comprehensive analysis of plastics and microplastic legislation worldwide. *Water, Air, and Soil Pollution, 229*, 345. https://doi.org/10.1007/s11270-018-4002-z

Liu, Z., Bai, Y., Ma, T., Liu, X., Wei, H., Meng, H., Fu, Y., Ma, Z., Zhang, L., & Zhao, J. (2022). Distribution and possible sources of atmospheric microplastic deposition in a valley basin city (Lanzhou, China). *Ecotoxicology and Environmental Safety, 233*, 113353. https://doi.org/10.1016/j.ecoenv.2022.113353

Lu, X., Wang, X., Liu, X., & Singh, V. P. (2023). Dispersal and transport of microplastic particles under different flow conditions in riverine ecosystem. *Journal of Hazardous Materials, 442*, 130033.

MOEFCC. (2021). Ministry of Environment, Forest and Climate Change, Notification number G.S.R.571(E) dated 12th August 2021. https://egazette.nic.in/WriteReadData/2022/233568.pdf.

Muhib, M. I., Uddin, M. K., Rahman, M. M., & Malafaia, G. (2023). Occurrence of microplastics in tap and bottled water, and food packaging: A narrative review on current knowledge. *Science of the Total Environment, 865*, 161274. https://doi.org/10.1016/j.scitotenv.2022.161274.

NASEM. (2022). *Reckoning with the U.S. Role in Global Ocean Plastic Waste*. The National Academies Press.

Osman, A. I., Hosny, M., Eltaweil, A. S., Omar, S., Elgarahy, A. M., Farghali, M., Yap, P.-S., Wu, Y.-S., Nagandran, S., Batumalaie, K., Gopinath, S. C. B., John, O. D., Sekar, M., Saikia, T., Karunanithi, P., Hatta, M. H. M., & Akinyede, K. A. (2023). Microplastic sources, formation, toxicity and remediation: A review. *Environmental Chemistry Letters, 21*, 2129–2169. https://doi.org/10.1007/s10311-023-01593-3

Pandey, D., Singh, A., Ramanathan, A., & Kumar, M. (2021). The combined exposure of microplastics and toxic contaminants in the floodplains of north India: A review. *Journal of Environmental Management, 279*, 111557. https://doi.org/10.1016/j.jenvman.2020.111557

Parolini, M., Antonioli, D., Borgogno, F., Gibellino, M. C., Fresta, J., Albonico, C., De Felice, B., Canuto, S., Concedi, D., & Romani, A. (2021). Microplastic contamination in snow from Western Italian alps. *International Journal of Environmental Research and Public Health, 18*, 768.

Patil, P. B., Maity, S., & Sarkar, A. (2022). Potential human health risk assessment of microplastic exposure: Current scenario and future perspectives. *Environmental Monitoring and Assessment, 194*, 898.

Prata, J. C., da Costa, J. P., Lopes, I., Duarte, A. C., & Rocha-Santos, T. (2020). Environmental exposure to microplastics: An overview on possible human health effects. *Science of the Total Environment, 702*, 134455. https://doi.org/10.1016/j.scitotenv.2019.134455

Primpke, S., Booth, A.M., Gerdts, G., Gomiero, A., Kögel, T., Lusher, A. L., Strand, J., Scholz-Böttcher, B. M., Galgani, F., & Provencher, J. F. (2022). Monitoring of microplastic pollution in the Arctic: Recent developments in polymer identification, quality assurance and control (QA/QC), and data reporting. *Arctic Science*.

Purwiyanto, A. I. S., Prartono, T., Riani, E., Naulita, Y., Cordova, M. R., & Koropitan, A. F. (2022). The deposition of atmospheric microplastics in Jakarta-Indonesia: The coastal urban area. *Marine Pollution Bulletin, 174*, 113195. https://doi.org/10.1016/j.marpolbul.2021.113195

Roblin, B., Ryan, M., Vreugdenhil, A., & Aherne, J. (2020). Ambient atmospheric deposition of anthropogenic microfibers and microplastics on the western periphery of Europe (Ireland). *Environmental Science and Technology, 54*, 11100–11108.

Rota, E., Bergami, E., Corsi, I., & Bargagli, R. (2022). Macro-and microplastics in the Antarctic environment: Ongoing assessment and perspectives. *Environments, 9*, 93.

Seay, J., & Ternes, M. E. (2022). A review of current challenges and legal advances in the global management of plastic waste. *Clean Technologies and Environmental Policy, 24*, 731–738. https://doi.org/10.1007/s10098-022-02289-y

Sharifinia, M., Bahmanbeigloo, Z. A., Keshavarzifard, M., Khanjani, M. H., & Lyons, B. P. (2020). Microplastic pollution as a grand challenge in marine research: A closer look at their adverse impacts on the immune and reproductive systems. *Ecotoxicology and Environmental Safety, 204*, 111109.

Sol, D., Solís-Balbín, C., Laca, A., Laca, A., & Díaz, M. (2023). A standard analytical approach and establishing criteria for microplastic concentrations in wastewater, drinking water and tap water. *Science of the Total Environment,* 899, 165356. https://doi.org/10.1016/j.scitotenv.2023.165356.

Sun, J., Peng, Z., Zhu, Z.-R., Fu, W., Dai, X., Ni, B.-J. (2022). The atmospheric microplastics deposition contributes to microplastic pollution in urban waters. *Water Research, 225*, 119116. https://doi.org/10.1016/j.watres.2022.119116.

Szewc, K., Graca, B., & Dołęga, A. (2021). Atmospheric deposition of microplastics in the coastal zone: Characteristics and relationship with meteorological factors. *Science of the Total Environment, 761*, 143272. https://doi.org/10.1016/j.scitotenv.2020.143272

Tadsuwan, K., & Babel, S. (2022). Unraveling microplastics removal in wastewater treatment plant: A comparative study of two wastewater treatment plants in Thailand. *Chemosphere, 307*, 135733. https://doi.org/10.1016/j.chemosphere.2022.135733

Talukdar, A., Bhattacharya, S., Bandyopadhyay, A., & Dey, A. (2023). Microplastic pollution in the Himalayas: Occurrence, distribution, accumulation and environmental impacts. *Science of the Total Environment, 874*, 162495. https://doi.org/10.1016/j.scitotenv.2023.162495

Thushari, G. G. N., & Senevirathna, J. D. M. (2020). Plastic pollution in the marine environment. *Heliyon, 6*, e04709–e04709. https://doi.org/10.1016/j.heliyon.2020.e04709

Truong, T.-N.-S., Strady, E., Kieu-Le, T.-C., Tran, Q.-V., Le, T.-M.-T., & Thuong, Q.-T. (2021). Microplastic in atmospheric fallouts of a developing Southeast Asian megacity under tropical climate. *Chemosphere, 272*, 129874. https://doi.org/10.1016/j.chemosphere.2021.129874

Tse, Y.-T., Chan, S. M., & Sze, E. T. (2022). Quantitative assessment of full size microplastics in bottled and tap water samples in Hong Kong. *International Journal of Environmental Research and Public Health, 19*, 13432.

UNEP. (2018a). Legal limits on single-use plastics and microplastics: A global review of national laws and regulations. https://www.unep.org/resources/publication/legal-limits-single-use-plastics-and-microplastics-global-review-national.

UNEP. (2018b). Plastics: A roadmap for sustainability. In *Single-Use Plast. A Roadmap Sustain.* https://www.unep.org/resources/report/single-use-plastics-roadmap-sustainability.

UNEP. (2022b). For people and planet: The UNEP strategy for 2022–2025. https://www.unep.org/resources/policy-and-strategy/people-and-planet-unep-strategy-2022-2025.

UNEP. (2022b). Fifth session of the United Nations Environment Assembly. https://www.unep.org/environmentassembly/unea5.

European Union. (2007). Regulation (EC) No 1907/2006 of the European Parliament and of the Council. https://eur-lex.europa.eu/LexUriServ/LexUriServ.do?uri=OJ:L:2007:136:0003:0280:en:PDF.

Vianello, A., Jensen, R. L., Liu, L., & Vollertsen, J. (2019). Simulating human exposure to indoor airborne microplastics using a Breathing Thermal Manikin. *Science and Reports, 9*, 8670. https://doi.org/10.1038/s41598-019-45054-w

Walker, T. R. (2021). (Micro)plastics and the UN sustainable development goals. *Current Opinion in Green and Sustainable Chemistry, 30*, 100497. https://doi.org/10.1016/j.cogsc.2021.100497.

Yang, L., Kang, S., Wang, Z., Luo, X., Guo, J., Gao, T., Chen, P., Yang, C., & Zhang, Y. (2022). Microplastic characteristic in the soil across the Tibetan Plateau. *Science of the Total Environment, 828*, 154518.

Yuan, Z., Li, H.-X., Lin, L., Pan, Y.-F., Liu, S., Hou, R., Xu, X.-R. (2022). Occurrence and human exposure risks of atmospheric microplastics: A review. *Gondwana Research, 108*, 200–212. https://doi.org/10.1016/j.gr.2022.02.001.

Zhang, Y., Wang, S., Olga, V., Xue, Y., Lv, S., Diao, X., Zhang, Y., Han, Q., & Zhou, H. (2022a). The potential effects of microplastic pollution on human digestive tract cells. *Chemosphere, 291*, 132714. https://doi.org/10.1016/j.chemosphere.2021.132714

Zhang, Y., Gao, T., Kang, S., Shi, H., Mai, L., Allen, D., & Allen, S. (2022b). Current status and future perspectives of microplastic pollution in typical cryospheric regions. *Earth-Science Rev, 226*, 103924.

Chapter 10
The Advancements and Detection Methodologies for Microplastic Detection in Environmental Samples

Dharaneesh Arunachalam Balasubramaniam, Rajapandiyan Panneerselvam, K. Akshaya, Ricky Rajamanickam, Gabriel Enrique De-la-Torre, and Rangabhashiyam Selvasembian

Abstract Microplastics (MPs) contamination has emerged as a significant environmental concern due to its extensive dispersion along with potential adverse effects on aquatic as well as terrestrial ecosystems. Microplastics' harmful effects have been seen to rise throughout the decades when they mix with other contaminants in a dynamic environmental setting. As a result, developing accurate, effective, and speedy analytical techniques for identifying MPs contamination has become a pressing issue. Understanding the origins, distribution, and implications of MPs requires reliable and efficient detection techniques in environmental samples. This chapter explores the methodologies and strategies for optical detection and identification of MPs in environmental samples, covering their potential, limitations, and the latest advances in destructive (thermal and GC-MS) and non-destructive (Fourier-transform infrared spectroscopy (FTIR) and Raman spectroscopy) detection techniques. By providing a brief overview of these detection methods, this chapter aims to inform further analysis and research efforts, evaluating their applicability across various sample matrices.

Keywords Microplastics · Contamination · Detection methods · Analytical techniques · Spectroscopy

---

D. A. Balasubramaniam
Department of Civil and Environmental Engineering, University of Windsor, Windsor, ON N9B 3P4, Canada

R. Panneerselvam
Department of Chemistry, SRM University AP, Amaravati 5222502, Andhra Pradesh, India

K. Akshaya · R. Rajamanickam · R. Selvasembian (✉)
Department of Environmental Science and Engineering, School of Engineering and Sciences, SRM University-AP, Amaravati 522240, Andhra Pradesh, India
e-mail: rambhashiyam@gmail.com

G. E. De-la-Torre
Grupo de Investigación de Biodiversidad, Medio Ambiente y Sociedad, Universidad San Ignacio de Loyola, Lima, Peru

A. K. Mishra et al. (eds.), *Microplastics*, https://doi.org/10.1007/978-981-97-6461-7_10

1 Introduction

The growing population causes environmental contamination, a significant stress for society. People contribute to environmental pollution by releasing harmful smoke, chemical fertilizers, and a variety of synthetic substances into the environment. Plastics are widely employed in modern human life because of their outstanding properties (e.g., resilience, lightweight materials, strength, and insulation) and affordability. It is commonly recognized that we generate massive volumes of plastic garbage every day, endangering the ecosystem in various ways. In 2024, global plastics production is expected to reach 502.5 million metric tons (Global Plastics Production, n.d.), representing an 8.5% increase from 2019 (Plastics Use by Application, n.d.). Plastics are categorized by their size: nano plastics (<100 nm), microplastics (<5 mm), mesoplastics (5–25 mm), and macroplastics (>25 mm) (Kasavan et al., 2021). Around 80% of the plastic waste in the world's oceans comes from land-based sources, whereas just 20% comes from marine sources (Li et al., 2016). Approximately 220 species of aquatic animals, including mussels, lobsters, seagulls, oysters, etc., have been shown to have microplastics (MPs) (Jinhui et al., 2019). Additionally, MPs are highly likely to be consumed or inhaled by various creatures, which will impair them because of their tiny sizes and difficulties degrading in the environment. Environmental pollution is primarily caused by MPs, which may be found in a wide range of components, forms, and morphologies. The percentage of MPs in the total mass of deposited plastic globally is predicted to be 13.2% by 2060. MPs may operate as a vector for the transportation of hydrophobic organic compounds and heavy metals in the environment by adsorbing these substances on their surface (Wu et al., 2020).

MPs, commonly detected in the environment, encompass diverse polymer types, including those derived from industrial processes for natural biopolymers like polylactic acid (PLA), polyhydroxybutyrate (PHB), and polyhydroxyalkanoate (PHA), along with conventional polymers such as polyethylene (PE), polyvinyl chloride (PVC), and polyethylene terephthalate (PET) (Ivleva, 2021). Tire wear particles (TWP), a form of MPs originating from styrene-butadiene rubber, might be inconspicuous due to their small size (Ivleva, 2021). Nevertheless, their impact on pollution is substantial. MPs primarily originate from microbeads present in products like shampoos, body scrubs, toothpaste, and glitter, as well as materials such as polyester, nylon, acrylic, car tyres, brake pads, and more extensive plastic (Singh & Kumar, 2024). Following their usage, a significant proportion of manufactured plastics (approximately 79%) is deliberately or inadvertently released into the environment (Geyer et al., 2017).

The pollution of aquatic ecosystems, including oceans, rivers, and lakes, with MPs has profound implications. Research indicates the presence of MPs on the ocean floor and within the organisms inhabiting the Earth (Botterell et al., 2019; Pereira et al., 2023). Figure 1 depicts the widespread accumulation of MPs in the environment through diverse mechanisms. Marine creatures, such as fish, seabirds, and aquatic mammals, face substantial challenges because they frequently confuse MPs with food and unwittingly swallow them when searching for food. This has severe repercussions, involving damage to internal systems, starvation, and mortality, disturbing the complicated equilibrium that exists in aquatic ecosystems. Furthermore, individuals who consume seafood or rely on food obtained from the marine system may face health hazards (Gunaalan et al., 2023). Another route for humans to be exposed to MPs in the natural environment is via packaged foods, inhaling particulates in the air, and skin contact (Gruber et al., 2023). MPs have also been detected in the placenta, blood, and in infants, indicating possible problems with prenatal immune development (Hu et al., 2021).

The scientific evidence offered on the impacts of MPs may be both persuasive and unsettling. Because these MPs are so prevalent in the environment as a whole, we need to pay close attention to identifying them in the samples present in the environment and working together to reduce their impact. MPs detection is also critical for understanding the environmental and health effects of plastic use, driving regulatory initiatives, encouraging research into the subject, and promoting public knowledge. While various detection techniques are currently under development, substantial challenges persist, including a heightened rate of detection errors and prolonged detection times, rendering the process challenging to execute. Researchers continue to explore rapid, cost-effective, efficient, reliable, and precise methods for detecting microplastics (Singh & Kumar, 2024).

As a result, the principal objectives of this chapter are to (i) highlight current advances in MPs detection in a variety of environmental systems, (ii) outline existing

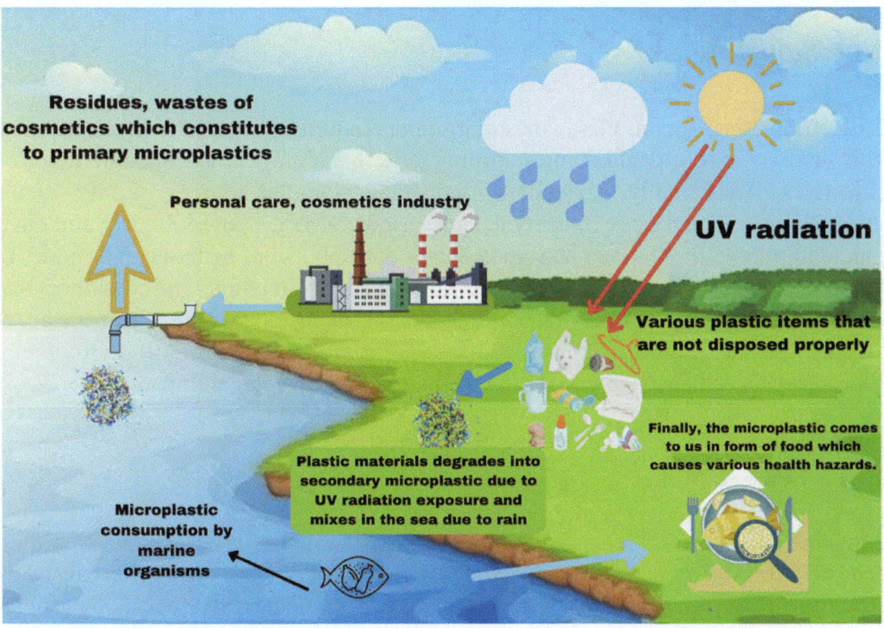

Fig. 1 Accumulation of microplastics in the environment

methods for identifying MPs in various environments, and (iii) analyze the benefits and drawbacks of approaches to detection so that researchers may determine which strategies are most suited for MPs detection based on the matrices used to discover MPs.

2 Sampling Techniques for MPs Detection

Analytical methodologies are necessary to quantify and chemically characterize microplastics. MPs detection approaches vary depending on their small size (micrometers to nanometers), shape, polymeric arrangement, and the existence of modifiers such as plasticizing agents, stabilization agents, dyes, and antioxidants (Hale, 2017; Ivleva, 2021). MPs' complexity stems from their primary as well as secondary states, hydrophobic nature, and surface electrical charge (Muhib et al., 2023). The existing approaches for detecting MPs rely mostly on its physical and chemical makeup. Microscopic approaches, thermal analysis combined with mass spectrometry, and pyrolysis gas chromatography techniques are all used for detection. Additional approaches are required to describe more specific features, such as weathering causes, deteriorating products, and adsorbed substances (chemicals). MPs detection consists of three steps: (a) sampling, (b) processing of samples (extraction process and purification process), and (c) detection.

2.1 Sampling

Collecting MPs from samples in the environment requires a comprehensive approach. This involves carefully choosing sampling sites that align with the specific research objectives. Additionally, it requires identifying potential sources of MPs in the selected areas. The process must consider various factors, both biological and environmental, including human activities, water currents, wind patterns, and more. To ensure the integrity of the samples, it is crucial to employ methods and containers that are free from contamination. This meticulous sampling strategy aims to provide accurate insights into the presence and distribution of microplastics in the environment, contributing to a better understanding of the broader ecological impact.

2.1.1 Soil Sampling

Soil serves as an initial reservoir for MPs, with a variety of processes causing their deposition on soil surfaces. These methods include atmospheric deposition caused by airborne movement, runoff from polluted waterways, and directly deposited particles from adjacent sources like highways or urban centers. The vertical dispersion of MPs inside the soil is controlled by the type of soil, utilization of land, regional

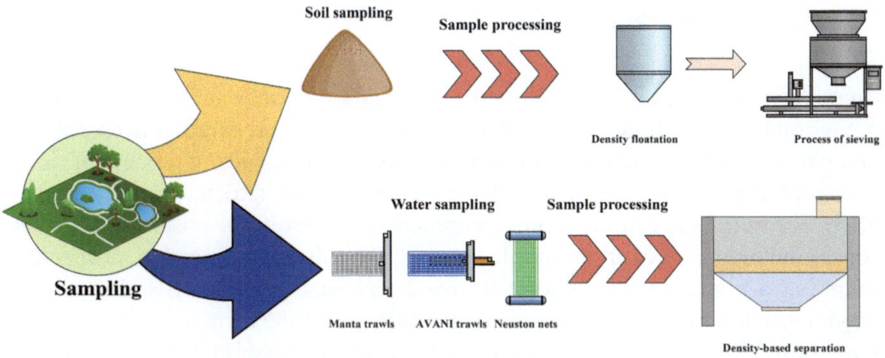

Fig. 2 Water and soil sampling techniques and sample processing for microplastic detection

weather, and the source of MPs contamination. Nonetheless, investigations have mostly examined the soil surface in the upper 10 cm (Möller et al., 2020; Piehl et al., 2018). The amount of soil sample obtained for MPs analysis is governed by a variety of factors, such as the objective of the study, the sampling technique, the projected MPs concentration level, and the detection ranges of the methods of analysis utilized. In the existing literature, reported quantities of soil samples for MPs analysis range from 50 g to 4 kg (Möller et al., 2020; Zhou et al., 2018). The sampling methods are further illustrated in Fig. 2, which highlights the soil and water sampling techniques used to detect MPs in their natural habitat.

2.1.2 Water Sampling

To acquire reliable and relevant findings, MPs from waterbodies must be collected and carried out carefully. MPs particles are often collected using sample collection equipment such as small-pore nets or sieves with meshes that are finer than the predicted size of MPs. Comparably, plankton tows with tiny mesh are commonly employed for surface water samples. Manta trawls (a type of fishing gear that drags a net through water) and AVANI trawls collect samples of drifting MPs on the ocean's surface (Eriksen et al., 2013, 2018). Neuston nets (Van Sebille et al., 2015) are most effective in more robust wave conditions, while manta trawls are utilized in gentler ocean conditions. A sampler designed for bottom sediment is employed to collect microplastics in sediment.

Numerous researchers have documented the procedures and elements that influence MPs sampling on ocean surfaces and sediments (Almuhtaram & Andrews, 2022; Briggs et al., 2019; Kovač Viršek et al., 2016; Löder & Gerdts, 2015; Prata et al., 2019). Undoubtedly, MPs sampling is influenced by a variety of elements in diverse aquatic habitats, including stream flow direction, flowing consistency, dimensions of the sampling area, seasonality and climatic fluctuations, and sampling frequency range (Singh & Kumar, 2024).

2.2 *Sample Processing*

Microplastics are generally small and varied, and they may be found in a variety of environmental matrices. As a result, rigorous sample preparation is essential for their examination. The density flotation process, filtering, and the process of sieving are essential steps for recovering MPs from soil samples (Rocha-Santos & Duarte, 2015). The initial stage in soil sample preprocessing is to remove observable particles, stones, and bigger items from samples of soil. The following procedure is drying the sample at a regulated temperature to avoid any heat-induced MPs changes. The bigger particles are then separated from the MPs by sifting them using various mesh dimensions. The density-based separation approach, which includes deposition in a solution of salt water or a dense liquid, isolates MPs from dirt particles by their densities (Constant et al., 2021; Monteiro & Pinto da Costa, 2022). Numerous studies in the literature have addressed the extraction of MPs from soil (Almuhtaram & Andrews, 2022; Möller et al., 2020; Prata et al., 2019).

Environmental samples, particularly water and dirt, may include organic debris that may cling to MPs. Eliminating this organic substance is critical for precisely measuring the mass and characteristics of the MPs (Prata et al., 2019). Figure 2 represents water and soil sampling methods, as well as sample processing methods essential for microplastic detection in diverse environmental conditions. MPs appear in a variety of sizes, from minute pieces to threads. Size fractionation, which splits particles into various size groups, was the most often utilized sample processing method. MPs in water samples are frequently maintained in sealed containers with fixative chemicals such as ethanol or isopropanol. The purpose of the fixative agent is to avert microbial development and deterioration of the MPs throughout preservation. Furthermore, the obtained samples were sieved to identify MPs. Several researchers examined the extraction of MPs from diverse sources (Elkhatib & Oyanedel-Craver, 2020; Kalaronis et al., 2022).

3 Visual Sorting and Inspection of Microplastics

Once the separation and purification processes are complete, the next step involves sorting the desired MPs from the rest of the matrix. Large plastics can be easily separated, but smaller ones require closer examination (Shim et al., 2017). They are typically observed under a dissection microscope after being dried in an oven at 60 °C. After being prepared, samples are carefully analyzed using instruments like stereomicroscopes, electron microscopes, or fluorescence microscopes (Kalaronis et al., 2022). The selection of these instruments depends on the characteristics of the microplastics in terms of size, shape, and quantity found in the sample. In addition, microscopy techniques like scanning electron microscopy (SEM) and transmission electron microscopy (TEM) provide high-resolution imaging capabilities, allowing for detailed morphological characterization and surface analysis of microplastics

(Girão, 2022). The visual inspection method has been extensively used for the characterization of microscopic particles due to its simplicity and ease of execution. These particles are manually separated depending on their size, color, and kind of plastic (fragments, granules, or beads) (Samanta et al., 2022).

SEM analyses things with electrons instead of light. Electrons are produced by an electron gun and when electrons meet with samples, X-rays are produced. These X-rays are then turned into pictures that can be used to examine the samples. Since it offers excellent clarity and high-definition photographs of the exterior surface of MPs, it facilitates differentiation between diverse organic components and synthetic MPs detected in an MPs sample. With the usage of energy-dispersive X-ray spectroscopy (EDX), even chemical composition can be analyzed. However, frequent usage of SEM–EDX is very expensive. The process of preparing samples for analysis takes more time. It also cannot distinguish between colored samples, and because of that, it needs an extra fluorescent dye (Cooper & Corcoran, 2010). Another equipment that slightly varies from SEM is polarized light microscopy, where polarized light is passed on to the sample instead of electrons. The waves resulting from the interaction of the light and the molecules are combined to form an image. It is fast in detecting MPs and has a simple working process, too. It can be used in the analysis of polyethylene terephthalate (PET), polypropylene (PP), and polyethylene-based MPs. But, it cannot detect opaque MPs (Sierra et al., 2020). An overview of these methods is presented in Fig. 3, which outlines various techniques used for identification and detection techniques employed.

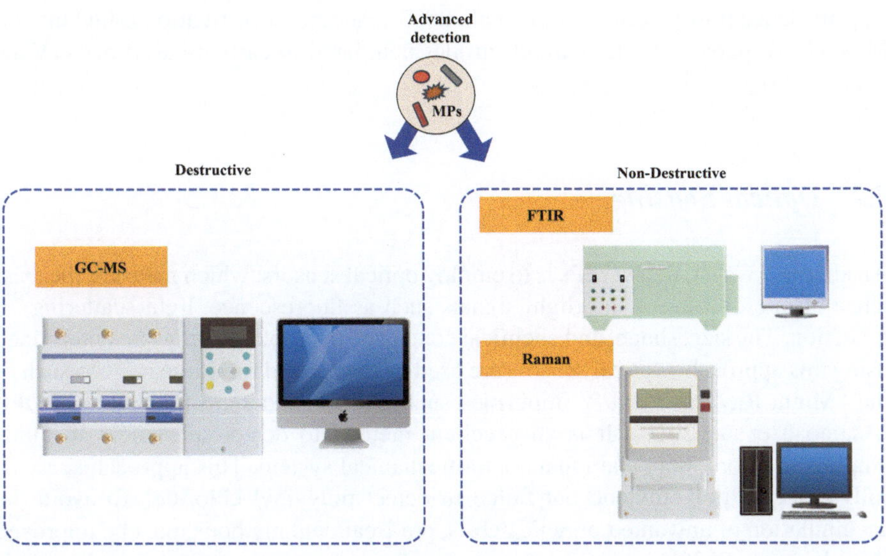

Fig. 3 Destructive and non-destructive methods in the detection of MPs

3.1 Electrochemical Impedance Spectroscopy

Impedance spectroscopy's minimal cross-sensitivity to biological materials establishes it as an effective technology for detecting and quantifying microplastics (MPs) within tested samples. (Colson & Michel, 2021) impedance spectroscopy experiments required the insertion of particle-spiked samples into a sampling funnel. Pristine water (degasification and filtration of an average particle size of 53 μm) was put into the funnel regularly to maintain a steady water level until the last particles were washed through. The flow was controlled by gravity. A trans-impedance amplifier and a lock-in amplifier were used to measure impedance across both receiving and transmitting electrodes at six frequencies at the same time. Post-processing revealed changes in impedance produced by particle movement via the flow cell. The procedure proceeded after the flow cell passage. After traversing the flow cell, the water-containing particles underwent filtration using a 53 μm sieve, and the microplastics were then counted through visual observation.

The precise differentiation between plastic and biological substances resulted in a minimal false detection rate of only 1%. Beyond essential identification, it examines the electrical characteristics of particles at a flow rate of 103 mL/min, obtaining a recovery rate of more than 90% for microplastics (MPs) ranging in size from 300 to 1000 μm. However, it was shown that the used approach has a decreased efficiency for particles smaller than 300 μm, which can be related to a lower recovery rate induced by the existence of deviations (Colson & Michel, 2021). This issue may be effectively addressed by using better electrodes with minimal gaps and identification by impedance flow cytometry (Du et al., 2023). Another investigation found that PS MPs adsorb poorly on platinum electrodes compared to carbon-based ones (Vidal et al., 2023).

3.2 Optical Sensing

An alternate way to detect MPs is to employ optical sensors, which measure the existence of microplastics using light signals such as fluorescence, light scattering, or reflection. The size, shape, and chemical content of microplastics may be determined using this approach. Optical sensors are used in specialized maritime regions such as the "Minta Ray" to identify impurities such as PET and semi-transparent LDPE (Asamoah et al., 2019). It is an excellent method to detect substances in fluids employing fluorophore stain in an opto-mechanical system. This approach successfully detected polyethylene but failed to detect polyvinyl chloride. To avoid the accumulation of unwanted organic debris, pre-treatment methods must be improved (Nicolai et al., 2022).

Another inexpensive mobile optical apparatus was developed and launched, which included a converged laser, diffraction, grating, and a charge-coupled device (CCD) sensor capable of detecting MPs at concentrations as low as 0.015% w/v (Iri et al.,

2021). Given the constraints of visual and optical microscopy methodologies for quantification and detection, a novel computational imaging technique that can recognize microplastics under UV light was developed. It was stated that when samples were displayed under UV light and dyed with NR, deviations of 10 and 8% were recorded, respectively. The approach employed in conjunction with image processing methods proved effective for detecting a wide range of solid and liquid MPs (Giardino et al., 2023).

Confirmation of potential microplastics identified through visual inspection often requires additional analytical techniques to verify their composition. Despite its low cost and simplicity, the approach cannot be recommended as a routine identification method due to the potential for underestimation and mistakes associated with manual or microscopic counting. Furthermore, the isolated MPs may include fragments, making the counting process more difficult. Vibrational spectroscopic techniques such as Fourier-transform infrared spectroscopy (FTIR) and Raman spectroscopy, are commonly used to verify the existence of plastic polymers and understand the chemical makeup of microplastic particles (Veerasingam et al., 2021). Therefore, it becomes necessary to employ other instrumental techniques like FT-IR and Raman spectroscopy for further identification.

4 Advancement in Microplastic Detection Methods

Advanced detection methods are categorized mainly into two types: destructive and non-destructive as depicted in Fig. 3. Destructive methods involve analyzing the sample by altering or damaging it, such as through techniques like pyrolysis-GC-MS, where the sample is heated to high temperatures for decomposition (Singh & Kumar, 2024). This process provides detailed information about the sample's chemical composition but renders it unsuitable for further analysis. In non-destructive means, we can further use the same sample for another type of analysis after finishing the analysis. Chemically, non-destructive FTIR (Fourier Transform Infrared Spectroscopy) is used. This approach involves exposing plastic samples to infrared radiation, which causes the chemical interactions between various molecules to vibrate, resulting in the formation of the appropriate spectra. The obtained spectrum is then compared with the reference spectrum that has already been analyzed for comparison (Murphy et al., 2016a). Three types of modes are present in the FT-IR method: transmission mode, reflectance, and attenuated total reflectance (ATR) modes. The uniqueness in cases of reflectance and ATR modes is that the step of sample preparation is not needed for the cases of opaque and thick MPs. The use of micro-FT-IR can reveal the composition of the polymer and the MPs abundance, which can help give clues as to the origin or the source where MPs was formed. The instrument setup for micro-FT-IR is expensive and demands skilled personnel due to its labor-intensive and time-consuming nature (Shim et al., 2017).

The focal plane array, an image sensor employed in combination with FTIR (FPA-FTIR), comprises arrays of light-sensing pixels positioned at the focal lens.

This configuration allows for the capture of high-definition images and ensures high detection efficiency. In analyzing samples from both the water treatment plant and the sludge, FPA-FTIR gives better explanations in both cases (Löder et al., 2015). However, it cannot identify stretched fiber images and has a limited diffraction span too. It becomes difficult to analyze particles with sizes less than 10–20 μm (Primpke et al., 2017).

Raman spectroscopy is another spectroscopy that has been employed for the analysis of MPs (Chakraborty et al., 2023). The laser beam has been forced onto the object, resulting in the formation of back-scattered lights of different wavelengths depending on the molecules or atoms present in the object, which further produces a unique spectrum for each polymer. The diameter of the laser beam is minimal, such that it allows the identification of MPs in the sizes of a few micrometers (Cole et al., 2013). While it offers finer spatial resolution compared to FT-IR, fluorescence can affect the analysis, and its cost is relatively high. Another type of FT-IR that can be used is by integrating thermogravimetric analysis with FT-IR. Here, heat is supplied, and the sample experiences a weight loss because of it. This method provides the thermal properties of the sample and helps to determine polyethylene, polypropylene, and polyamide. However, this method is destructive and expensive (Sierra et al., 2020).

5 Discussion on Various Detection Methods

5.1 Fourier Transform Infrared Spectroscopy

FT-IR is a non-destructive spectroscopic method frequently used for MPs analysis. The chemical composition of the plastic substance can be obtained from the generated spectrum by comparing it to existing reference spectra recorded in a library (Murphy et al., 2016b). As a result, it is critical to create a library that comprises a non-typical assortment of polymers as a reference from multiple plastic sources to guarantee that plastic particles may be analyzed adequately. Despite its broad spectrum of features in detecting MPs, FT-IR is also known as laborious (Harrison et al., 2012). The complexity of this method might be alleviated by including automated particle identification in a picture analyzer. This will undoubtedly minimize operational strain and improve data accuracy.

FPA-FTIR explains the properties of a polymeric-based sample collected from both the wastewater treatment plant and the sludge better than other analytical techniques. It has been claimed that high-resolution pictures of specimens may be produced utilizing IR transmission as opposed to IR reflection (Mintenig et al., 2017). However, there are certain drawbacks to this analytical approach, including the difficulties in detecting stretched fiber images, the limited high-resolution diffraction range of 10 μm–1000 cm, and the challenges in studying particulates less than 10–20 μm in diameter (Li et al., 2018).

5.2 Microfluidics

Microfluidics has been dubbed the "latest environmental savior" due to its effectiveness in detecting MPs (Ece et al., 2023). Several techniques, including a PMMA (Polymethyl Methacrylate)-oriented integrated microfluidic approach, are shown for quantifying MPs in river water samples. According to (Zhang et al., 2023a, b), the approach is straightforward, low-cost, and requires few accessories. Another examination recovered and subsequently identified MPs ranging in size from 1 to 10 μm using DC electrophoretic pressures. Resistance voltage monitoring was used to identify polystyrene in a simple broad PDMS medium outfitted with a set of copper microwire electrodes and coupled to a DC power supply meter. Decreasing the flow rate via microfluidic channels aided in removing and identifying MPs (Zabihihesari et al., 2023). Another investigation (Zhang et al., 2023a, b) employed a microfluidic approach in conjunction with a microfluidic chip to detect Nile Red (NR) dyed particles passing via microchannels. This technique is reported to have a wide range of environmental applications. It has also been reported that NR must be diluted 100 times since less than that causes particle aggregation, which causes obstruction in microfluidics channels (Mesquita et al., 2022).

5.3 Pyrolysis Gas Chromatography

Heat pyrolysis and pyrolysis gas chromatography/mass spectrometry, often referred to as pyrolysis GC-MS, are two analytical procedures involving pyrolysis. In these methods, components are typically broken down, and the resulting products are adsorbed onto a solid-phase apparatus before being characterized using GC-MS. Because it provides data on particle molecular characterization, Py-GC-MS is recognized as one of the most efficient techniques for measuring MPs. This approach was reported to have a standard deviation of under 10% and effectively identified all polymers except for PU (Polyurethane), PET, and methylene diphenyl diisocyanate (MDI) (Matsueda et al., 2021). Since the prior sample limit was 0.5 mg, the thermal pyrolysis technique has a benefit over Py-GC-MS technology. A different technique of degradation is thermal extraction with thermogravimetric analysis (TGA) combined with the use of thermal desorption gas chromatography-mass spectrometry (TDS-GC-MS), in which samples are burnt at 600 °C, but this results in a reduction in the distribution of sizes and the melting of polymeric particles prior to degradation. To tackle this problem, a thermal degradation technique using GC-MS as a thermo-analytical method with sample volumes up to 20 mg is given. This approach effectively identifies PP, PE, and PS (Dümichen et al., 2017). Another research used mass spectrometry in tandem to increase the Py-GC-MS approach's selectivity and degree of sensitivity. The updated approach revealed a detection threshold of 15–70 ng and improved the classification of polyethylene and polypropylene (Albignac et al., 2023). For materials with a wide range of particle sizes, FTIR, LDIR, or

Raman spectroscopy are favored despite their analytical challenges. However, due to its quicker analytical time and simple sample separation based on size, the Py-GC-MS technique with minimal sample pre-treatment steps is preferred for enhanced characterization (Fischer & Scholz-Böttcher, 2019).

5.4 Raman Spectroscopy

Among the various techniques used to detect, characterize, and analyze micro and nanoparticles, including microscopy, Py-GC-MS, FTIR, and Raman spectroscopy. Raman imaging stands out as one of the most effective approaches. It can identify particles as small as 20 μm. Additionally, Raman imaging is frequently combined with methods such as principal component analysis to enhance signal detection in imaging (Luo et al., 2023). Surface-enhanced Raman spectroscopy (SERS) was developed as a result of higher sensitivity in Raman spectroscopy, and it is presently utilized to identify Polystyrene particles of size 350 nm and polyethylene particles of range 1–4 μm (Kim et al., 2024). The critical component is the fabrication of SERS substrates, which brings about signal enhancement via electromagnetic and chemical enhancement (Yin et al., 2021). It was revealed that the spectroscopic method yielded stable PS signals but not PE indications. Polystyrene MPs may be detected at concentrations as low as 6.5 μm (Mikac et al., 2023). Raman spectroscopy, in conjunction with spatial heterodyne apparatus, is a novel way of detecting MPs. A system of optics that collects Raman signals from samples and is linked to SHS (Spatial Heterodyne Spectrometer). SHS has a high ratio of signal to noise, which was shown to be insufficient in the opening of a typical dispersive Raman spectrometer (Xue et al., 2022). Table 1 summarizes the advantages and disadvantages of the detection of MPs from the samples using analytical techniques.

6 Conclusion

MPs pollution needs systematic and reliable detection methods in order to reveal its existence, origins, and environmental issues. As worldwide concern over MPs pollution grows, detection in the natural environment becomes critical. Detecting MPs in samples from the environment requires a multifaceted strategy that includes techniques ranging from classic visual evaluation to innovative spectroscopy methods and imaging technologies. All methods have advantages and shortcomings, and when used together, they can provide an improved understanding of MPs contamination. As the subject evolves, improved techniques for detection will be vital to effective MPs research and administration. Microscopic technologies show intricate morphological traits, enhancing our understanding of particle attributes and function. Visual identification, being the primary identifying method, has limits in terms of precision as well

Table 1 Advantages and disadvantages of detection methods of MPs

Identification method	Advantages	Disadvantages	References
FTIR	The non-destructive method analyses chemical compositions and can analyze 1 μm size particles	Time-consuming method and labor intensive	Harrison et al. (2012)
Microfluidics	Low cost and requires fewer accessories. Large MPs can be removed effectively	Microfluidics, in conjunction with Nile Red, produces particle aggregation, which leads to occlusion	Mesquita et al. (2022)
Raman Spectroscopy	The non-destructive method reduces false negative data	The instrument is expensive, and the process is time-consuming	Cole et al. (2013)
Thermogravimetric Analysis	Polymer and chemical composition can be determined simultaneously	Detection facility for several plastic polymers is a limited, destructive method	Sierra et al. (2020)
Pyrolysis Gas Chromatography	PY-GC-MS is one of the most efficient methods to detect MPs and has a standard deviation of <10%. It can be used to detect most of the polymers	Destructive method and cannot be able to detect PU, PET, and MDI	Matsueda et al. (2021)

as effectiveness. Advanced methods, including Fourier Transform Infrared Spectroscopy and Raman spectroscopy, provide molecular data on MPs constitution but face problems in complicated matrices. Py-GC-MS has demonstrated itself as a viable technology for testing MPs, offering a unique method of detecting polymer plastics via thermally degraded intermediates. Prospective advancement may encompass the integration of Machine Learning and Artificial Intelligence (AI) for the automated recognition and classification of microplastics (MPs). Moreover, creating mobile and field-deployable solutions is expected to facilitate instantaneous monitoring, thereby strengthening our capability to tackle the problem of MPs contamination. Developments in robotics and automation offer the ability to simplify the detection process, allowing for the quick examination of huge quantities of samples.

7 Declaration of Competing Interest

The authors declare they have no known competing interests.

Acknowledgements The authors acknowledge the financial support provided by the SRM University-AP under the grant: SRMAP/URG/SEED/2023–2024/021

Author Contributions Dharaneesh Arunachalam Balasubramaniam: Methodology, Conceptualization, Writing, original draft preparation and visualization.

Rajapandiyan Panneerselvam: Writing and Review & Editing.

K. Akshaya: Writing and Review & Editing.

Ricky Rajamanickam: Review & Editing and visualization.

Rangabhashiyam Selvasembian: Methodology, Supervision, Conceptualization, Writing—Review & Editing.

References

Albignac, M., de Oliveira, T., Landebrit, L., Miquel, S., Auguin, B., Leroy, E., Maria, E., Mingotaud, A. F., & ter Halle, A. (2023). Tandem mass spectrometry enhances the performances of pyrolysis-gas chromatography-mass spectrometry for microplastic quantification. *Journal of Analytical and Applied Pyrolysis, 172*, 105993. https://doi.org/10.1016/J.JAAP.2023.105993

Almuhtaram, H., & Andrews, R. C. (2022). Sampling microplastics in water matrices: A need for standardization. *ACS ES and T Water, 2*(8), 1276–1278. https://doi.org/10.1021/ACSESTWATER.2C00236/ASSET/IMAGES/MEDIUM/EW2C00236_0001.GIF

Asamoah, B. O., Kanyathare, B., Roussey, M., & Peiponen, K. E. (2019). A prototype of a portable optical sensor for the detection of transparent and translucent microplastics in freshwater. *Chemosphere, 231*, 161–167. https://doi.org/10.1016/J.CHEMOSPHERE.2019.05.114

Botterell, Z. L. R., Beaumont, N., Dorrington, T., Steinke, M., Thompson, R. C., & Lindeque, P. K. (2019). Bioavailability and effects of microplastics on marine zooplankton: A review. *Environmental Pollution, 245*, 98–110. https://doi.org/10.1016/J.ENVPOL.2018.10.065

Briggs, E., de Moura, E. A. B., Furusawa, H. A., Cotrim, M. E. B., Oguzie, E. E., & Lugao, A. B. (2019). Microplastics: A novel method for surface water sampling and sample extraction in Elechi Creek, Rivers State, Nigeria. *Minerals, Metals and Materials Series* (pp. 269–281). https://doi.org/10.1007/978-3-030-05749-7_27/COVER.

Chakraborty, I., Banik, S., Biswas, R., Yamamoto, T., Noothalapati, H., & Mazumder, N. (2023). Raman spectroscopy for microplastic detection in water sources: a systematic review. *International Journal of Environmental Science and Technology, 20*(9), 10435–10448. https://doi.org/10.1007/s13762-022-04505-0

Cole, M., Lindeque, P., Fileman, E., Halsband, C., Goodhead, R., Moger, J., & Galloway, T. S. (2013). Microplastic ingestion by zooplankton. *Environmental Science & Technology, 47*(12), 6646–6655. https://doi.org/10.1021/ES400663F

Colson, B. C., & Michel, A. P. M. (2021). Flow-through quantification of microplastics using impedance spectroscopy. *ACS Sensors, 6*(1), 238–244. https://doi.org/10.1021/ACSSENSORS.0C02223/ASSET/IMAGES/LARGE/SE0C02223_0005.JPEG

Constant, M., Billon, G., Breton, N., & Alary, C. (2021). Extraction of microplastics from sediment matrices: Experimental comparative analysis. *Journal of Hazardous Materials, 420*, 126571. https://doi.org/10.1016/J.JHAZMAT.2021.126571

Cooper, D. A., & Corcoran, P. L. (2010). Effects of mechanical and chemical processes on the degradation of plastic beach debris on the island of Kauai, Hawaii. *Marine Pollution Bulletin, 60*(5), 650–654. https://doi.org/10.1016/J.MARPOLBUL.2009.12.026

Du, H., Chen, G., & Wang, J. (2023). Highly selective electrochemical impedance spectroscopy-based graphene electrode for rapid detection of microplastics. *Science of the Total Environment, 862*, 160873. https://doi.org/10.1016/J.SCITOTENV.2022.160873

Dümichen, E., Eisentraut, P., Bannick, C. G., Barthel, A. K., Senz, R., & Braun, U. (2017). Fast identification of microplastics in complex environmental samples by a thermal degradation method. *Chemosphere, 174*, 572–584. https://doi.org/10.1016/J.CHEMOSPHERE.2017.02.010

Ece, E., Hacıosmanoğlu, N., & Inci, F. (2023). Microfluidics as a Ray of Hope for Microplastic Pollution. *Biosensors, 13*(3), 332. https://doi.org/10.3390/BIOS13030332.

Elkhatib, D., & Oyanedel-Craver, V. (2020). A critical review of extraction and identification methods of microplastics in wastewater and drinking water. *Environmental Science and Technology, 54*(12), 7037–7049. https://doi.org/10.1021/ACS.EST.9B06672/SUPPL_FILE/ES9B06672_SI_001.PDF

Eriksen, M., Liboiron, M., Kiessling, T., Charron, L., Alling, A., Lebreton, L., Richards, H., Roth, B., Ory, N. C., Hidalgo-Ruz, V., Meerhoff, E., Box, C., Cummins, A., & Thiel, M. (2018). Microplastic sampling with the AVANI trawl compared to two neuston trawls in the Bay of Bengal and South Pacific. *Environmental Pollution, 232*, 430–439. https://doi.org/10.1016/J.ENVPOL.2017.09.058

Eriksen, M., Mason, S., Wilson, S., Box, C., Zellers, A., Edwards, W., Farley, H., & Amato, S. (2013). Microplastic pollution in the surface waters of the Laurentian Great Lakes. *Marine Pollution Bulletin, 77*(1–2), 177–182. https://doi.org/10.1016/J.MARPOLBUL.2013.10.007

Fischer, M., & Scholz-Böttcher, B. M. (2019). Microplastics analysis in environmental samples-recent pyrolysis-gas chromatography-mass spectrometry method improvements to increase the reliability of mass-related data. *Analytical Methods, 11*(18), 2489–2497. https://doi.org/10.1039/c9ay00600a

Geyer, R., Jambeck, J. R., & Law, K. L. (2017). Production, use, and fate of all plastics ever made. *Science Advances, 3*(7). https://doi.org/10.1126/SCIADV.1700782/SUPPL_FILE/1700782_SM.PDF.

Giardino, M., Balestra, V., Janner, D., & Bellopede, R. (2023). Automated method for routine microplastic detection and quantification. *Science of the Total Environment, 859*, 160036. https://doi.org/10.1016/J.SCITOTENV.2022.160036

Girão, A. V. (2022). SEM/EDS and optical microscopy analysis of microplastics. In *Handbook of microplastics in the environment* (pp. 57–78). Springer. https://doi.org/10.1007/978-3-030-39041-9_7

Global plastics production. (n.d.). Retrieved January 19, 2024, from https://ourworldindata.org/grapher/global-plastics-production.

Gruber, E. S., Stadlbauer, V., Pichler, V., Resch-Fauster, K., Todorovic, A., Meisel, T. C., Trawoeger, S., Hollóczki, O., Turner, S. D., Wadsak, W., Vethaak, A. D., & Kenner, L. (2023). To waste or not to waste: Questioning potential health risks of micro- and nanoplastics with a focus on their ingestion and potential carcinogenicity. *Exposure and Health, 15*(1), 33–51. https://doi.org/10.1007/s12403-022-00470-8

Gunaalan, K., Nielsen, T. G., Rodríguez Torres, R., Lorenz, C., Vianello, A., Andersen, C. A., Vollertsen, J., & Almeda, R. (2023). Is zooplankton an entry point of microplastics into the marine food web? *Environmental Science and Technology, 57*(31), 11643–11655. https://doi.org/10.1021/ACS.EST.3C02575/ASSET/IMAGES/LARGE/ES3C02575_0004.JPEG

Hale, R. C. (2017). Analytical challenges associated with the determination of microplastics in the environment. *Analytical Methods, 9*(9), 1326–1327. https://doi.org/10.1039/C7AY90015E

Harrison, J. P., Ojeda, J. J., & Romero-González, M. E. (2012). The applicability of reflectance micro-Fourier-transform infrared spectroscopy for the detection of synthetic microplastics in marine sediments. *Science of the Total Environment, 416*, 455–463. https://doi.org/10.1016/J.SCITOTENV.2011.11.078

Hu, J., Qin, X., Zhang, J., Zhu, Y., Zeng, W., Lin, Y., & Liu, X. (2021). Polystyrene microplastics disturb maternal-fetal immune balance and cause reproductive toxicity in pregnant mice. *Reproductive Toxicology, 106*, 42–50. https://doi.org/10.1016/J.REPROTOX.2021.10.002

Iri, A. H., Shahrah, M. H. A., Ali, A. M., Qadri, S. A., Erdem, T., Ozdur, I. T., & Icoz, K. (2021). Optical detection of microplastics in water. *Environmental Science and Pollution Research, 28*(45), 63860–63866. https://doi.org/10.1007/S11356-021-12358-2/TABLES/1

Ivleva, N. P. (2021). Chemical analysis of microplastics and nanoplastics: Challenges, advanced methods, and perspectives. *Chemical Reviews, 121*(19), 11886–11936. https://doi.org/10.1021/ACS.CHEMREV.1C00178/ASSET/IMAGES/LARGE/CR1C00178_0015.JPEG

Jinhui, S., Sudong, X., Yan, N., Xia, P., Jiahao, Q., & Yongjian, X. (2019). Effects of microplastics and attached heavy metals on growth, immunity, and heavy metal accumulation in the yellow seahorse, Hippocampus Kuda Bleeker. *Marine Pollution Bulletin, 149*, 110510. https://doi.org/10.1016/J.MARPOLBUL.2019.110510

Kalaronis, D., Ainali, N. M., Evgenidou, E., Kyzas, G. Z., Yang, X., Bikiaris, D. N., & Lambropoulou, D. A. (2022). Microscopic techniques as means for the determination of microplastics and nanoplastics in the aquatic environment: A concise review. *Green Analytical Chemistry, 3*, 100036. https://doi.org/10.1016/J.GREEAC.2022.100036

Kasavan, S., Yusoff, S., Rahmat Fakri, M. F., & Siron, R. (2021). Plastic pollution in water ecosystems: A bibliometric analysis from 2000 to 2020. *Journal of Cleaner Production, 313*, 127946. https://doi.org/10.1016/J.JCLEPRO.2021.127946

Kim, J. Y., Koh, E. H., Yang, J. Y., Mun, C., Lee, S., Lee, H., Kim, J., Park, S. G., Kang, M., Kim, D. H., & Jung, H. S. (2024). 3D plasmonic gold nanopocket structure for SERS machine learning-based microplastic detection. *Advanced Functional Materials, 34*(2), 2307584. https://doi.org/10.1002/adfm.202307584

Kovač Viršek, M., Palatinus, A., Koren, Š, Peterlin, M., Horvat, P., & Kržan, A. (2016). Protocol for microplastics sampling on the sea surface and sample analysis. *Journal of Visualized Experiments: Jove, 118*, 55161. https://doi.org/10.3791/55161

Li, J., Liu, H., & Paul Chen, J. (2018). Microplastics in freshwater systems: A review on occurrence, environmental effects, and methods for microplastics detection. *Water Research, 137*, 362–374. https://doi.org/10.1016/J.WATRES.2017.12.056

Li, W. C., Tse, H. F., & Fok, L. (2016). Plastic waste in the marine environment: A review of sources, occurrence and effects. *Science of the Total Environment, 566–567*, 333–349. https://doi.org/10.1016/J.SCITOTENV.2016.05.084

Löder, M. G. J., & Gerdts, G. (2015). Methodology used for the detection and identification of microplastics—A critical appraisal. Marine Anthropogenic Litter, 201–227. https://doi.org/10.1007/978-3-319-16510-3_8/FIGURES/7.

Löder, M. G. J., Kuczera, M., Mintenig, S., Lorenz, C., & Gerdts, G. (2015). Focal plane array detector-based micro-Fourier-transform infrared imaging for the analysis of microplastics in environmental samples. *Environmental Chemistry, 12*(5), 563–581. https://doi.org/10.1071/EN14205

Luo, Y., Awoyemi, O. S., Naidu, R., & Fang, C. (2023). Detection of microplastics and nanoplastics released from a kitchen blender using Raman imaging. *Journal of Hazardous Materials, 453*, 131403. https://doi.org/10.1016/J.JHAZMAT.2023.131403

Matsueda, M., Mattonai, M., Iwai, I., Watanabe, A., Teramae, N., Robberson, W., Ohtani, H., Kim, Y. M., & Watanabe, C. (2021). Preparation and test of a reference mixture of eleven polymers with deactivated inorganic diluent for microplastics analysis by pyrolysis-GC–MS. *Journal of Analytical and Applied Pyrolysis, 154*, 104993. https://doi.org/10.1016/J.JAAP.2020.104993

Mesquita, P., Gong, L., & Lin, Y. (2022). A low-cost microfluidic method for microplastics identification: Towards continuous recognition. *Micromachines, 13*(4), 499. https://doi.org/10.3390/MI13040499.

Mikac, L., Rigó, I., Himics, L., Tolić, A., Ivanda, M., & Veres, M. (2023). Surface-enhanced Raman spectroscopy for the detection of microplastics. *Applied Surface Science, 608*, 155239. https://doi.org/10.1016/J.APSUSC.2022.155239

Mintenig, S. M., Int-Veen, I., Löder, M. G. J., Primpke, S., & Gerdts, G. (2017). Identification of microplastic in effluents of waste water treatment plants using focal plane array-based micro-Fourier-transform infrared imaging. *Water Research, 108*, 365–372. https://doi.org/10.1016/J.WATRES.2016.11.015

Möller, J. N., Löder, M. G. J., & Laforsch, C. (2020). Finding microplastics in soils: A review of analytical methods. *Environmental Science and Technology, 54*(4), 2078–2090. https://doi.org/10.1021/ACS.EST.9B04618/SUPPL_FILE/ES9B04618_SI_001.PDF

Monteiro, S. S., & Pinto da Costa, J. (2022). Methods for the extraction of microplastics in complex solid, water and biota samples. *Trends in Environmental Analytical Chemistry, 33*, e00151. https://doi.org/10.1016/J.TEAC.2021.E00151

Muhib, M. I., Uddin, M. K., Rahman, M. M., & Malafaia, G. (2023). Occurrence of microplastics in tap and bottled water, and food packaging: A narrative review on current knowledge. *Science of the Total Environment, 865*, 161274. https://doi.org/10.1016/J.SCITOTENV.2022.161274

Murphy, F., Ewins, C., Carbonnier, F., & Quinn, B. (2016a). Wastewater Treatment Works (WwTW) as a source of microplastics in the aquatic environment. *Environmental Science & Technology, 50*(11), 5800–5808. https://doi.org/10.1021/ACS.EST.5B05416

Murphy, F., Ewins, C., Carbonnier, F., & Quinn, B. (2016b). Wastewater Treatment Works (WwTW) as a source of microplastics in the aquatic environment. *Environmental Science and Technology, 50*(11), 5800–5808. https://doi.org/10.1021/acs.est.5b05416

Nicolai, E., Pizzoferrato, R., Li, Y., Frattegiani, S., Nucara, A., & Costa, G. (2022). A new optical method for quantitative detection of microplastics in water based on real-time fluorescence analysis. *Water (switzerland), 14*(20), 3235. https://doi.org/10.3390/W14203235/S1

Pereira, R., Rodrigues, S. M., Silva, D., Freitas, V., Almeida, C. M. R., & Ramos, S. (2023). Microplastic contamination in large migratory fishes collected in the open Atlantic Ocean. *Marine Pollution Bulletin, 186*, 114454. https://doi.org/10.1016/J.MARPOLBUL.2022.114454

Piehl, S., Leibner, A., Löder, M. G. J., Dris, R., Bogner, C., & Laforsch, C. (2018). Identification and quantification of macro- and microplastics on an agricultural farmland. *Scientific Reports, 8*(1), 1–9. https://doi.org/10.1038/s41598-018-36172-y.

Plastics use by application. (n.d.). Retrieved January 19, 2024, from https://stats.oecd.org/viewhtml.aspx?datasetcode=PLASTIC_USE_10&lang=en.

Prata, J. C., da Costa, J. P., Duarte, A. C., & Rocha-Santos, T. (2019). Methods for sampling and detection of microplastics in water and sediment: A critical review. *TrAC Trends in Analytical Chemistry, 110*, 150–159. https://doi.org/10.1016/J.TRAC.2018.10.029

Primpke, S., Lorenz, C., Rascher-Friesenhausen, R., & Gerdts, G. (2017). An automated approach for microplastics analysis using focal plane array (FPA) FTIR microscopy and image analysis. *Analytical Methods, 9*(9), 1499–1511. https://doi.org/10.1039/C6AY02476A

Rocha-Santos, T., & Duarte, A. C. (2015). A critical overview of the analytical approaches to the occurrence, the fate and the behavior of microplastics in the environment. *TrAC Trends in Analytical Chemistry, 65*, 47–53. https://doi.org/10.1016/J.TRAC.2014.10.011

Samanta, P., Dey, S., Kundu, D., Dutta, D., Jambulkar, R., Mishra, R., Ghosh, A. R., & Kumar, S. (2022). An insight on sampling, identification, quantification and characteristics of microplastics in solid wastes. *Trends in Environmental Analytical Chemistry, 36*, e00181. https://doi.org/10.1016/j.teac.2022.e00181

Shim, W. J., Hong, S. H., & Eo, S. E. (2017). Identification methods in microplastic analysis: A review. *Analytical Methods, 9*(9), 1384–1391. https://doi.org/10.1039/C6AY02558G

Sierra, I., Chialanza, M. R., Faccio, R., Carrizo, D., Fornaro, L., & Pérez-Parada, A. (2020). Identification of microplastics in wastewater samples by means of polarized light optical microscopy. *Environmental Science and Pollution Research, 27*(7), 7409–7419. https://doi.org/10.1007/S11356-019-07011-Y/FIGURES/6

Singh, B., & Kumar, A. (2024). Advances in microplastics detection: A comprehensive review of methodologies and their effectiveness. *TrAC Trends in Analytical Chemistry, 170*, 117440. https://doi.org/10.1016/J.TRAC.2023.117440

Van Sebille, E., Wilcox, C., Lebreton, L., Maximenko, N., Hardesty, B. D., Van Franeker, J. A., Eriksen, M., Siegel, D., Galgani, F., & Law, K. L. (2015). A global inventory of small floating plastic debris. *Environmental Research Letters, 10*(12), 124006. https://doi.org/10.1088/1748-9326/10/12/124006

Veerasingam, S., Ranjani, M., Venkatachalapathy, R., Bagaev, A., Mukhanov, V., Litvinyuk, D., Mugilarasan, M., Gurumoorthi, K., Guganathan, L., Aboobacker, V. M., & Vethamony, P. (2021). Contributions of Fourier transform infrared spectroscopy in microplastic pollution research: A review. *Critical Reviews in Environmental Science and Technology, 51*(22), 2681–2743. https://doi.org/10.1080/10643389.2020.1807450

Vidal, J. C., Midón, J., Vidal, A. B., Ciomaga, D., & Laborda, F. (2023). Detection, quantification, and characterization of polystyrene microplastics and adsorbed bisphenol A contaminant using electroanalytical techniques. *Microchimica Acta, 190*(6), 203. https://doi.org/10.1007/s00604-023-05780-5

Wu, M., Yang, C., Du, C., & Liu, H. (2020). Microplastics in waters and soils: Occurrence, analytical methods and ecotoxicological effects. *Ecotoxicology and Environmental Safety, 202*, 110910. https://doi.org/10.1016/J.ECOENV.2020.110910

Xue, Q., Wang, N., Yang, H., Yang, J., & Bai, H. (2022). Detection of microplastics based on spatial heterodyne Raman spectroscopy. *Spectrochimica Acta Part a: Molecular and Biomolecular Spectroscopy, 283*, 121712. https://doi.org/10.1016/J.SAA.2022.121712

Yin, R., Ge, H., Chen, H., Du, J., Sun, Z., Tan, H., & Wang, S. (2021). Sensitive and rapid detection of trace microplastics concentrated through Au-nanoparticle-decorated sponge on the basis of surface-enhanced Raman spectroscopy. *Environmental Advances, 5*, 100096. https://doi.org/10.1016/j.envadv.2021.100096

Zabihihesari, A., Khalili, A., Farshchi-Heydari, M. J., Eilaghi, A., & Rezai, P. (2023). Simple microfluidic device for simultaneous extraction and detection of microplastics in water using DC electrical signal. *New Journal of Chemistry, 47*(19), 9050–9060. https://doi.org/10.1039/d2nj06268b

Zhang, M., Wang, X., Zhang, Y., & Fan, Y. (2023a). Integrated sample processing and counting microfluidic device for microplastics analysis. *Analytica Chimica Acta, 1261*, 341237. https://doi.org/10.1016/J.ACA.2023.341237

Zhang, Y., Zhang, M., & Fan, Y. (2023b). Assessment of microplastics using microfluidic approach. *Environmental Geochemistry and Health, 45*(3), 1045–1052. https://doi.org/10.1007/S10653-022-01262-4/FIGURES/7

Zhou, Q., Zhang, H., Fu, C., Zhou, Y., Dai, Z., Li, Y., Tu, C., & Luo, Y. (2018). The distribution and morphology of microplastics in coastal soils adjacent to the Bohai Sea and the Yellow Sea. *Geoderma, 322*, 201–208. https://doi.org/10.1016/J.GEODERMA.2018.02.015

Chapter 11
Advancements in Microplastic Removal Techniques and Sustainable Solutions for Plastic Reduction

Dharaneesh Arunachalam Balasubramaniam, Anushka Upamali Rajapaksha, Meththika Vithanage, Digvijay Kumar, Ricky Rajamanickam, and Rangabhashiyam Selvasembian

Abstract The release of microplastics into the ecosystem with wastewater treatment facilities is a growing problem across the world. Wastewater treatment Plants (WWTPs) discharge microplastics in terrestrial and aquatic systems, mostly from the fabric, laundry, and cosmetics sectors. Despite substantial studies on microplastics (MPs) in the natural environment, removal tactics, and WWTP management plans that emphasize their environmental impact, not much is understood concerning MPs' destiny and behavior during diverse treatment procedures. MPs are affected by treatment methods in varied ways due to their variable physical and chemical properties, resulting in differential removal efficacy. MPs recovered from WWTPs can build in soil and affect ecosystems on land. Few researches have looked at the cost, energy consumption, and alternatives of large-scale microplastic cleanup using contemporary treatment technologies. To protect aquatic and terrestrial environments from microplastic pollution, targeted and cost-effective management strategies must close knowledge gaps. This chapter summarizes recent advances in microplastic removal methods and their efficiencies. Classical treatment method, electrocoagulation method, magnetic extraction, biological process, membrane filtration, pulse clarification, and metal organic frameworks are discussed for microplastic removal. To minimize MPs, alternatives to plastics and severe limitations, such as microplastic waste conversion, should be addressed. MPs should also be managed by policy implementation and awareness.

D. A. Balasubramaniam
Department of Civil and Environmental Engineering, University of Windsor, Windsor, ON N9B 3P4, Canada

A. U. Rajapaksha · M. Vithanage
Ecosphere Resilience Research Centre, Faculty of Applied Sciences, University of Sri Jayewardenepura, Nugegoda 10250, Sri Lanka

D. Kumar · R. Rajamanickam · R. Selvasembian (✉)
Department of Environmental Science and Engineering, School of Engineering and Sciences, SRM University-AP, Amaravati 522240, Andhra Pradesh, India
e-mail: rambhashiyam@gmail.com

A. K. Mishra et al. (eds.), *Microplastics*, https://doi.org/10.1007/978-981-97-6461-7_11

Keywords Emerging contaminants · Plastic pollution · WWTPs · Sustainable economy

1 Introduction

The plastics manufacturing began in the 1950s in the United States of America by DuPont de Nemours, Inc. Up to now, the virgin plastics production has been evaluated at 8300 million metric tons. Plastics are categorized by their size: nano plastics (<100 nm), MPs (<5 mm), mesoplastics (5–25 mm), and macroplastics (>25 mm) (Kasavan et al., 2021). Around 80% of the plastic waste in the world's oceans comes from land-based sources, whereas just 20% comes from marine sources (Li et al., 2016). Approximately 220 different species of aquatic animals, including mussels, lobsters, seagulls, oysters, etc., have been shown to have MPs (Jinhui et al., 2019). Additionally, MPs have a high likelihood of being consumed or inhaled by a variety of creatures, which will impair them because of their tiny sizes and difficulties degrading in the environment. Environmental pollution is largely caused by MPs, which may be found in a wide range of components, forms, and morphologies. The percentage of MPs in the total mass of deposited plastic in the globe is predicted to be 13.2% by 2060 (Sharma et al., 2021). MPs may operate as a vector for the transportation of hydrophobic organic compounds and heavy metals in the environment by adsorbing these substances on their surface (Wu et al., 2020).

Developing effective methods to eliminate MPs from the environment is critical to avoiding the numerous health risks connected with their contamination. Although several common treatment procedures may be employed to eliminate MPs from water waste, their effectiveness remains a difficulty. Furthermore, in order for these technologies to be financially sustainable, they must be made widely available at low cost. The physical elimination of MPs using autonomous micromotors has still to be validated for practical applications. Electrocoagulation, or EC, approaches have yielded encouraging outcomes. However, the power need is a significant problem for applications that are large in scale (Perren et al., 2018). Different types of filtration systems may be effective in removing MPs. These devices need to be capable enough of being incorporated into local WWTPs, allowing MPs to be eliminated with other contaminants in the water. Furthermore, because the origins and destinations of MPs are diverse, no single approach will ever be capable to deliver a comprehensive solution. Depending on the origin of MPs, procedures such as biological trapping and filtering may be utilized.

This study found that MPs constitute a substantial hazard to the environments due to insufficient treatment and continued release into the environment. This study found that MPs constitute a substantial hazard to the environments due to insufficient treatment and continued release into the environment. Few articles have addressed the pretreatment of MPs, removal techniques in WWTPs, related difficulties, and their possible application as solutions for averting environmental threats, despite the fact that numerous reviews and research studies have thoroughly examined this subject and emphasized the environmental effects. In light of these facts, this study sheds light on the way MPs react with various treatment technologies due to their different chemical and physical characteristics, leading to differing removal efficiency. It also goes beyond examining current methods for the pretreatment processes, elimination, and administration of MPs in WWTPs. This thorough analysis of the literature offers a deeper understanding of the issues and possibilities for efficient microplastic management. This research emphasizes the significance of assessing the environmental effects of MPs extracted from wastewater treatment facilities, in contrast to many other investigations that have concentrated only on removal.

Regardless of the metadata on the detection, assimilation, identification and quantitative determination of MPs can be found in a wide range of articles (Pivokonsky et al., 2018; Reddy et al., 2006; Xiong et al., 2019), more extensive findings on MPs that include their emergence, transmission, sampling, pre-treatment, and classification in various environmental system, along with ecological impacts on marine and soil life forms including humans, are still required.. Therefore, this chapter aims to focus on discussing various sources of microplastic contamination and critically explore the treatment methods. Such as membrane filtration, electrocoagulation, pulse clarification, magnetic extraction, biological process, classical treatment methods, and metal organic frameworks. This brief study also emphasized the removal efficiency of each method, their advantages and disadvantages in detail. Microplastic mitigation methods such as Reducing the usage of plastics, education and awareness, improvement of waste disposal methods, avoiding littering, and recycling concepts were also discussed in depth.

2 Sources of Microplastics

2.1 Sources of Microplastics in Terrestrial Area

2.1.1 Plastic Mulching

Mulch film, composed of polyvinyl chloride (PVC) and polyethylene (PE), has emerged as a prominent practice in global agricultural production due to its considerable monetary advantage (Huang et al., 2020). Plastic mulch covers approximately 20 million hectares of cropland worldwide, with China accounting for the greater part. The extensive use of plastic mulching in croplands shows that it is one of the world's leading sources of MPs (Huang et al., 2020; Weber & Opp, 2020). Substantial bits of plastic mulch particles that remain on the farmland's soil are naturally exposed to UV light, generating photocatalytic degradation and fragility (Astner et al., 2019; Huang et al., 2020; Li et al., 2020). The integration of MPs into subsoil is owing to the hard surface layer of the MPs, which has been heavily worn and hence allows clay minerals to easily adhere (Horton et al., 2017; Li et al., 2020). In Europe and North America, between 0.11 million and 0.73 million tons of MPs are deposited into

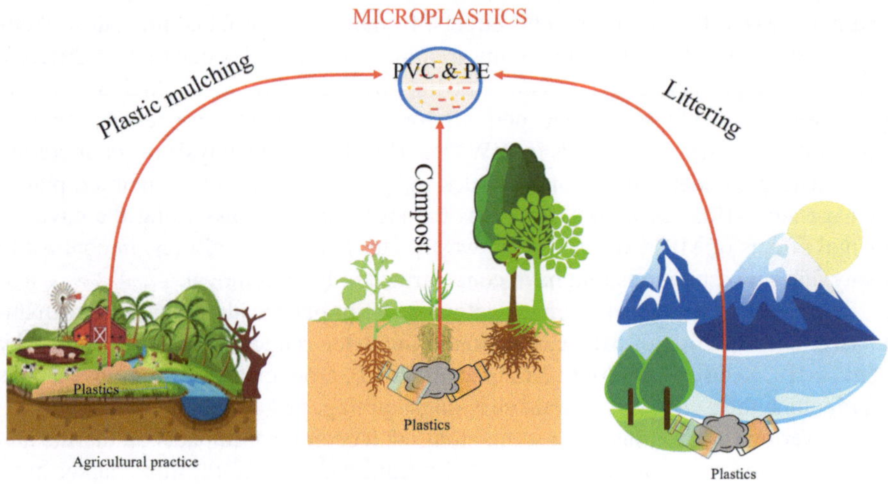

Fig. 1 Sources of Microplastics in Terrestrial Area

agricultural areas each year, respectively (Nizzetto et al., 2016), which was supposed to be greater than the global annual deposit of microplastic in the ocean. Because of its widespread usage, insufficiency of understanding in plastic waste management among farmer groups, and an absence of adequate waste management in mulching plastics, plastic mulching can be a substantial contributor of microplastic formation in the terrestrial environment as depicted in Fig. 1 (Qadeer et al., 2021).

2.1.2 Littering

Because of the recent advanced development of plastics and unforeseen or ineffective management framework, there is a wide diversity and spectrum of microplastic waste in the soil ecosystem. A large amount of plastic waste has the potential to produce MPs in aquatic habitats (Akdogan & Guven, 2019; Bläsing & Amelung, 2018; Kumar et al., 2020). A predicted 6.3 billion tons of plastic waste were created globally over a 65-year period beginning in 1950, with 4.97 billion tons destined for landfills and natural ecosystems (Geyer et al., 2017). The most prevalent form of microplastic particles is that of fibers, whereas the most prominent polymers in a litter are polyethylene, polypropylene, and polystyrene (Tata et al., 2020). Many variables, including soil features (e.g., soil fissures, porosity, and aggregation), lifeforms in soil (e.g., fungus, bacterium, vegetation, and other organisms), and anthropogenic impacts such as cropping and irrigation, can impact the lateral and vertical movement of MPs in soils (Al-Jaibachi et al., 2019; Li et al., 2020). Although, there is presently no descriptive approach to the quantitative estimation of MPs present in the soil as a result of garbage disposal.

2.1.3 Compost

Composting soil can potentially offer a good conduit for MPs to enter the soil. Typically, biodegradable material is used as fertilizer on farms after being composted and otherwise treated to reuse minerals, micronutrients, and other elements. However, it has also been noted that composts made of organic waste, in particular, include plastics as a result of ineffective garbage disposal and management (Bläsing & Amelung, 2018). Hazardous trace metals are detected in MPs because they were used in the manufacturing process of the polymers and can wind up in compost. Additionally, MPs present in compost can act as a platform for the harmful adsorption of metals, allowing them to be transferred to croplands and absorbed by plants (Rolsky et al., 2020; Vithanage et al., 2021). The surface structure and chemical contents are critical characteristics for determining the degree of deterioration in MPs (Li et al., 2020). (Wei et al., 2019) reported that Bisphenol-A and radical oxidants released by MPs of Polyethylene and Poly Vinyl Chloride might diminish the population of activated sludge anaerobic digesting microorganisms. It has been shown that organic fertilizers made from biological waste matter include MPs, where the sizes of particles are higher than 1 mm (Weithmann et al., 2018).

2.2 Sources of Microplastics in Water

2.2.1 Sewage Sludge

MPs contamination occurs as a result of the use of sewage sludge and stormwater. MPs from drainage enter wastewater treatment systems via a variety of routes as depicted in Fig. 2 (Gao et al., 2020; Jian et al., 2020). The discharge of masterbatch (plastics) from the plastic production facility, the processing of microbeads from personal care and intimate hygiene products, the laundering of garments fabrics that release plastic and polymer fibers, and the production of MPs from synthetic rubber tyres from automobiles all transmitted to sewage systems (Gao et al., 2020). MPs are subjected to physical and complex chemical deteriorating processes in the Wastewater treatment facility, which might lead to the adsorption of hazardous contaminants (Li et al., 2018). The sewage treatment techniques used will have an impact on the concentration of MPs in WWTP effluent. However, studies have revealed that tertiary treatment in some WWTPs did not further lower the amount of microplastic particles in the effluent (Mintenig et al., 2017; Sun et al., 2019). The scarcity of sewage treatment plants, exacerbated by anarchic urban development centered on the coast, adds to the deterioration of both the plants, hence explains the presence of MPs indisputably (Ouali et al., 2018). Table 1 illustrates the MP concentration in the sewage sludge from various places across the world. These data suggest that, sewage sludge is one of the contributors to the source of MPs in water. A portion of them are released from the drainage system, but the majority of MPs are segregated throughout the sewage disposal method and eventually enter in sewage sludge.

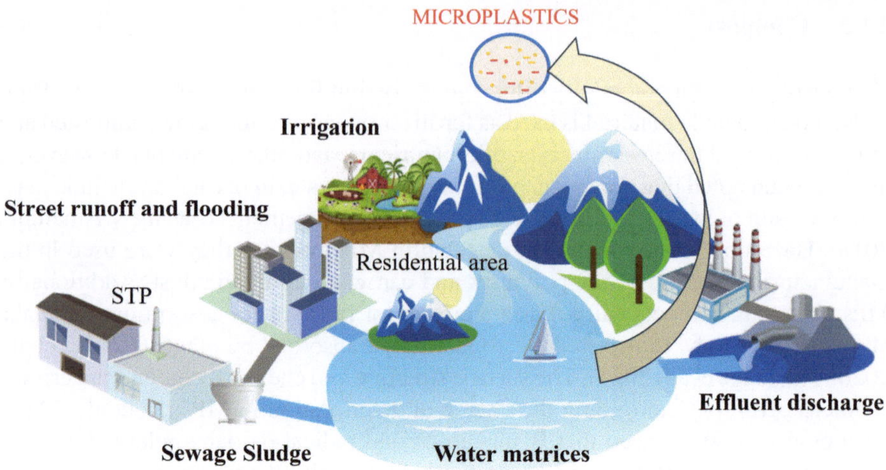

Fig. 2 Sources of MPs in Water

Table 1 Concentration of MPs in different Wastewater Treatment plants

Flowrate (m^3)/day	Type of sludge	Concentration of MPs (Per kg of sludge)	Microplastic size range	Location	References
300,000	Dry sludge Heated dry sludge	$(133.00 \pm 59.00) \times 10^3$ $(101.00 \pm 19.00) \times 10^3$	36–472 μm 29–222μm	Madrid, Spain	Edo et al. (2020)
80,000	Primary sludge	67×10^3	(1)300–5000 μm	Falconara Marittima, Italy	Pittura et al. (2021)
	Activated sludge	5.3×10^3			
	Dewatered sludge	4.74×10^3			
20,000	Dewatered sludge	$(240.3 \pm 31.4) \times 10^3$	60–4200 μm	Wuhan City, China	Liu et al. (2021)
190,000	Activated sludge	7.91 ± 0.44	1.5–1000 μm	New South Wales, Australia	Raju et al. (2020)
1,600,000	Primary sludge	$(14.9 \pm 6.30) \times 10^3$	>1 μm	Vancouver, Canada	Gies et al. (2018)
	Secondary sludge	$(4.40 \pm 2.80) \times 10^3$			

2.2.2 Irrigation

Numerous studies have confirmed that agronomic irrigation water contains MPs (Di & Wang, 2018; Jian et al., 2020). Despite the fact that a considerable number of MPs may be eliminated during the sewage treatment method, high quantities of MPs still remain in the filtered sewage (Ziajahromi et al., 2017). In the practice of "puddling," a tractor-driven rotary plough is used to plough a field that has been soaked with irrigation water. This technique has the ability to easily release soil-accumulated MPs, particularly microcapsules, into irrigation water (Katsumi et al., 2021). (Katsumi et al., 2020) reported that, the concentration of MPs in the irrigation season (9423–74,869 pieces/m^2) is far higher than the non-irrigation season (77–1477 pieces/m^2). High quantities of MPs have been shown to be present in rivers, lakes, reservoirs, and groundwater thus far, according to thorough and comprehensive research (Koelmans et al., 2019; Yang et al., 2021). As a result of irrigation, the MPs that are present in the water sources will be carried to the surface and get deposited.

2.2.3 Street Runoff and Flooding

Street runoff and floods, in addition to deliberate irrigation, are key channels for the transport and accumulation of macroplastics and MPs (Bläsing & Amelung, 2018). Street runoff and floods can result in improper disposal of garbage near roadways and tire degradation (Kumar et al., 2020). Several of them have been inherently MPs, while others are progressively changed into MPs as a result of persistent environmental exposures (Yang et al., 2021).

(Werbowski et al., 2021) investigated the Stormwater discharge from 12 San Francisco Bay Area locations which was then tested for MPs and anthropogenic particles following six storm events between December 2016 and November 2018. The land usage and size of the sites emptying into San Francisco Bay vary. During storms, composite depth-integrated samples were collected, with 3–8 samples taken over the hydrograph. For collection, an ISCO sampler or a container and winch mechanism were employed. For quality control, field blanks and copies were used. The San Francisco Estuary Institute prepared the samples before shipping them to the University of Toronto for laboratory investigation, which included the injection of isopropyl alcohol to suppress algae development. The research also includes a second case study of a rain garden to evaluate its efficiency in decreasing MPs in stormwater runoff.

The findings revealed the presence of MPs and other man-made particles in all samples. Anthropogenic particles were consistently seen and recognized based on morphology, color, size, and kind. The particle concentrations in each sample ranged from 1.1 to 24.6 particles per liter of stormwater. MPs made up around 39% of all particles which was examined using spectroscopy. This highlights that stormwater is an important source of MPs.

3 Methods for Treating Microplastic Affected Samples

Many study findings have shown the widespread presence of MPs in marine and freshwater habitats. Urban wastewater treatment plant (WWTP) discharges are a major source of MPs in the marine environment (Gies et al., 2018; Mintenig et al., 2017; Ziajahromi et al., 2017). While WWTPs have not been specifically intended or optimized for the deliberate elimination of MPs (Carr et al., 2016; Lares et al., 2018; Ziajahromi et al., 2017). The removal effectiveness of MPs in WWTPs will likely differ in accordance with the method of treatment and the physiochemical characteristics of the polymer in consideration. Although several reviews have been published recently about the processing of plastics in wastewater (Sun et al., 2019; Zhang et al., 2020), only a few computed the effectiveness of removal for MPs for pertinent treatment methods based on the body of research that was accessible that are often encountered in wastewater.

3.1 Classical Treatment Methods

Numerous wastewater treatment facilities implement coagulation and agglomeration methodologies to foster the formation of larger particles from pollutants, thereby enhancing the efficiency of their subsequent removal processes (Hu et al., 2012; Lee et al., 2012). These procedures use additional coagulants, Fe- and Al-based salts, and uptake-complexation mechanisms to bind minute particles by the development of solid bonds among suspended particles (Chorghe et al., 2017). These mechanisms employ Fe- and Al-based salts as well as some other coagulants to attach small particles through uptake-complexation pathways triggered by the exchange of ligands, resulting in the formation of strong chemical bonds between suspended particles.

The removal of polyethylene MPs employing the coagulants of iron and aluminum along with the process of ultrafiltration was studied by Ariza-Tarazona et al. (2019). Various concentrations of Fe^{3+} and Al^{3+} were experimented out and it is concluded that the salt coagulant of Al^{3+} has better performance than Fe^{3+}. The pH of the solution plays a major role in affecting the removal efficiency. Increase in pH reduces the removal efficiency considerably especially for MPs with a diameter less than 0.5 mm. For the above drawback, the usage of polyacrylamide (PAM), an augmenting coagulating agent, significantly improved the removal effectiveness of smaller MPs compared to macro particles. A surge of 61% in removal efficiency is reported when using PAM for the removal of smaller MPs, while only a 14% increase is observed in the larger MPs. The removal of MPs was also investigated using the same approach but with a $FeCl_3.6H_2O$ coagulation agent. Neutral PH favors the removal of smaller MPs and when the PH is substantially increased, the removal efficiency is reported to be increased considerably (Ariza-Tarazona et al., 2019). With traditional treatment methods showing a positive impact on the removal of MPs, extensive research should be done to improve the mechanisms and to cover the existing drawbacks in the current method employed.

3.2 Electrocoagulation Method

Due to treatment bias, contemporary treatment facilities can eliminate the overwhelming bulk of MPs (>90%), however, some MPs can still elude the treatment plant and end out in the ocean with the wastewater (Ziajahromi et al., 2017). Clearly, a creative, economical, and energy-efficient solution is essential to substitute for the current three-step treatment process and stop the scraping of MPs in wastewater (Perren et al., 2018). An efficient three-step wastewater management procedure that doesn't depend on the chemicals or microorganisms used in traditional activated sludge or general chemistry coagulation is provided by electrocoagulation, a type of electrochemical technology (Behbahani et al., 2011). Electrocoagulation is a sophisticated process wherein electrodes generate cations in the presence of an electric field. Environmental friendliness, cheap operational cost, energy effectiveness, sludge reduction, simplicity of automation, and cost efficiency are all perks of electrochemical processes (Murthy & Parmar, 2011).

Water/wastewater treatment has shown success with electrocoagulation technology. The two primary flocculation processes are charge neutralization and salvage flocculation, while pH range and applied voltage are also important characteristics. Because Al^{3+} and Fe^{3+} have a positive electrical charge throughout hydrolysis, the features of flocs play a prominent part in MP removal during the coagulation process (Perren et al., 2018). The removal effectiveness and processes of 4 main MPs by EC process utilizing Al and Fe anodes were reported. According to the existing data, the impact of the aluminum anode electrocoagulation method on MPs removal was notably significant (>98.6%). Recent studies also showed that the electrocoagulation process had a greater impact on removing fiber MPs from wastewater (>99%). In the field of MPs research in the past, fiber MPs were commonly disregarded (Shen et al., 2022). Electrocoagulation technology also has several issues that cannot be overlooked, which may hinder the technology's continued growth in microplastic removal, primarily for the four factors listed below (1) Strong conductivity is required. (2) The anode must be replaced on a constant schedule. (3) Metal ions that persist. (4) Passivation of anodes (Perren et al., 2018).

Addressing these issues may further the chances of deploying the technology for wider use.

3.3 Magnetic Extraction

Recovering MPs (MPs) of small dimensions, specifically 150 μm, presents a unique challenge. The physical removal of these smaller MPs is intricate, and the efficacy of existing extraction methods is subject to scrutiny (Grbic et al., 2019). To complement current methods and facilitate the extraction of tiny MPs, researchers created and showed an effective method for removing and isolating MP particles from a variety of environmental matrices (Mai et al., 2018; Perren et al., 2018; Talvitie et al., 2017). Fe nanoparticles were selected because of their significant surface-to-volume ratio, relatively inexpensive, and ferromagnetic characteristics (Reinhard, 1987). To provide a proof of concept, spike and recovery tests were done on three different fractions utilizing this magnetic extraction technique: (1) Big MPs (from 1 to 8 mm), (2) intermediate MPs (200 μm–1 mm), and (3) small MPs (<20 μm).

Ultimately, a prototype of the system was demonstrated over a wide variety of MP sizes, along with the lowest size fraction. This is significant considering researchers are continuing to analyze MPs in the ecosystem at decreasing sizes (Enders et al., 2017). As a result, studies have focused in effectively extracting tiny MPs (Mintenig et al., 2018). Toward the authors' knowledge (Grbic et al., 2019), the lowest size variations of polyethylene and polystyrene investigated utilizing density methods are 200–400 μm and 80–309 μm (Quinn et al., 2017), correspondingly, with both attaining >90% recovery. Digestion yielded a 93.6% removal of PS spheres of 10 μm (Claessens et al., 2013). An electrostatic separator was investigated on parched PE and PS up to 63 μm (Fuller & Gautam, 2016), as well as pressured fluid extraction on HDPE of 50 μm, with optimistic recoveries (>80%). The performance of such tiny MPs (20 m) is unclear, despite the fact that remarkable recovery percentages have been seen with bigger MPs employing a variety of techniques (Grbic et al., 2019) reported that 93% recovery rates for smaller MPs, 84% recovery rates for medium MPs and 92% recovery rates for larger MPs. No other single removal method had an impressive recovery rate considering all 3 size fractions of MPs.

There are still limits to magnetic extraction methods. The stated outcomes for intermediate MPs are, on aggregate, poorer than those for small and big MPs. This might be due to the medium-sized fraction's reduced surface-to-volume ratio. The recovery rate for sediments might even be decreased due to soil debris inhibiting the nanoparticles from contacting the polymer (Grbic et al., 2019). Furthermore, if lipophilic chemicals or lifeforms (for example, fat content in fish samples) were found in the soil sample, indiscriminate interaction of nanoparticles likely diminishes their efficiency. As a result, once the specimen has been cleaned, magnetic extraction is an excellent follow-up to density separation or digestion. Furthermore, this approach appears to be quite suitable for rapid extraction in clean specimens (Wang et al., 2015). Magnetic extraction reportedly causes highly delicate MPs to break. In future research, the magnetism strength might be tailored to the polymers of relevance, using a considerably weaker force of magnetism for MPs prone to breaking. However, consistent collection might be done to reduce magnet interaction. Ultimately, the approach was evaluated under constrained settings, and the extraction of MPs was primitive in comparison to that of a constructed apparatus. Nonetheless, the proof-of-concept trials suggest that we can get high recoveries.

3.4 Biological Process

PET (polyethylene terephthalate) is a widely used material. Because of its impact resilience, low weight, and exceptional clarity, it is mostly used in bottles (Andrady & Neal, 2009). PET use has risen throughout the years due to its superior structural qualities and durability. PET trash has now become a significant source of plastic pollution as the usage has increased. As a result, many ways of treating or recycling PET have been studied. Engineering procedures for PET segregation and chemical mechanisms for PET breakdown, for example, were documented, leading to a rise in the net PET recycled content (Choudhary et al., 2019; Sinha et al., 2010; Welle, 2011). Nevertheless, primarily clean PET materials free of leftover foodstuffs or residues can be processed by the methods, hence the recycling process has fallen dramatically (Taniguchi et al., 2019). Presently, the majority of non-recyclable PET trash is burnt, which emits a significant quantity of air pollution such as CO_2 (Chen et al., 2020a, b; Lin & Yang, 2007). As a result, biological solutions for plastic pollution have gained popularity. Scientists have confirmed 27 enzymes can destroy synthetic composites of polymeric materials, with the majority of the enzymes being Esterase and Lipase (Danso et al., 2018; Müller et al., 2005). Furthermore, in several circumstances, plastic breakdown is not accomplished by a singular enzyme; rather, numerous enzymes and biochemical functions must be employed (Kong et al., 2019; Tokiwa et al., 2009; Yoshida et al., 2016). Ideonella sakaiensis 201-F6, a bacteria that can breakdown PET and utilize it as a single source of carbon, was discovered in PET waste in 2016 (Yoshida et al., 2016). Researchers have identified two enzymes, which exhibits PET degrading ability, namely, PETase—PET hydrolase and MHETase-mono(2-hydroxyethyl) terephthalic acid hydrolase (Austin et al., 2018; Yoshida et al., 2016). Non-toxic monomers are produced as a result of these enzymes hydrolyzing PET. Many investigations have been conducted since PETase's inception in trying to confirm its framework to improve its action (Austin et al., 2018; Han et al., 2017; Joo et al., 2018). PETase synthesis in bacterial systems was also investigated for its possible utility in the biological recycling of PET (Chen et al., 2020a, b; Huang et al., 2018). For the synthesis of PETase, bacterial systems have several advantages: fast growth pace, cheap cost, ease of modification, well-established genomic technologies, and so on. Bacteria, on the other hand, may be regarded as pollution due to endotoxins or the necessity for a rich carbon supply for development. Furthermore, if bacteria grows rapidly and spreads into the ecosystem, it poses a concern (Kim et al., 2020).

Microalgae is far more appropriate for ecological-oriented applications since it does not demand carbon sources amid photoautotrophic circumstances and does not typically contain endotoxins (Yan et al., 2016). Chlamydomonas reinhardtii is a photosynthetic unicellular microbe with several benefits as a model species. C. reinhardtii is acceptable for ecologically sustainable applications since it is generally recognised as safe (GRAS) (Harris, 2001; Kim et al., 2020). PET MPs scrapped from PET bottles are exposed to PETase enzyme. The conversion time for this process was generally reported to be 4 weeks and a conversion percentage was found to be 35.17%.

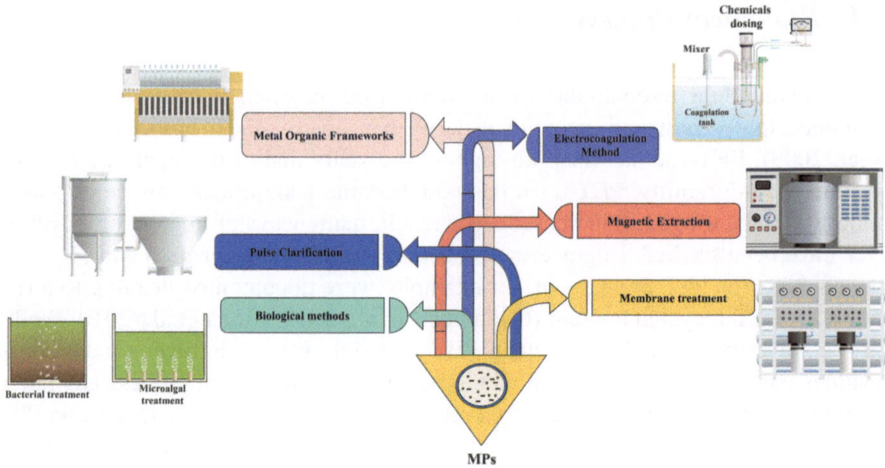

Fig. 3 Methods for treating microplastic from the samples

One of the primary reasons for this low conversion rate was due to the coated pigment found in the PET sample bottles (Kim et al., 2020).

These findings imply that microalgae might be used as a biological remediation technique for plastic pollution, particularly in aquatic and terrestrial ecosystems. Despite the lower conversion rate of this research, it holds out hope for a successful future in the elimination of MPs from the ecosystem. This unique method does not extract MPs rather it provides a permanent solution to convert the MPs into something less harmful. Figure 3 provides the overview of the treatment methods available for treating MPs.

3.5 Membrane Filtration

Due to its low power consumption, increased removal rate, and small plant footprint, ultrafiltration (UF) is a practical water treatment alternative for producing high-quality drinking water at a low cost (Jambeck et al., 2015; Moslehyani et al., 2019). Presently, one of the main water treatment procedures utilized in contemporary water treatment facilities is ultrafiltration, which shows a significant removal of organic contaminants when combined with the coagulation phase. However, such techniques are not well-suited for removing the MPs that persist in the resultant effluents (Talvitie et al., 2017). Nonetheless, the alarming concentrations of microplastic within freshwater ecosystems need a thorough examination of the characteristics of microplastic elimination during UF and coagulation procedures, especially as both are water treatment methods required for the production of drinkable water (XIA et al., 2007). (Ma et al., 2013) highlighted how polyethylene (PE) may be removed from water using ultrafiltration and coagulation techniques with the use of an activator with an iron-based activator. PE is by far the most prevalent plastic pollutant in water, and because of its density, which is very similar to that of water (0.92–0.97 g/cm^3), it can be challenging to remove utilizing wastewater treatment systems. Following coagulation, there was a poor removal efficiency of PE fragments (less than 15%), showing that the solo coagulation procedure was unsuccessful for microplastic elimination. Conversely, when polyacrylamide (PAM) was introduced to improve coagulation capability, the maximum removal of tiny PE (d 0.5 mm) rose dramatically from 13 to 91%.

Finally, while UF may be used to completely remove PE fragments in general, further research is required to fully comprehend how the existence of plastic particles influences the development of cake layers and, ultimately, fouling.

Dynamic Membrane Solution has recently emerged as an appealing innovation for treating wastewater (Saleem et al., 2017). This approach is focused on the production of a cake layer (DM) that works as a supplementary membrane/barrier formed when particulates as well as other foulants in waste effluent are processed through a supportive membrane. MBR's purpose in MP treatment is to reduce solution diversity by organic matter microbial degradation, allowing for MP filtration and subsequent treatment. In theory, the process starts as soon as a flow that has been previously treated reaches the bioreactor, which is where the function of organic material biodegradation is carried out. The resulting combined liquor is then pushed through a semi-crossflow filtering system for extraction. The MP is accumulated in the retentate flow as a result of the membrane technology (Talvitie et al., 2017). In comparison to other sophisticated treatment procedures, the employed MBR demonstrated a significant enhancement in MP elimination (99%), superior final effluent purity, and a great chance for limiting the number of process steps by substituting traditional secondary clarifiers (Talvitie et al., 2017).

3.6 Pulse Clarification

Another method for removing MPs from wastewater treatment facilities as well as additional treatment facilities is pulse clarification. It has been discovered that pulse clarification and filtering may remove MPs with an 85% removal efficiency (Sarkar et al., 2021). The procedure entails confining microscopic fragments in the coagulated sludge blanket. The pulse keeps the sludge layer from constricting, minimizing trapping. Finally, screening adds to the removal of MPs. However, there are numerous inherent disadvantages to filtering techniques, including: (1) the method is slow, with impending pore obstruction; (2) the process runs under pressure, thereby rendering it costly; and (3) membrane replenishment is an intensive procedure. Furthermore, while normal wastewater treatment procedures can help remove MPs to some extent, none of these procedures are exclusively focused on MP removal or breakdown.

3.7 *Metal–Organic Frameworks*

Metal–organic frameworks (MOFs) are heterogeneous chemical substances or structures formed by the combination of metals with ligands that are organic in nature. Chemical components with a substantial surface area, porosity, and diverse properties have been proven to help in the trapping of a variety of pollutants (Gnanasekaran et al., 2021; Wan et al., 2022; Zhang et al., 2019).

To be successful at trapping MPs, the material needs to possess a porous surface, an adequate structure for trapping the contaminants, and an elevated degree of resilience. MOFs' outstanding properties make them an ideal material for adsorption for MPs absorption. Several zirconium MOF-based foams have been developed for eliminating simulated MPs from saltwater or aquatic environments. Interpenetrated pores, outstanding uniformity, and exceptional endurance were identified in these materials. The UiO-66-OH@MF-3 was claimed to function better than others, with an effective removal rate of $95.5 \pm 1.2\%$ and a greater recycling efficiency during large-quantity filtration trials. This research also offered a laboratory-scale autonomous solar-powered filtration device. As a result, the excellent performance of MOF foam components paired with the unique notion of automated filtration mechanisms may aid in the development of creative techniques for eliminating MPs (Chen et al., 2020a, b).

Other studies have shown the practicality of MOFs for facilitating the elimination of MPs, among which is ZIF-67 for the elimination of PS MPs from an aqueous environment (Wan et al., 2022). For the elimination of MPs adsorbing dye pollutants from textile effluent PSF/MIL-100 (Fe)) was used in this case (Gnanasekaran et al., 2021). Table 2 summarizes the advantages and disadvantages of MPs treatment.

4 Microplastic Mitigation: Proactive Environmental Controls

4.1 *Reducing the Usage of Plastics*

After efficient production, reducing wasteful consumption is a wise choice, but it may be difficult due to concerns about food security and inconvenient availability (Beitzen-Heineke et al., 2017). However, it is still possible to minimize unnecessary packaging (such as double packaging) or choose eco-friendly alternatives. Long-term strategies for reducing plastic usage include official (in schools) and informal (news, clean-ups) teaching about the impact on the environment of customer perception. This may encourage people to choose products without microbeads, which could be facilitated by accurate labeling (Chang, 2015; Santos et al., 2005). Organizations would be required to adjust their products in response to heightened consumer interest in alternatives devoid of plastics. Yet, these possibilities would be restricted, necessitating the implementation of strategic development approaches. Enforcement

Table 2 Comparison of different MP treatment methods

Removal method	Advantages	Disadvantages	Mean removal efficiency (%)	References
Traditional treatment process (coagulation and flocculation)	Efficient in removing smaller MPs	Highly uneconomical	WTP 1—70% WTP 2—81% WTP 3—83%	Pivokonsky et al. (2018)
Magnetic extraction	Fe nanoparticles-low cost, higher specific surface area, and ferromagnetic properties	This process causes brittle MPs to fragment—nanoplastics	Large MPs: (1 mm-8 mm) PP—96% ± 6 PVC—92% ± 7 PRT—74% ± 9 Medium MPs:(200 μm–1 mm) [PP, PU, PVC, PS & PE] RO water sample—59–100% Sediment sample—49–90% Small MPs: (<20 μm) PE—96% ± 13% PS—88% ± 17%	Grbic et al. (2019)
Biological process (Green algae—*C. reinhadtii*)	*C. reinhadtii* biological agent used has high growth, low cost and ease of manipulation	Low conversion rate	35.17%	Kim et al. (2020)
Membrane seperation	Membrane technology has strong selectivity and separation	Membrane fouling and high cost	90%	Ziajahromi et al. (2017)
Electrocoagulation	It doesn't rely on chemical compounds or microorganisms that are utilised in other procedures.	Anode passivation	99%	Perren et al. (2018)
Pulse clarification	Substantial removal efficiency	Slow and expensive method	85%	Sarkar et al. (2021)
Metal organic framework	Greater removal efficiency, suitable for removing MPs from saltwater and freshwater aquatic environments	Micron and nano meter sized MPs were difficult to remove. No data on testing under various conditions	95.5 ± 1.2%	Chen et al. (2020a, b)

WTP indicates wastewater treatment plants

PP—Polypropylene, PVC—Polyvinyl Chloride, PET—Polyethene Terephthalate, PE—Polyethylene, PS—Polystyrene measures, such as implementing taxes on consumption, imposing restrictions on advertisements, and banning single-use items, can be combined with voluntary initiatives by industries, known as corporate social responsibility (CSR) (Landon-Lane, 2018).

4.2 Education and Awareness

The internet may be used as a tool for education since social media currently has a bigger influence than traditional sources of information (Jansen et al., 2008) and because 80% of search queries attempt to retrieve information (Dabbagh & Kitsantas, 2012). When examining the trends for the keywords "MPs" and "microbeads" on search engines, blogs, and social media, the general public increased understanding of "MPs" as well as the behavior pattern consisting of news releases by media outlets that are followed by social media sharing and search engine triggers could be observed and appropriate actions could be taken to make the crisis reach the publics ear. This provides a different path for unique learning and awareness of this ecological issue (Miller, 2010).

Additionally, education and awareness programs should be successful in changing people's behavior over the long term (Santos et al., 2005), for example, by lowering littering through moral duties rather than ineffectual littering punishments (Grasmick et al., 1991). The usage of hazardous materials must be limited, littering reduced, and recycling rates increased. These practical actions might be pushed by leveraging consumer online activity. However, the existence of plastic-free alternatives is necessary for consumption decrease (Santos et al., 2005).

4.3 Improving Waste Disposal Methods

Waste management is built on the principle of the four R's: Reducing, Reusing, Recycling, and Recovery. Even if the objective is to minimize and recycle waste by intervening with manufacturing and utilization, a suitable Comprehensive Waste Management System should be employed to manage waste as a resource (Schneider & Ragossnig, 2015; Thompson et al., 2009). Packaging is infrequently used outside of high-valued goods such as electronics and automobiles (Liu et al., 2018), since it is challenging to reuse and needs recovery, organizing, and refilling. Therefore, only final waste, such as ash, must be landfilled; anything else that cannot be retrieved should only be used as feedstock or in material recovery. The environment will include less plastics thanks to effective solid waste management, which also slows down the formation of MPs (Wu et al., 2017).

However, implementing a comprehensive solid waste management system is time-consuming and costly. It is probable that emerging countries who do not have waste management systems will take an extended period to implement such complicated systems. Managing trash is a primary priority in these conditions to avoid hazards to public health and the formation of marine litter. As a result, landfills and waste incinerators can be used as principal ways of eliminating waste, gradually giving way to more ecologically friendly techniques. Marine trash is a global issue that needs global cooperation and response. Thus, global collaboration may aid in this change by exchanging information and sponsoring waste management infrastructure in developing countries.

4.4 Avoid Littering

Another contributing reason to marine litter is the inappropriate disposal of waste, frequently referred to as littering. Penalties have been used to curb littering up to this point. But, they are ineffective since they need continual monitoring (Burgess et al., 1971). Although social penalties such as shame (self-inflicted) or humiliation (socially enforced) can reduce conduct, they may be paired with positive behavior, such as offering incentives for appropriate waste disposal or emphasizing the individual's moral duty (Grasmick et al., 1991). Positive incentives such as tax rebates or deposit-refund programs are instances that might be used to increase recycling rates (Bing et al., 2014).

4.5 Recycling

The sophisticated recycling process of plastics comprises the following steps: collecting garbage that has been segregated by customers or facilities; sorting recyclables and eliminating impurities; grounding and sorting by polymer and color; and extruding every type of polymer and color into pellets; as well as selling recycled pellets to the manufacturing industry (Bing et al., 2014; Ignatyev et al., 2014; Schneider & Ragossnig, 2015; Walker & Xanthos, 2018). Waste polymer segregation is extremely difficult and may have an impact on the final quality of recovered plastics. While secondary recycling (scaling back) creates lower grade plastic to be utilized in less demanding processes from compromised plastics, such as household garbage, better-grade plastics are produced through primary recycling (closed-loop) from clean materials often produced by manufacturers. Recycled plastics should ideally be employed in robust and long-lasting applications (Gu & Ozbakkaloglu, 2016; Kazemi Najafi, 2013; Poulikakos et al., 2017; Walker & Xanthos, 2018).

Recycled plastics should ideally be employed in robust and long-lasting applications (Ignatyev et al., 2014; Walker & Xanthos, 2018). In order to enhance the qualities of concrete (Gu & Ozbakkaloglu, 2016) and asphalt (Poulikakos et al., 2017), leftover plastic can be shredded. Mixed polymers or blends of polymers and materials other than plastics (like wood) have the potential to form durable and cost-effective thermoset composites. These composites can find applications, such as in the construction of railroad sleepers (Ferdous et al., 2015). Composites, which cannot be further recycled, may lessen the requirement for polymer isolation and are much more robust for external use than their traditional ones (such as wood). Recycling is the ideal waste management strategy given the environmental consequences (Arena et al., 2003; Chilton et al., 2010). As an illustration, the recycling of PET and PE necessitates only 50% of the energy expended in the production of new polymers (Arena et al., 2003). The capacity to replace virgin polymers is limited, and excessive organic contamination may make burning more advantageous than recycling (Lazarevic et al., 2010). The environmental implications associated with recycling often arise from the utilization of non-renewable energy, shipping, and the incorporation of fillers and preservatives (Gu et al., 2017). Although, recycling is still a top focus in waste treatment. Recycling benefits the environment in a variety of ways, including the conservation of resources and energy, the reduction of pollutant emissions, the elimination of the need for landfills, the stimulation of local economies through job creation, and the reduction of commodity imports. Recycling is therefore more expensive than other options, but it also contributes to society and is not costly overall. For these reasons, nations ought to make an effort to increase their recyclability rate. By enhancing primary recycling, levying prices on unprocessed plastics, or requiring the utilization of recycled plastics in all goods, as was previously said, recycling may also become financially sustainable. However, not all plastics can be recycled; mixed, tainted, and deteriorated polymers can be utilized as biomass resources or in waste heat recovery but are not suitable for recycling.

5 Conclusions and Future Recommendations

Wastewater treatment facilities serve as both sources and destinations for MPs, posing a notable concern when released into the environment, especially aquatic ecosystems. The most studied treatment methods are membrane filtration, electro coagulation, pulse clarification, magnetic extraction, biological process, classical treatment methods, and metal organic frameworks of all the methods discussed, electrocoagulation has the higher removal efficiency (99%) but anode passivation limits its usage in a large scale. Similarly, magnetic extraction has a comparable higher removal efficiency (96%) but causes brittle MPs to fragment. Novel treatment methods such as biological process (Green algae—*C. reinhadtii*) provides a cost-effective solution but has a lower removal rate (35%). Pulse clarification (95%), membrane filtration (90%), metal organic framework (95%), and classical treatment method (83%) are some other methods with impressive removal efficiencies but have a certain disadvantage while going for a full-scale implementation. One single method isn't effective to remove MPs altogether and we need to come up with a combination of methods to overcome this scenario. Microplastic mitigation methods such as reducing the usage of plastics, education and awareness, improvement of waste disposal methods, avoiding littering and recycling concepts are recommended to reduce MP generation and contamination in the environment.

MPs have unique physical and chemical characteristics, making it crucial to study how they interact with different treatment techniques. This understanding will enhance removal efficiency and lessen MPs' environmental impact. Evaluating the environmental impact of MPs recovered from WWTPs is crucial. Research on the accumulation of MPs in soil and their influence on terrestrial ecosystems is crucial for developing mitigation strategies and environmental protection. The treatment procedure's affordability and consumption of energy are also important factors to consider. Analyzing the expense, consumption of energy, and choices of implementing modern microplastic elimination methods is crucial for wastewater management bodies to arrive at sound choices about their adoption. Further study is needed to develop and evaluate innovative microplastic removal methods. Finally, to reduce microplastic production, it's important to create and employ alternatives to plastics, as well as implement strict regulations and waste management measures.

6 Declaration of Competing Interest

The authors declare they have no known competing interests.

Acknowledgements The authors acknowledge the financial support provided by the SRM University-AP under the grant: SRMAP/URG/SEED/2023-2024/021

Author Contributions Dharaneesh Arunachalam Balasubramaniam: Writing and original draft preparation.

Anushka Upamali Rajapaksha: Writing and Review & Editing.

Meththika Vithanage: Writing and Review & Editing.

Digvijay Kumar: Writing and Review & Editing.

Ricky Rajamanickam: Review & Editing and visualization.

Rangabhashiyam Selvasembian: Methodology, Supervision, Conceptualization, Writing—Review & Editing.

References

Akdogan, Z., & Guven, B. (2019). Microplastics in the environment: A critical review of current understanding and identification of future research needs. *Environmental Pollution, 254*, 113011. https://doi.org/10.1016/J.ENVPOL.2019.113011

Al-Jaibachi, R., Cuthbert, R. N., & Callaghan, A. (2019). Examining effects of ontogenic microplastic transference on Culex mosquito mortality and adult weight. *Science of the Total Environment, 651*, 871–876. https://doi.org/10.1016/J.SCITOTENV.2018.09.236

Andrady, A. L., & Neal, M. A. (2009). Applications and societal benefits of plastics. *Philosophical Transactions of the Royal Society B: Biological Sciences, 364*(1526), 1977. https://doi.org/10.1098/RSTB.2008.0304

Arena, U., Mastellone, M. L., & Perugini, F. (2003). Life cycle assessment of a plastic packaging recycling system. *International Journal of Life Cycle Assessment, 8*(2), 92–98. https://doi.org/10.1007/BF02978432/METRICS

Ariza-Tarazona, M. C., Villarreal-Chiu, J. F., Barbieri, V., Siligardi, C., & Cedillo-González, E. I. (2019). New strategy for microplastic degradation: Green photocatalysis using a protein-based porous N-TiO 2 semiconductor. *Ceramics International, 45*(7), 9618–9624. https://doi.org/10.1016/J.CERAMINT.2018.10.208

Astner, A. F., Hayes, D. G., O'Neill, H., Evans, B. R., Pingali, S. V., Urban, V. S., & Young, T. M. (2019). Mechanical formation of micro- and nano-plastic materials for environmental studies in agricultural ecosystems. *Science of the Total Environment, 685*, 1097–1106. https://doi.org/10.1016/J.SCITOTENV.2019.06.241

Austin, H. P., Allen, M. D., Donohoe, B. S., Rorrer, N. A., Kearns, F. L., Silveira, R. L., Pollard, B. C., Dominick, G., Duman, R., Omari, K. El, Mykhaylyk, V., Wagner, A., Michener, W. E., Amore, A., Skaf, M. S., Crowley, M. F., Thorne, A. W., Johnson, C. W., Lee Woodcock, H., … Beckham, G. T. (2018). Characterization and engineering of a plastic-degrading aromatic polyesterase. *Proceedings of the National Academy of Sciences of the United States of America, 115*(19), E4350–E4357. https://doi.org/10.1073/PNAS.1718804115.

Behbahani, M., Moghaddam, M. R. A., & Arami, M. (2011). Techno-economical evaluation of fluoride removal by electrocoagulation process: Optimization through response surface methodology. *Desalination, 271*(1–3), 209–218. https://doi.org/10.1016/J.DESAL.2010.12.033

Beitzen-Heineke, E. F., Balta-Ozkan, N., & Reefke, H. (2017). The prospects of zero-packaging grocery stores to improve the social and environmental impacts of the food supply chain. *Journal of Cleaner Production, 140*, 1528–1541. https://doi.org/10.1016/J.JCLEPRO.2016.09.227

Bing, X., Bloemhof-Ruwaard, J. M., & Van Der Vorst, J. G. A. J. (2014). Sustainable reverse logistics network design for household plastic waste. *Flexible Services and Manufacturing Journal, 26*(1–2), 119–142. https://doi.org/10.1007/S10696-012-9149-0/FIGURES/5

Bläsing, M., & Amelung, W. (2018). Plastics in soil: Analytical methods and possible sources. *Science of the Total Environment, 612*, 422–435. https://doi.org/10.1016/J.SCITOTENV.2017.08.086

Burgess, R. L., Clark, R. N., & Hendee, J. C. (1971). An experimental analysis of anti-litter procedures. *Journal of Applied Behavior Analysis, 4*(2), 71. https://doi.org/10.1901/JABA.1971.4-71

Carr, S. A., Liu, J., & Tesoro, A. G. (2016). Transport and fate of microplastic particles in wastewater treatment plants. *Water Research, 91*, 174–182. https://doi.org/10.1016/J.WATRES.2016.01.002

Chang, M. (2015). Reducing microplastics from facial exfoliating cleansers in wastewater through treatment versus consumer product decisions. *Marine Pollution Bulletin, 101*(1), 330–333. https://doi.org/10.1016/J.MARPOLBUL.2015.10.074

Chen, Y. J., Chen, Y., Miao, C., Wang, Y. R., Gao, G. K., Yang, R. X., Zhu, H. J., Wang, J. H., Li, S. L., & Lan, Y. Q. (2020a). Metal–organic framework-based foams for efficient microplastics removal. *Journal of Materials Chemistry A, 8*(29), 14644–14652. https://doi.org/10.1039/D0TA04891G

Chen, Z., Wang, Y., Cheng, Y., Wang, X., Tong, S., Yang, H., & Wang, Z. (2020b). Efficient biodegradation of highly crystallized polyethylene terephthalate through cell surface display of bacterial PETase. *Science of the Total Environment, 709*, 136138. https://doi.org/10.1016/J.SCITOTENV.2019.136138

Chilton, T., Burnley, S., & Nesaratnam, S. (2010). A life cycle assessment of the closed-loop recycling and thermal recovery of post-consumer PET. *Resources, Conservation and Recycling, 54*(12), 1241–1249. https://doi.org/10.1016/J.RESCONREC.2010.04.002

Chorghe, D., Sari, M. A., & Chellam, S. (2017). Boron removal from hydraulic fracturing wastewater by aluminum and iron coagulation: Mechanisms and limitations. *Water Research, 126*, 481–487. https://doi.org/10.1016/J.WATRES.2017.09.057

Choudhary, K., Sangwan, K. S., & Goyal, D. (2019). Environment and economic impacts assessment of PET waste recycling with conventional and renewable sources of energy. *Procedia CIRP, 80*, 422–427. https://doi.org/10.1016/J.PROCIR.2019.01.096

Claessens, M., Van Cauwenberghe, L., Vandegehuchte, M. B., & Janssen, C. R. (2013). New techniques for the detection of microplastics in sediments and field collected organisms. *Marine Pollution Bulletin, 70*(1–2), 227–233. https://doi.org/10.1016/J.MARPOLBUL.2013.03.009

Dabbagh, N., & Kitsantas, A. (2012). Personal learning environments, social media, and self-regulated learning: A natural formula for connecting formal and informal learning. *The Internet and Higher Education, 15*(1), 3–8. https://doi.org/10.1016/J.IHEDUC.2011.06.002

Danso, D., Schmeisser, C., Chow, J., Zimmermann, W., Wei, R., Leggewie, C., Li, X., Hazen, T., & Streit, W. R. (2018). New insights into the function and global distribution of Polyethylene Terephthalate (PET)-degrading bacteria and enzymes in marine and terrestrial metagenomes. *Applied and Environmental Microbiology, 84*(8). https://doi.org/10.1128/AEM.02773-17.

Di, M., & Wang, J. (2018). Microplastics in surface waters and sediments of the Three Gorges Reservoir, China. *Science of the Total Environment, 616–617*, 1620–1627. https://doi.org/10.1016/J.SCITOTENV.2017.10.150

Edo, C., González-Pleiter, M., Leganés, F., Fernández-Piñas, F., & Rosal, R. (2020). Fate of microplastics in wastewater treatment plants and their environmental dispersion with effluent and sludge. *Environmental Pollution (Barking, Essex : 1987), 259*. https://doi.org/10.1016/J.ENVPOL.2019.113837.

Enders, K., Lenz, R., Beer, S., & Stedmon, C. A. (2017). Extraction of microplastic from biota: Recommended acidic digestion destroys common plastic polymers. *ICES Journal of Marine Science, 74*(1), 326–331. https://doi.org/10.1093/ICESJMS/FSW173

Ferdous, W., Manalo, A., Van Erp, G., Aravinthan, T., Kaewunruen, S., & Remennikov, A. (2015). Composite railway sleepers—Recent developments, challenges and future prospects. *Composite Structures, 134*, 158–168. https://doi.org/10.1016/J.COMPSTRUCT.2015.08.058

Fuller, S., & Gautam, A. (2016). A procedure for measuring microplastics using pressurized fluid extraction. *Environmental Science and Technology, 50*(11), 5774–5780. https://doi.org/10.1021/ACS.EST.6B00816/ASSET/IMAGES/LARGE/ES-2016-00816B_0006.JPEG

Gao, D., Li, X. Y., & Liu, H. T. (2020). Source, occurrence, migration and potential environmental risk of microplastics in sewage sludge and during sludge amendment to soil. *Science of the Total Environment, 742*, 140355. https://doi.org/10.1016/J.SCITOTENV.2020.140355

Geyer, R., Jambeck, J. R., & Law, K. L. (2017). Production, use, and fate of all plastics ever made. *Science Advances, 3*(7). https://doi.org/10.1126/SCIADV.1700782/SUPPL_FILE/1700782_SM.PDF.

Gies, E. A., LeNoble, J. L., Noël, M., Etemadifar, A., Bishay, F., Hall, E. R., & Ross, P. S. (2018). Retention of microplastics in a major secondary wastewater treatment plant in Vancouver, Canada. *Marine Pollution Bulletin, 133*, 553–561. https://doi.org/10.1016/J.MARPOLBUL.2018.06.006

Gnanasekaran, G., Arthanareeswaran, G., & Mok, Y. S. (2021). A high-flux metal-organic framework membrane (PSF/MIL-100 (Fe)) for the removal of microplastics adsorbing dye contaminants from textile wastewater. *Separation and Purification Technology, 277*, 119655. https://doi.org/10.1016/J.SEPPUR.2021.119655

Grasmick, H. G., Bursik, R. J., & Kinsey, K. A. (1991). Shame and embarrassment as deterrents to noncompliance with the law *23*(2), 233–251. https://doi.org/10.1177/0013916591232006.

Grbic, J., Nguyen, B., Guo, E., You, J. B., Sinton, D., & Rochman, C. M. (2019). Magnetic extraction of microplastics from environmental samples. *Environmental Science and Technology Letters, 6*(2), 68–72. https://doi.org/10.1021/ACS.ESTLETT.8B00671/SUPPL_FILE/EZ8B00671_SI_001.PDF

Gu, F., Guo, J., Zhang, W., Summers, P. A., & Hall, P. (2017). From waste plastics to industrial raw materials: A life cycle assessment of mechanical plastic recycling practice based on a real-world case study. *Science of the Total Environment, 601–602*, 1192–1207. https://doi.org/10.1016/J.SCITOTENV.2017.05.278

Gu, L., & Ozbakkaloglu, T. (2016). Use of recycled plastics in concrete: A critical review. *Waste Management (New York, N.Y.), 51*, 19–42. https://doi.org/10.1016/J.WASMAN.2016.03.005.

Han, X., Liu, W., Huang, J. W., Ma, J., Zheng, Y., Ko, T. P., Xu, L., Cheng, Y. S., Chen, C. C., & Guo, R. T. (2017). Structural insight into catalytic mechanism of PET hydrolase. *Nature Communications, 8*(1). https://doi.org/10.1038/S41467-017-02255-Z

Harris, E. H. (2001). Chlamydomonas as a model organism. *Annual Review of Plant Physiology and Plant Molecular Biology, 52*, 363–406. https://doi.org/10.1146/ANNUREV.ARPLANT.52.1.363

Horton, A. A., Walton, A., Spurgeon, D. J., Lahive, E., & Svendsen, C. (2017). Microplastics in freshwater and terrestrial environments: Evaluating the current understanding to identify the knowledge gaps and future research priorities. *Science of the Total Environment, 586*, 127–141. https://doi.org/10.1016/J.SCITOTENV.2017.01.190

Huang, X., Cao, L., Qin, Z., Li, S., Kong, W., & Liu, Y. (2018). Tat-independent secretion of polyethylene terephthalate hydrolase PETase in bacillus subtilis 168 mediated by its native signal peptide. *Journal of Agricultural and Food Chemistry, 66*(50), 13217–13227. https://doi.org/10.1021/ACS.JAFC.8B05038/SUPPL_FILE/JF8B05038_SI_001.PDF

Huang, Y., Liu, Q., Jia, W., Yan, C., & Wang, J. (2020). Agricultural plastic mulching as a source of microplastics in the terrestrial environment. *Environmental Pollution, 260*, 114096. https://doi.org/10.1016/J.ENVPOL.2020.114096

Hu, C., Liu, H., Chen, G., & Qu, J. (2012). Effect of aluminum speciation on arsenic removal during coagulation process. *Separation and Purification Technology, 86*, 35–40. https://doi.org/10.1016/J.SEPPUR.2011.10.017

Ignatyev, I. A., Thielemans, W., & Vander Beke, B. (2014). Recycling of polymers: A review. *Chemsuschem, 7*(6), 1579–1593. https://doi.org/10.1002/CSSC.201300898

Jambeck, J. R., Geyer, R., Wilcox, C., Siegler, T. R., Perryman, M., Andrady, A., Narayan, R., & Law, K. L. (2015). Marine pollution. Plastic waste inputs from land into the ocean. *Science (New York, N.Y.), 347*(6223), 768–771. https://doi.org/10.1126/SCIENCE.1260352.

Jansen, B. J., Booth, D. L., & Spink, A. (2008). Determining the informational, navigational, and transactional intent of Web queries. *Information Processing & Management, 44*(3), 1251–1266. https://doi.org/10.1016/J.IPM.2007.07.015

Jian, M., Zhang, Y., Yang, W., Zhou, L., Liu, S., & Xu, E. G. (2020). Occurrence and distribution of microplastics in China's largest freshwater lake system. *Chemosphere, 261*, 128186. https://doi.org/10.1016/J.CHEMOSPHERE.2020.128186

Jinhui, S., Sudong, X., Yan, N., Xia, P., Jiahao, Q., & Yongjian, X. (2019). Effects of microplastics and attached heavy metals on growth, immunity, and heavy metal accumulation in the yellow seahorse, Hippocampus Kuda Bleeker. *Marine Pollution Bulletin, 149*, 110510. https://doi.org/10.1016/J.MARPOLBUL.2019.110510

Joo, S., Cho, I. J., Seo, H., Son, H. F., Sagong, H. Y., Shin, T. J., Choi, S. Y., Lee, S. Y., & Kim, K. J. (2018). Structural insight into molecular mechanism of poly(ethylene terephthalate) degradation. *Nature Communications, 9*(1), 1–12. https://doi.org/10.1038/s41467-018-02881-1.

Kasavan, S., Yusoff, S., Rahmat Fakri, M. F., & Siron, R. (2021). Plastic pollution in water ecosystems: A bibliometric analysis from 2000 to 2020. *Journal of Cleaner Production, 313*. https://doi.org/10.1016/J.JCLEPRO.2021.127946.

Katsumi, N., Kusube, T., Nagao, S., & Okochi, H. (2020). The role of coated fertilizer used in paddy fields as a source of microplastics in the marine environment. *Marine Pollution Bulletin, 161*, 111727. https://doi.org/10.1016/J.MARPOLBUL.2020.111727

Katsumi, N., Kusube, T., Nagao, S., & Okochi, H. (2021). The input–output balance of microplastics derived from coated fertilizer in paddy fields and the timing of their discharge during the irrigation season. *Chemosphere, 279*, 130574. https://doi.org/10.1016/J.CHEMOSPHERE.2021.130574

Kazemi Najafi, S. (2013). Use of recycled plastics in wood plastic composites—A review. *Waste Management, 33*(9), 1898–1905. https://doi.org/10.1016/J.WASMAN.2013.05.017

Kim, J. W., Park, S. B., Tran, Q. G., Cho, D. H., Choi, D. Y., Lee, Y. J., & Kim, H. S. (2020). Functional expression of polyethylene terephthalate-degrading enzyme (PETase) in green microalgae. *Microbial Cell Factories, 19*(1), 1–9. https://doi.org/10.1186/S12934-020-01355-8/TABLES/1

Koelmans, A. A., Mohamed Nor, N. H., Hermsen, E., Kooi, M., Mintenig, S. M., & De France, J. (2019). Microplastics in freshwaters and drinking water: Critical review and assessment of data quality. *Water Research, 155*, 410–422. https://doi.org/10.1016/J.WATRES.2019.02.054

Kong, H. G., Kim, H. H., Chung, J. H., Jun, J. H., Lee, S., Kim, H. M., Jeon, S., Park, S. G., Bhak, J., & Ryu, C. M. (2019). The Galleria mellonella hologenome supports microbiota-independent metabolism of long-chain hydrocarbon beeswax. *Cell Reports, 26*(9), 2451-2464.e5. https://doi.org/10.1016/J.CELREP.2019.02.018

Kumar, M., Xiong, X., He, M., Tsang, D. C. W., Gupta, J., Khan, E., Harrad, S., Hou, D., Ok, Y. S., & Bolan, N. S. (2020). Microplastics as pollutants in agricultural soils. *Environmental Pollution, 265*, 114980. https://doi.org/10.1016/J.ENVPOL.2020.114980

Landon-Lane, M. (2018). Corporate social responsibility in marine plastic debris governance. *Marine Pollution Bulletin, 127*, 310–319. https://doi.org/10.1016/J.MARPOLBUL.2017.11.054

Lares, M., Ncibi, M. C., Sillanpää, M., & Sillanpää, M. (2018). Occurrence, identification and removal of microplastic particles and fibers in conventional activated sludge process and advanced MBR technology. *Water Research, 133*, 236–246. https://doi.org/10.1016/J.WATRES.2018.01.049

Lazarevic, D., Aoustin, E., Buclet, N., & Brandt, N. (2010). Plastic waste management in the context of a European recycling society: Comparing results and uncertainties in a life cycle perspective. *Resources, Conservation and Recycling, 55*(2), 246–259. https://doi.org/10.1016/J.RESCONREC.2010.09.014

Lee, K. E., Morad, N., Teng, T. T., & Poh, B. T. (2012). Development, characterization and the application of hybrid materials in coagulation/flocculation of wastewater: A review. *Chemical Engineering Journal, 203*, 370–386. https://doi.org/10.1016/J.CEJ.2012.06.109

Lin, Y. H., & Yang, M. H. (2007). Catalytic conversion of commingled polymer waste into chemicals and fuels over spent FCC commercial catalyst in a fluidised-bed reactor. *Applied Catalysis b: Environmental, 69*(3–4), 145–153. https://doi.org/10.1016/J.APCATB.2006.07.005

Liu, W., Zhang, J., Liu, H., Guo, X., Zhang, X., Yao, X., Cao, Z., & Zhang, T. (2021). A review of the removal of microplastics in global wastewater treatment plants: Characteristics and mechanisms. *Environment International, 146*, 106277. https://doi.org/10.1016/J.ENVINT.2020.106277

Liu, Z., Adams, M., & Walker, T. R. (2018). Are exports of recyclables from developed to developing countries waste pollution transfer or part of the global circular economy? *Resources, Conservation and Recycling, 136*, 22–23. https://doi.org/10.1016/J.RESCONREC.2018.04.005

Li, W. C., Tse, H. F., & Fok, L. (2016). Plastic waste in the marine environment: A review of sources, occurrence and effects. *Science of the Total Environment, 566–567*, 333–349. https://doi.org/10.1016/J.SCITOTENV.2016.05.084

Li, W., Wufuer, R., Duo, J., Wang, S., Luo, Y., Zhang, D., & Pan, X. (2020). Microplastics in agricultural soils: Extraction and characterization after different periods of polythene film mulching in an arid region. *Science of the Total Environment, 749*, 141420. https://doi.org/10.1016/J.SCITOTENV.2020.141420

Li, X., Chen, L., Mei, Q., Dong, B., Dai, X., Ding, G., & Zeng, E. Y. (2018). Microplastics in sewage sludge from the wastewater treatment plants in China. *Water Research, 142*, 75–85. https://doi.org/10.1016/J.WATRES.2018.05.034

Mai, L., Bao, L. J., Shi, L., Wong, C. S., & Zeng, E. Y. (2018). A review of methods for measuring microplastics in aquatic environments. *Environmental Science and Pollution Research International, 25*(12), 11319–11332. https://doi.org/10.1007/S11356-018-1692-0

Ma, J., Wang, Z., Zou, X., Feng, J., & Wu, Z. (2013). Microbial communities in an anaerobic dynamic membrane bioreactor (AnDMBR) for municipal wastewater treatment: Comparison of bulk sludge and cake layer. *Process Biochemistry, 48*(3), 510–516. https://doi.org/10.1016/J.PROCBIO.2013.02.003

Miller, J. D. (2010). Adult science learning in the internet era. *Curator: The Museum Journal, 53*(2), 191–208. https://doi.org/10.1111/J.2151-6952.2010.00019.X.

Mintenig, S. M., Bäuerlein, P. S., Koelmans, A. A., Dekker, S. C., & Van Wezel, A. P. (2018). Closing the gap between small and smaller: Towards a framework to analyse nano- and microplastics in aqueous environmental samples. *Environmental Science: Nano, 5*(7), 1640–1649. https://doi.org/10.1039/C8EN00186C

Mintenig, S. M., Int-Veen, I., Löder, M. G. J., Primpke, S., & Gerdts, G. (2017). Identification of microplastic in effluents of waste water treatment plants using focal plane array-based micro-Fourier-transform infrared imaging. *Water Research, 108*, 365–372. https://doi.org/10.1016/J.WATRES.2016.11.015

Moslehyani, A., Ismail, A. F., Matsuura, T., Rahman, M. A., & Goh, P. S. (2019). Recent progresses of Ultrafiltration (UF) membranes and processes in water treatment. In *Membrane Separation Principles and Applications: From Material Selection to Mechanisms and Industrial Uses* (pp. 85–110). https://doi.org/10.1016/B978-0-12-812815-2.00003-X.

Müller, R. J., Schrader, H., Profe, J., Dresler, K., & Deckwer, W. D. (2005). Enzymatic degradation of poly(ethylene terephthalate): Rapid hydrolyse using a hydrolase from T. fusca. *Macromolecular Rapid Communications, 26*(17), 1400–1405. https://doi.org/10.1002/MARC.200500410.

Murthy, Z. V. P., & Parmar, S. (2011). Removal of strontium by electrocoagulation using stainless steel and aluminum electrodes. *Desalination, 282*, 63–67. https://doi.org/10.1016/J.DESAL.2011.08.058

Nizzetto, L., Futter, M., & Langaas, S. (2016). Are agricultural soils dumps for microplastics of urban origin? *Environmental Science and Technology, 50*(20), 10777–10779. https://doi.org/10.1021/ACS.EST.6B04140/ASSET/IMAGES/LARGE/ES-2016-04140A_0001.JPEG

Ouali, N., Belabed, B. E., & Chenchouni, H. (2018). Modelling environment contamination with heavy metals in flathead grey mullet Mugil cephalus and upper sediments from north African coasts of the Mediterranean Sea. *Science of the Total Environment, 639*, 156–174. https://doi.org/10.1016/J.SCITOTENV.2018.04.377

Perren, W., Wojtasik, A., & Cai, Q. (2018). Removal of microbeads from wastewater using electrocoagulation. *ACS Omega, 3*(3), 3357–3364. https://doi.org/10.1021/ACSOMEGA.7B02037/ASSET/IMAGES/AO-2017-02037A_M013.GIF

Pittura, L., Foglia, A., Akyol, Ç., Cipolletta, G., Benedetti, M., Regoli, F., Eusebi, A. L., Sabbatini, S., Tseng, L. Y., Katsou, E., Gorbi, S., & Fatone, F. (2021). Microplastics in real wastewater treatment schemes: Comparative assessment and relevant inhibition effects on anaerobic processes. *Chemosphere, 262*, 128415. https://doi.org/10.1016/J.CHEMOSPHERE.2020.128415

Pivokonsky, M., Cermakova, L., Novotna, K., Peer, P., Cajthaml, T., & Janda, V. (2018). Occurrence of microplastics in raw and treated drinking water. *Science of the Total Environment, 643*, 1644–1651. https://doi.org/10.1016/J.SCITOTENV.2018.08.102

Poulikakos, L. D., Papadaskalopoulou, C., Hofko, B., Gschösser, F., Cannone Falchetto, A., Bueno, M., Arraigada, M., Sousa, J., Ruiz, R., Petit, C., Loizidou, M., & Partl, M. N. (2017). Harvesting the unexplored potential of European waste materials for road construction. *Resources, Conservation and Recycling, 116*, 32–44. https://doi.org/10.1016/J.RESCONREC.2016.09.008

Qadeer, A., Ajmal, Z., Usman, M., Zhao, X., & Chang, S. (2021). Agricultural plastic mulching as a potential key source of microplastic pollution in the terrestrial ecosystem and consequences. *Resources, Conservation and Recycling, 175*, 105855. https://doi.org/10.1016/j.resconrec.2021.105855

Quinn, B., Murphy, F., & Ewins, C. (2017). Validation of density separation for the rapid recovery of microplastics from sediment. *Analytical Methods, 9*(9), 1491–1498. https://doi.org/10.1039/C6AY02542K

Raju, S., Carbery, M., Kuttykattil, A., Senthirajah, K., Lundmark, A., Rogers, Z., Suresh, S. C. B., Evans, G., & Palanisami, T. (2020). Improved methodology to determine the fate and transport of microplastics in a secondary wastewater treatment plant. *Water Research, 173*, 115549. https://doi.org/10.1016/J.WATRES.2020.115549

Reddy, M. S., Basha, S., Adimurthy, S., & Ramachandraiah, G. (2006). Description of the small plastics fragments in marine sediments along the Alang-Sosiya ship-breaking yard, India. *Estuarine, Coastal and Shelf Science, 68*(3–4), 656–660. https://doi.org/10.1016/J.ECSS.2006.03.018

Reinhard, G. (1987). Surface characterization of iron and steel prior to coating. *Progress in Organic Coatings, 15*(2), 125–148. https://doi.org/10.1016/0033-0655(87)80003-1

Rolsky, C., Kelkar, V., Driver, E., & Halden, R. U. (2020). Municipal sewage sludge as a source of microplastics in the environment. *Current Opinion in Environmental Science & Health, 14*, 16–22. https://doi.org/10.1016/J.COESH.2019.12.001

Saleem, M., Alibardi, L., Cossu, R., Lavagnolo, M. C., & Spagni, A. (2017). Analysis of fouling development under dynamic membrane filtration operation. *Chemical Engineering Journal, 312*, 136–143. https://doi.org/10.1016/J.CEJ.2016.11.123

Santos, I. R., Friedrich, A. C., Wallner-Kersanach, M., & Fillmann, G. (2005). Influence of socio-economic characteristics of beach users on litter generation. *Ocean & Coastal Management, 48*(9–10), 742–752. https://doi.org/10.1016/J.OCECOAMAN.2005.08.006

Sarkar, D. J., Das Sarkar, S., Das, B. K., Praharaj, J. K., Mahajan, D. K., Purokait, B., Mohanty, T. R., Mohanty, D., Gogoi, P., Kumar, V. S., Behera, B. K., Manna, R. K., & Samanta, S. (2021). Microplastics removal efficiency of drinking water treatment plant with pulse clarifier. *Journal of Hazardous Materials, 413*. https://doi.org/10.1016/J.JHAZMAT.2021.125347

Schneider, D. R., & Ragossnig, A. M. (2015). Recycling and incineration, contradiction or coexistence? *Waste Management & Research : THe Journal of the International Solid Wastes and Public Cleansing Association, ISWA, 33*(8), 693–695. https://doi.org/10.1177/0734242X15593421

Sharma, S., Basu, S., Shetti, N. P., Nadagouda, M. N., & Aminabhavi, T. M. (2021). Microplastics in the environment: Occurrence, perils, and eradication. *Chemical Engineering Journal, 408*, 127317. https://doi.org/10.1016/j.cej.2020.127317

Shen, M., Zhang, Y., Almatrafi, E., Hu, T., Zhou, C., Song, B., Zeng, Z., & Zeng, G. (2022). Efficient removal of microplastics from wastewater by an electrocoagulation process. *Chemical Engineering Journal, 428*, 131161. https://doi.org/10.1016/J.CEJ.2021.131161

Sinha, V., Patel, M. R., & Patel, J. V. (2010). Pet waste management by chemical recycling: A review. *Journal of Polymers and the Environment, 18*(1), 8–25. https://doi.org/10.1007/S10924-008-0106-7

Sun, J., Dai, X., Wang, Q., van Loosdrecht, M. C. M., & Ni, B. J. (2019). Microplastics in wastewater treatment plants: Detection, occurrence and removal. *Water Research, 152*, 21–37. https://doi.org/10.1016/J.WATRES.2018.12.050

Talvitie, J., Mikola, A., Koistinen, A., & Setälä, O. (2017). Solutions to microplastic pollution—Removal of microplastics from wastewater effluent with advanced wastewater treatment technologies. *Water Research, 123*, 401–407. https://doi.org/10.1016/J.WATRES.2017.07.005

Taniguchi, I., Yoshida, S., Hiraga, K., Miyamoto, K., Kimura, Y., & Oda, K. (2019). Biodegradation of PET: Current status and application aspects. *ACS Catalysis, 9*(5), 4089–4105. https://doi.org/10.1021/ACSCATAL.8B05171

Tata, T., Belabed, B. E., Bououdina, M., & Bellucci, S. (2020). Occurrence and characterization of surface sediment microplastics and litter from North African coasts of Mediterranean Sea: Preliminary research and first evidence. *Science of the Total Environment, 713*, 136664. https://doi.org/10.1016/J.SCITOTENV.2020.136664

Thompson, R. C., Moore, C. J., Saal, F. S. V., & Swan, S. H. (2009). Plastics, the environment and human health: Current consensus and future trends. *Philosophical Transactions of the Royal Society B: Biological Sciences, 364*(1526), 2153. https://doi.org/10.1098/RSTB.2009.0053

Tokiwa, Y., Calabia, B. P., Ugwu, C. U., & Aiba, S. (2009). Biodegradability of plastics. *International Journal of Molecular Sciences, 10*(9), 3722–3742. https://doi.org/10.3390/IJMS10093722.

Vithanage, M., Ramanayaka, S., Hasinthara, S., & Navaratne, A. (2021). Compost as a carrier for microplastics and plastic-bound toxic metals into agroecosystems. *Current Opinion in Environmental Science & Health, 24*, 100297. https://doi.org/10.1016/J.COESH.2021.100297

Walker, T. R., & Xanthos, D. (2018). A call for Canada to move toward zero plastic waste by reducing and recycling single-use plastics. *Resources, Conservation and Recycling, 133*, 99–100. https://doi.org/10.1016/J.RESCONREC.2018.02.014

Wang, S. K., Stiles, A. R., Guo, C., & Liu, C. Z. (2015). Harvesting microalgae by magnetic separation: A review. *Algal Research, 9*, 178–185. https://doi.org/10.1016/J.ALGAL.2015.03.005

Wan, H., Wang, J., Sheng, X., Yan, J., Zhang, W., & Xu, Y. (2022). Removal of polystyrene microplastics from aqueous solution using the metal–organic framework material of ZIF-67. *Toxics, 10*(2), 70. https://doi.org/10.3390/TOXICS10020070.

Weber, C. J., & Opp, C. (2020). Spatial patterns of mesoplastics and coarse microplastics in floodplain soils as resulting from land use and fluvial processes. *Environmental Pollution, 267*, 115390. https://doi.org/10.1016/J.ENVPOL.2020.115390

Weithmann, N., Möller, J. N., Löder, M. G. J., Piehl, S., Laforsch, C., & Freitag, R. (2018). Organic fertilizer as a vehicle for the entry of microplastic into the environment. *Science Advances, 4*(4). https://doi.org/10.1126/SCIADV.AAP8060/ASSET/B5C67CCE-780A-440E-9135-8EAE418A098E/ASSETS/GRAPHIC/AAP8060-F2.JPEG.

Wei, W., Huang, Q. S., Sun, J., Dai, X., & Ni, B. J. (2019). Revealing the mechanisms of polyethylene microplastics affecting anaerobic digestion of waste activated sludge. *Environmental Science and Technology, 53*(16), 9604–9613. https://doi.org/10.1021/ACS.EST.9B02971/ASSET/IMAGES/LARGE/ES9B02971_0005.JPEG

Welle, F. (2011). Twenty years of PET bottle to bottle recycling—An overview. *Resources, Conservation and Recycling, 55*(11), 865–875. https://doi.org/10.1016/j.resconrec.2011.04.009

Werbowski, L. M., Gilbreath, A. N., Munno, K., Zhu, X., Grbic, J., Wu, T., Sutton, R., Sedlak, M. D., Deshpande, A. D., & Rochman, C. M. (2021). Urban stormwater runoff: A major pathway for anthropogenic particles, black rubbery fragments, and other types of microplastics to urban receiving waters. *ACS ES&T Water, 1*(6), 1420–1428. https://doi.org/10.1021/ACSESTWATER.1C00017

Wu, M., Yang, C., Du, C., & Liu, H. (2020). Microplastics in waters and soils: Occurrence, analytical methods and ecotoxicological effects. *Ecotoxicology and Environmental Safety, 202*, 110910. https://doi.org/10.1016/J.ECOENV.2020.110910

Wu, W. M., Yang, J., & Criddle, C. S. (2017). Microplastics pollution and reduction strategies. *Frontiers of Environmental Science and Engineering, 11*(1), 1–4. https://doi.org/10.1007/S11783-017-0897-7/METRICS

Xia, S. J., Liu, Y. N., Li, X., & Yao, J. J. (2007). Drinking water production by ultrafiltration of Songhuajiang River with PAC adsorption. *Journal of Environmental Sciences, 19*(5), 536–539. https://doi.org/10.1016/S1001-0742(07)60089-8

Xiong, X., Wu, C., Elser, J. J., Mei, Z., & Hao, Y. (2019). Occurrence and fate of microplastic debris in middle and lower reaches of the Yangtze River—From inland to the sea. *Science of the Total Environment, 659*, 66–73. https://doi.org/10.1016/J.SCITOTENV.2018.12.313

Yang, L., Zhang, Y., Kang, S., Wang, Z., & Wu, C. (2021). Microplastics in soil: A review on methods, occurrence, sources, and potential risk. *Science of the Total Environment, 780*, 146546. https://doi.org/10.1016/J.SCITOTENV.2021.146546

Yan, N., Fan, C., Chen, Y., & Hu, Z. (2016). The potential for microalgae as bioreactors to produce pharmaceuticals. *International Journal of Molecular Sciences, 17*(6), 962. https://doi.org/10.3390/IJMS17060962.

Yoshida, S., Hiraga, K., Takehana, T., Taniguchi, I., Yamaji, H., Maeda, Y., Toyohara, K., Miyamoto, K., Kimura, Y., & Oda, K. (2016). A bacterium that degrades and assimilates poly(ethylene terephthalate). *Science (New York, N.Y.), 351*(6278), 1196–1199. https://doi.org/10.1126/SCIENCE.AAD6359.

Zhang, J., Wang, L., Halden, R. U., & Kannan, K. (2019). Polyethylene terephthalate and polycarbonate microplastics in sewage sludge collected from the United States. *Environmental Science and Technology Letters, 6*(11), 650–655. https://doi.org/10.1021/ACS.ESTLETT.9B00601/ASSET/IMAGES/LARGE/EZ9B00601_0002.JPEG

Zhang, X., Chen, J., & Li, J. (2020). The removal of microplastics in the wastewater treatment process and their potential impact on anaerobic digestion due to pollutants association. *Chemosphere, 251*, 126360. https://doi.org/10.1016/J.CHEMOSPHERE.2020.126360

Ziajahromi, S., Neale, P. A., Rintoul, L., & Leusch, F. D. L. (2017). Wastewater treatment plants as a pathway for microplastics: Development of a new approach to sample wastewater-based microplastics. *Water Research, 112*, 93–99. https://doi.org/10.1016/J.WATRES.2017.01.042

Chapter 12
An Advanced Approach of MOF-Mediated Microplastic Degradation After Confiscating Microplastics by MOFs

Pratibha Chauhan and Ajay Kumar Mishra

Abstract Microplastic pollution poses a substantial risk to aquatic ecosystems and human health, necessitating innovative strategies for its removal and degradation. The study proposes an advanced approach utilizing Metal–Organic Frameworks (MOFs) for the efficient confiscation and subsequent degradation of microplastics. Through a comprehensive review of relevant literature, this paper discusses the characteristics of MOFs tailored for microplastic adsorption. Moreover, it explores the mechanisms underlying MOF-mediated microplastic capture and degradation, highlighting the role of interaction sites within the MOF. The proposed method offers a promising path for lessening the environmental impact of microplastic pollution.

Keywords Absorption microplastics · MOF · Removal · Application · Pollution

1 Introduction

Plastics in the form of nano and microplastics are extremely harmful and challenging to remove from the ecosystem. Nanoplastic or microplastic comes into the ecosystem through the direct release or degradation of plastic items by different processes (https://iupac.org/event/1st-international-conference-on-noncovalent-interactions-icni/). These could cause serious health problems and prove to be fatal if taken above the proposed limits. The problem multiplies many times by their efficiency of absorption. Microplastics are good absorbers at the surface, which leads

P. Chauhan (✉) · A. K. Mishra
Academy of Nanotechnology and Waste Water Innovations, Johannesburg, South Africa
e-mail: sendtoprati@gmail.com

A. K. Mishra
Department of Chemistry, University of the Western Cape, Cape Town, South Africa

Department of Chemistry, Durban University of Technology, Steve Biko Road, Durban 4001, South Africa

A. K. Mishra et al. (eds.), *Microplastics*, https://doi.org/10.1007/978-981-97-6461-7_12

them to absorb harmful heavy metals, Pharma debris, and pathogens from the water, assembling the pool of harmful substances (Anand et al., 2023).

If we look at the mechanism of Microplastic pollution spread, it is chiefly by absorption/ adsorption processes. Nano and microplastics have a high surface area, which enables them for effective adsorption and absorption (Singh et al., 2021). Then the efficient solution of this problem can come from including the phenomena of absorption simply, in the process of complete removal. Only the materials that have higher absorption capacity can solve this problem; such materials that come to mind first- are metal–organic frameworks, highly applied materials in the current decade (López-Magano et al., 2020). In recent years, metal–organic framework has been used in different forms, shapes, and sizes. MOFs in the form of membranes could be used for the purpose of absorbing microplastic and consequential other pollutants from the water systems. Thus, the very idea of applying metal–organic frameworks for the degradation of microplastics and capturing microplastics in MOF foam, is not only interesting and advanced approach to address the issue of microplastic pollution but also a very applied and genuine method to eradicate the problem.

The first step is the application of MOFs for confiscating microplastics in MOF foam (Chen et al., 2020; Mehmood et al., 2023; Zhao et al., 2022). This approach is very useful for the concentration and removal of microplastics from water. The process makes it easier to target microplastic species precisely and process them. MOF foam is an advanced material that combines the properties of MOFs with the porous structure of foam material, where MOF foam can behave as an efficient absorbent (Gnanasekaran & Mok, 2021; Lin et al., 2021). For their high surface area and porosity of MOFs are efficient to absorb microplastics from the water bodies. MOF can be developed, but forming MOF foam material stable and effective in real world in environmental conditions is a critical challenge. Moreover, the environmental sustainability and cost effectiveness of MOF foam for large-scale production is also a matter of consideration for the real-life application of this approach, it is important to develop MOF-Foam that can integrate with existing water treatment processes and infrastructure (Mukherjee et al., 2024). It is important That MOF-Foam can selectively capture and degrade different types of microplastics which vary in size and composition. Creating MOF-foam materials that are stable and effective in real-world environmental conditions is a critical challenge. Scalability For practical implementation, the production of MOF-foam and MOFs on a large scale must be cost-effective and environmentally sustainable. The environmental impact of MOF-foam production, as well as the fate of MOFs and their byproducts in the environment, should be carefully assessed. The developed MOF-mediated degradation and MOF-foam capture systems should be designed to integrate with existing water treatment processes and infrastructure. Ensuring that MOFs can selectively capture and degrade different types of microplastics, which vary in size and composition, is important. The electrostatic attraction between the polar groups on the microplastic and the MOF can lead to the adsorption or binding of the microplastics onto the MOF's surface or within its pores (Chen et al., 2020). This interaction can be useful in capturing and removing microplastics from a solution or environment (Fig. 1).

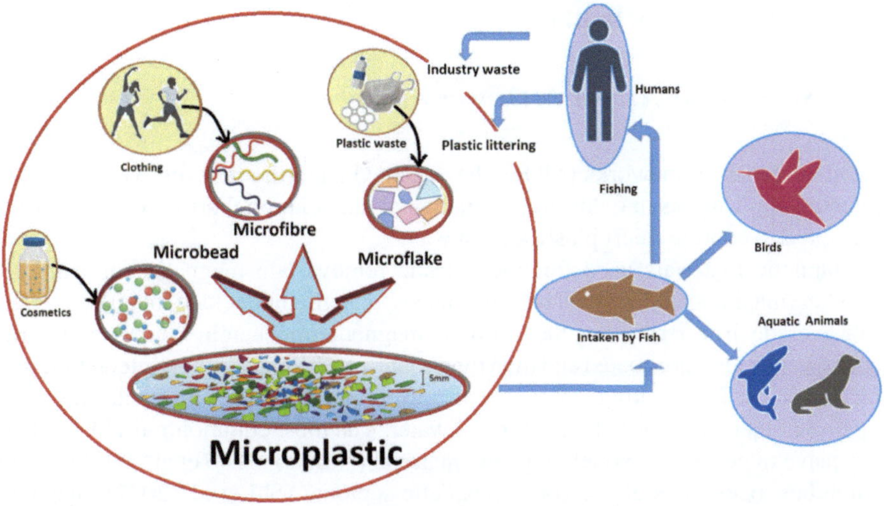

Fig. 1 The schematics of the life cycle of different types of microplastic in an ecosystem

The second step is MOF-mediated microplastic degradation- MOFs are recently invented and designed for specific actions as catalytic centers, using a precise metal complex. For the requirement of degradation of microplastic degradation process, it mainly involves MOFs as a catalyst to break down chemical bonds and transforming them into neutral non-harmful forms. Apart from their chemical constitution, MOFs physical properties of high surface area and tunability are very useful to capture microplastics. MOF provides numerous exposed active sites within MOFs. The combination of MOF-mediated degradation and MOF-foam for microplastic capture represents a holistic strategy for easing the environmental impact of microplastics. This advanced approach shows promise for addressing the issue of microplastic pollution, but further research and development are needed to optimize the materials, techniques, and processes involved. As technology evolves, innovative solutions like MOF-mediated microplastic degradation and MOF-foam capture can play a significant role in combatting microplastic pollution in a sustainable and effective manner. Therefore, the combination of MOF-mediated degradation And MOF form for microplastic capture. Represents a holistic strategy for minimizing the environmental impact of microplastics. However, there are several challenges.

2 MOFs as Adsorbents

2.1 State of the Art for MOFs as Adsorbents

Metal–Organic Frameworks (MOFs) have gained attention in recent years for their potential applications in fields of environmental remediation. Various materials were attempted to capture microplastics from water.

Magnetic materials used for microplastic removal are magnetic nanoparticles or other magnetically responsive substances that might include magnetic materials include iron oxides in the form of magnetite/maghemite and ferrites. The magnetic particles are coated and then functionalized to enhance their interaction with microplastics. This coating can be adapted to selectively attract microplastics while repelling other substances present in the water. The most common reusable and not expensive materials are novel magnetic material (NMAs) (Zhao et al., 2022) carbon nanotubes, (Céspedes et al., 2004) magnetic sepiolite (Shi et al., 2022), magnetic nanocomposites, magn (Pasanen et al., 2023) magnetic biochar, (Ye et al., 2020) and other magnetic systems (Dedeler; Atzori et al., 2015; Cao et al., 2023). The magnetic materials attract microplastics, because of its iron content. These magnetic materials are dispersed in water, which forms the suspension, and then it is introduced to the water containing microplastics. When an external magnetic field is applied to the water, the magnetic materials become magnetized and exert a magnetic force on the microplastics (Atzori et al., 2015). This force allows the magnetic particles to attract and capture microplastics. This separation process is achieved by applying a magnetic field gradient, causing the magnetic particles with attached microplastics to move towards a collection plate (Cao et al., 2023). When the microplastics are captured by the magnetic materials, these can be separated from the water using a magnetic field. The accumulated plastics can be recovered from magnetic materials for proper disposal or recycling. They studied potential adsorption processes, and indulged kinetic forces, interactions, and complexation processes. They found that these NMAs were four times more effective than normal adsorbents (Zhao et al., 2022). There have been histories related to Mechanism reusability and adsorbs and ability of magnetic material for removing polystyrene nano plastic from the aqueous ecosystem.

2.2 Establishment of MOF-Membrane Filter

MOFs are crystalline materials with high porosity and tunable structures, offering them suitable systems for adsorption and further catalysis. MOF-membranes have been used, which have extra advantages over conventional membranes because of high functionality, surface area, and porosity (Gnanasekaran et al., 2021; Kathuria et al., 2013; Kumar Dey et al., 2023; Li et al., 2018, 2020; Mohana et al., 2022; Peng & Yang, 2020; Peng et al., 2017). The technique of using Metal–Organic Framework

Membrane is to filter water in its juvenile state. The research that has been conducted by now, using metal–organic framework and a membrane to filter water encounters different types of troubles. Despite the great advantages of employing metal–organic framework, these membranes face many challenges. There has been research that mainly details all the challenges of using metal–organic framework membranes, along with different matrix systems, to better facilitate them for microplastic removal (Mohana et al., 2022).

2.3 Establishment of MOF-Foam Filter

The combination of MOFs with foams can enhance the efficiency of microplastic capture and degradation processes. It's been shown that MOFs can absorb Microplastics from 72 to 99.9%, despite their drawback of occurring in powder form and their instability. The problem could be solved by using some matrix material. For supporting MOFs. There has been a review article investigating the long-term usability with effective regeneration for reuse of MOF-foam (Chen et al., 2016, 2020). Other studies mainly focus on filtering microplastics out of water bodies with the help of MOFs (He et al., 2014; Lin et al., 2014). Although, the field requires more theoretical treatments for the mechanism of action, facilitated by metal–organic framework for the effective capture of microplastics.

2.4 Mechanisms of Microplastic Confiscation Within MOFs

Several mechanisms have been proposed for how MOFs can aid in microplastic degradation. One approach is physical adsorption, where microplastics are captured within the pores of MOFs, leading to their isolation from the environment. Another approach involves catalytic degradation, where MOFs with suitable metal sites can facilitate the breakdown of plastic polymers through chemical reactions such as hydrolysis, oxidation, or reduction.

2.5 Physical Adsorption

See Fig. 2.

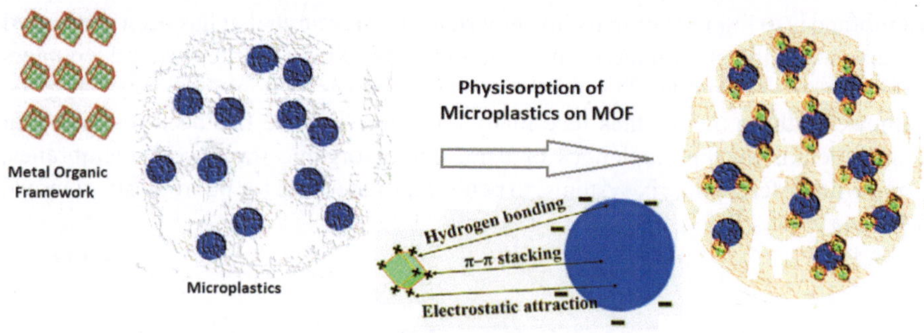

Fig. 2 A schematic of physical adsorption with indulged non-bonded interactions

2.6 Steps in Physical Adsorption

2.6.1 Selection of MOF

Interaction of microplastics with MOF-microplastics are introduced to the MOF material by mixing them in a solution or suspending them in water. The MOF and microplastics are brought into contact, allowing the microplastics to come into proximity with the MOF's porous structure. The choice of MOF is critical for efficient physisorption. MOFs are known for their high surface area, tunable pore sizes, and structural diversity (Chen et al., 2023). The selection of an appropriate MOF depends on the characteristics of the microplastics to be removed. The selection of MOF includes the following criteria. The pore size of the MOF should be large enough to accommodate the microplastic particles but small enough to prevent them from escaping. Different MOFs have varying pore sizes, so selecting one with an appropriate size is essential (Cacho-Bailo et al., 2016).

2.6.2 MOF Characteristics

MOFs with higher surface areas provide more binding sites for the microplastic particles, enhancing their adsorption capacity. The MOF is typically exposed to a solution or water system containing microplastics. The interaction between the MOF and microplastics starts as the particles encounter the MOF (Ploetz et al., 2020).

2.6.3 Adsorption Capacity

The adsorption capacity of the MOF refers to the maximum amount of microplastics that the MOF can adsorb. The capacity depends on the MOF's surface area, pore size, and the specific microplastic type. The rate at which microplastics are adsorbed onto the MOF is influenced by factors like the concentration of microplastics in the solution, temperature, and the specific MOF used (Devi & Kumar, 2023).

MOFs can often be regenerated by desorbing the adsorbed microplastics and then reused. Regeneration can be achieved through changes in environmental conditions or by specific desorption techniques. Researchers can analyze the adsorbed microplastics to determine their types, quantities, and physical characteristics. Techniques such as spectroscopy, microscopy, and chemical analysis are commonly used for this purpose (Nijem et al., 2012).

Once the MOF has adsorbed enough microplastics, it can be separated from the solution, and the adsorbed microplastics can be recovered for further analysis or disposal. In some cases, the MOF can be regenerated, either by changing environmental conditions (e.g., pH or temperature) or by applying specific techniques to release the captured microplastics. Regeneration is an essential aspect to ensure the MOF's reusability (Chen et al., 2023).

Researchers may analyze the adsorbed microplastics to determine the types and quantities captured. Techniques such as spectroscopy, microscopy, and chemical analysis can be used. The effectiveness of MOFs for capturing microplastics depends on various factors, including the MOF's properties, the nature of the microplastics, and the environmental conditions (Shekhah et al., 2011). Researchers continue to explore this field to develop more efficient MOFs and understand the challenges and potential applications of MOFs in mitigating microplastic pollution.

3 Primary Forces Driving the Physisorption

The primary forces driving the physisorption of microplastics onto MOFs are van der Waals forces. These are weak, attractive forces that arise from fluctuations in electron distribution within molecules. Although individually weak, van der Waals forces can be significant when large MOF surfaces are involved. Additionally, other non-covalent interactions, such as dipole–dipole interactions and hydrogen bonding, may contribute to the physisorption of microplastics on the MOF surface, depending on the MOF's chemical properties.

3.1 Non-bonding Forces

Non-bonding forces are the pillars of the process of capturing microplastics and nanoplastics within MOFs. The size scale of nanoplastics and microplastics should be similar to metal–organic frameworks. Absorbing materials should be matching with physical and chemical properties to remove microplastics from the medium. Apart from that there should be some non-bonding interactions that bind microplastics irreversible until it gets degraded or removed for further processing.

Non-bonding forces have a crucial role in absorbing microplastics within MOF-foam. MOF foams are three-dimensional, porous materials and are increasingly being explored for applications in adsorbing and removing pollutants like microplastics from water or other environmental sources. Non-bonded contacts are the critical forces for upholding the three-dimensional, long-range order, of encapsulated "target guest microplastic moieties," within the host-sponge framework (Ratera & Veciana, 2011; Sanna et al., 2015). These non-bonding interactions are the necessary forces, for proceeding each and every step of confiscating microplastics (Ning et al., 2016; Politzer & Murray, 2019). Beginning from the construction of "MOF-architect" (Bradshaw et al., 2014; Lei et al., 2018; Zhu et al., 2018) "sponge solvent-exchange" (Mondal et al., 2020) "sponge-activation," and effective and efficient "guest uptake step" (Brunet et al., 2017) all are principally dependent on these non-bonding forces, in the MOF method (Politzer & Murray, 2019). Therefore, these are the solitary reasons to keep guest microplastic moieties within the framework, and to prevent them from eluding back. Even the very weak interactions have profound importance in molecular retention in MOF (Daze & Hof, 2016). The perforation of all microplastics in MOF is largely determined by these weak interactions. The electrostatic interactions are the main cause of all types of interactions that are divided into types of intermolecular forces within crystal packing (Rzęsikowska et al., 2019). A prime example is hydrogen and halogen interactions (Varadwaj et al., 2019) dipole–dipole, aromatic stacking, dipole-induced dipole interactions, and hydrophobic solvent interactions (Alkorta et al., 2020). Ultimately, the influence of these interactions is dependent on other molecular properties. Atomic charges, induced polarizability, atomic distances, molecular geometry, and dielectric geometry of microplastic moiety and solvent, are other factors Influencing inter-molecular forces (Mondal et al., 2020). This orchestral harmony is required to balance target guest–host MOF supramolecular infrastructure.

3.1.1 Importance of Non-bonded Contacts for Confiscating Microplastics in MOF

Indeed, numerous non-covalent bonds perform collectively. A regularly arranged guest microplastic moiety within the sponge is the resultant symphonic action of all non-bonding interactions. The orchestral organization of these interactions provides order to target guest moieties, in the constrained atmosphere within the sponge (Chauhan et al., 2013). The relative counterpart of all non-bonding interactions is presented here for each non-bonding interaction. Protein and ligand interactions are very much like supramolecular guest–host complexes. H-bonding is the most predominat interaction after solvent. A π–π stacking interactions, C-Hσ…π and cation…π interactions are major stabilizing forces for guest moieties in sponge framework. The encapsulation of guests—into hosts; ranging from small synthetic cryptands and cavitands (Cao et al., 2023) to complex MOFs/COFs and HOFs, to large supramolecular proteins—for organic and inorganic guest moieties is well documented (Kathuria et al., 2013; Kumar Dey et al., 2023; Li et al., 2018, 2020; Mohana et al., 2022). The encapsulation of organic guest moieties by cyclodextrins—synthetic hosts, or albumins and globulin in very polar a solvent like water, as well as of ionic guest as catalytic antibodies or enzymes for substrates; follow the same trend of affinity to host, under the very general Principle of non-bonded contacts (Chauhan & Mishra, 2023; Chauhan et al., 2022; Peng & Yang, 2020; Peng et al., 2017).

3.1.2 A Vital Force of All Contacts

On considering, the overall underlying Principle, of all interactions that are accountable for all types of crystal packing, from simple crystal packing to advanced supramolecular crystal packing of MOF, needed for the MOF method, all require a barrier of electrostatic potential, to get initialized (Hobza, 2012).

Electrostatic potential works everywhere, for literally every chemical reaction, occurring in our body or in any living organism or outside in nature (Fig. 4). All biochemical reactions, beginning from TCA—Tricarboxylic acid cycle, Kreb's cycle, and Citric acid cycle, the second stage of cellular respiration, all require this three-step transfer of electrons from one atom to another of microplastic moieties, in the presence of oxygen to get energy from organic microplastics. In plant systems too, photosynthesis is primarily governed by the difference in electrostatic potential. Nerve functioning in our body, all are dependent on the electrostatic potential created (Scheiner, 2018).

3.1.3 The Kinetic Aspect of Non-bonding Forces

The amount of energy released during non-covalent bonding is minimal in the range of 1–10 kcal/mol (https://courses.lumenlearning.com/boundless-chemistry/chapter/kinetic-molecular-theory/). This amount is ~1/100th part of ionic and covalent bonding (https://courses.lumenlearning.com/physics/chapter/13-4-kinetic-theory-atomic-and-molecular-explanation-of-pressure-and-temperature/). At NTP, or at 25 °C temperature, the average kinetic energy of microplastic moieties is ~0.6 kcal/mol. Therefore, at physiologic room temperature, 25–36 °C, molecules have enough energy to form non-covalent bonds or interactions with surrounding (Kathuria et al., 2013; Lodish et al., 2000). A single non-covalent bond, would have very less stabilized energy, and thus would have transient existence for its individuality. But the overall play of all types of interaction would have enough energy as a sum of all energies of all non-bonding interactions, to stabilize a microplastic moiety in any atmosphere in the solvent in co-crystals or as supramolecular complexation (Rest et al., 2015). The simplified expression of the average kinetic energy is as follows—

Ionic bonds are the strongest, then covalent bonding. Among non-bonded contacts, H-bonding is considered to be a very strong interaction, able to carry out phase change in compounds, where possible relative to its analogous counterparts (https://iupac.org/event/1st-international-conference-on-noncovalent-interactions-icni/). For crystallographic studies or crystal studies, it is the prime force that could be observed reversing structure from one polymorph to another (Sharma & Sastry, 2015). Dipole–dipole bond interaction is a relatively feeble force but causes relative stability in complexes. Van der Waals forces work everywhere in crystal formation, and are very tangible forces, not to be liable for any noticeable change in crystal structure.

4 NBIs Are Necessary for MOF-Microplastic Structure Integrity

Sponge's physical features such as, surface properties, absorption characteristics, and internal atmosphere too, are largely determined by non-bonding forces involved in sponge (Liu et al., 2014). Non-covalent interactions, NBI, is a dynamic equilibrium between opposite forces of attractive and repulsive forces (Zohar et al., 2019) occurring simultaneously between atoms of microplastic moieties. At equilibrium, the net force or potential energy is zero. At every point, the movement of atoms is the consequence of net attractive and repulsive forces acting simultaneously. When the attractive forces balance the repulsive forces, the bond becomes stable. On increasing temperature, basically this balance is disturbed. This disturbs all NBIs as a result all MOF's frameworks collapse, resulting in deformed MOF. This collapse may also result when an unfavorable solvent is inserted. It increases repulsive forces, so the MOF framework collapses. Fundamentally, subatomic particles are the governing factors for underlying forces between microplastic moieties at any level from very atomic model to supramolecular infinite networks that govern guest–host complexation in metal–organic frameworks too. The behavior of an atom or microplastic moiety is governed by the number of subatomic particles, electrons, protons, and neutrons. Therefore basically, behind every kind of interaction there is electrostatic force that works.

4.1 Electrostatic Attraction

If the microplastics contain charged or polar groups on their surfaces, these groups can interact with oppositely charged or polar sites on the MOF foam. This results in electrostatic attraction, which helps in the adsorption of microplastics onto the MOF foam. The Coulombic interactions between charged species or the dipole–dipole interactions between polar groups can be strong enough to retain microplastics within the foam structure. Electrostatic force as a major governing force for non-covalent interactions (Novoa, 1999) is apparent by computational methods for interaction analysis, surface generation softwares, etc., where, an electrostatic force is treated, quantum mechanically by Classical columbic theory and Hellmann–Feynman theorem (Silverman, 1984; Ventra & Pantelides, 2000) and by Schrödinger equation in quantum mechanics (Politzer & Murray, 2019). Electrostatic forces, which arise due to the presence of charged or polar functional groups, can affect the interaction between microplastics and MOF foam. Altogether, non-covalent interactions primarily arise from electrostatic potential. Electrostatic potential originates from subatomic charged particles, electrons, and protons that work like Coulomb's point charges. These opposite charges, or uneven distribution of charge in a microplastic moiety, generate a potential gradient. Generated potential among microplastic moieties, render them to react or to be encapsulated.

The chemical nature of elements is largely determined by inherent electron-pulling power or electronegativity and electro-affinity properties of elements. Electronegativity is a charge-shifting property toward one atom. Charge shifting causes, one of the atoms to be electron deficient and the other one, electron rich. Uneven distribution of charges causes various interactions to take place among different microplastic moieties in any medium. Multiple dipoles work under Coulomb's law of attraction. Attraction and repulsion forces, at one equilibrium point, finally stabilize these dipoles in certain orientations (Bauzá et al., 2018; Kim et al., 2015). This force is significant for guest recognition in supramolecular entity with distance r^{-1} and spreading throughout the microplastic moiety, not only a net charge but also a charge distribution over the whole microplastic moiety. Consequently, many positive and negative terms contribute to the Columbic electrostatic energy of a ligand–receptor complex (https://www.sciencedirect.com/topics/chemistry/electrostatic-interaction).

$$E_{el} = \sum \frac{q_l q_r}{4\pi \varepsilon_0 \varepsilon r}$$

ε_0 - permittivity of free space,
ε - dielectric constant of solvent,
$q_l q_r$ - the charges.

Though, the electrostatic energy is not applied for all charge distribution throughout the microplastic moiety and neighboring microplastic moiety in a medium. To describe the dispersion of electric potential in any solution in the direction normal to point charge, Poisson–Boltzmann rule is used. The equation for three-dimensional electrostatic potential of ions in a diffused layer (Mauro, 1962) is the same.

Poisson–Boltzmann equation (BPE) is broadly applied for electrostatic potential, wherever electrons create electric potential. In the electro-chemistry field, it is applied as the Gouy–Chapman equation, in solution chemistry, it is functional as Debey–Huckel theory, in the colloidal chemistry branch, and it is used as Derjaguin–Laandau–Verwey–Overbeek equation.

4.2 Ion–Dipole Interactions

In addition to simple electrostatic attractions, ion–dipole interactions can also come into play if the MOF foam contains charged sites (such as metal ions) and the microplastics have polar functional groups. Ion–dipole interactions occur between charged ions and polar microplastic moieties, contributing to the binding of microplastics within the MOF foam. Microplastics are composed of polymers and have charged groups or polar regions on their surfaces. While MOFs are porous materials with metal nodes and organic linkers, containing charged groups within their structure. Polar Groups on Microplastics have polar functional groups on their surfaces or within their polymer structure. These polar groups include hydroxyl (-OH), carbonyl (C = O), or other polar moieties. MOF structure's polar groups serve as budding sites for ion–dipole interactions. The ion–dipole interaction occurs when the polar groups on the microplastic surface or within the MOF structure interact with charged ions or polar molecules in the surrounding environment. The induced dipole in the polar molecule and the ion's charge creates an attractive force between them. This force is essentially an electrostatic attraction between the positive end of the induced dipole and the negative charge of the ion, or vice versa. The magnitude of this force is proportional to the product of the ion's charge and the magnitude of the dipole moment of the polar molecule. The interaction potential energy between the ion and the polar molecule decreases as they come closer together due to their electrostatic attraction. This potential energy reaches a minimum when they are at an optimal distance, and it increases as they move farther apart. The strength of the ion–dipole interaction between microplastics and MOFs depends on the nature of the polar groups involved, their charge or dipole moment, and the distance between the microplastics and the MOF. Stronger ion–dipole interactions will lead to more effective adsorption and binding of microplastics.

4.3 Magnetic Interaction

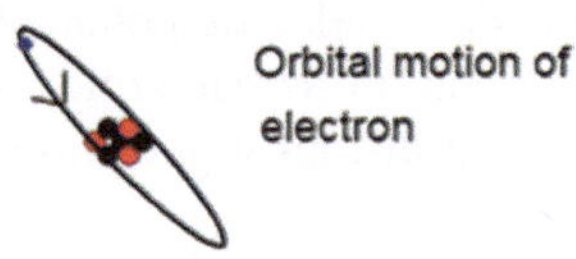

$$\left|\vec{L}\right| = mvr$$

$$|L| = \sqrt{l(l+1)}\hbar$$

where, l = 0, 1, 2, 3, … n − 1

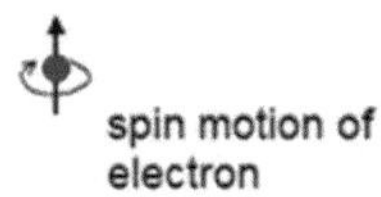

Quantized spin angular moment

$$S = \hbar\sqrt{s(s+1)} = \frac{h}{4\pi}\sqrt{n(n+2)}$$

In terms of energy, magnetic energy is a big counterpart for total non-covalent interaction energy (https://www.britannica.com/science/crystal). Magnetism influences structure, orientation, and crystal packing in supramolecular complexes (Baskett & Lahti, 2005; Seber et al., 2013). Many times magnetism causes polymorphism (Novoa & Deumal, 1997) and self-assembly of molecules (Pilkington & Decurtins, 2003) in crystal packing. There is a strong correlation between structure and magnetism (Deumal et al., 1999; Yue et al., 2008) in crystal growth (Seber et al., 2013). Bohr's model, assigned to account for a periodic model, is solely based on magnetism. Based on magnetism, it describes the positions of electrons (orbitals) within an atom. Electrons have two types of motion, orbital and spin motion in an atom (Scaramucci et al., 2012). Consequently, a moving charge particle generates a magnetic field in a perpendicular direction. Two magnetic fields generated by orbital rotation and spin motion, have some resultant magnetic moment (Reiff et al., 2009). Russell Saunders' coupling scheme calculates the resultant magnetic moment for transition metal complexes (Figgis & Lewis, 1964; Mokrousov et al., 2007). John–teller distortion also calculates distortions in metal complexes (Atzori et al., 2015; Mokrousov et al., 2007). John–Teller distortion also calculates distortions in metal complexes (Atzori et al., 2015). Orientations of microplastic moiety has a big counterpart from magnetism (Kim & Kee, 2016). These orientations are largely like crystal packing models. They are like small magnets with north and south ends directing from south to north. Oppositely aligned magnets tend to be canceled to give specific models like antiparallel π-stacking (Novoa, 1999). Conclusively, the net magnetism cause microplastic moieties to crystallize in certain way. It is accountable for arrangement of microplastics in crystal packing (Deumal et al., 1999). Since the magnetic force is always perpendicular to the velocity of a charged particle, the particle will undergo circular motion. Different forces would result in different amounts and directions of motion (Fig. 3). In any crystal packing these forces need to be nullified or balanced. Only a few crystals have magnetic fields.

4.4 Exchange Attraction- Repulsion Interaction

$$\text{Repulsive forces} \rightleftharpoons \text{Attraction forces}$$

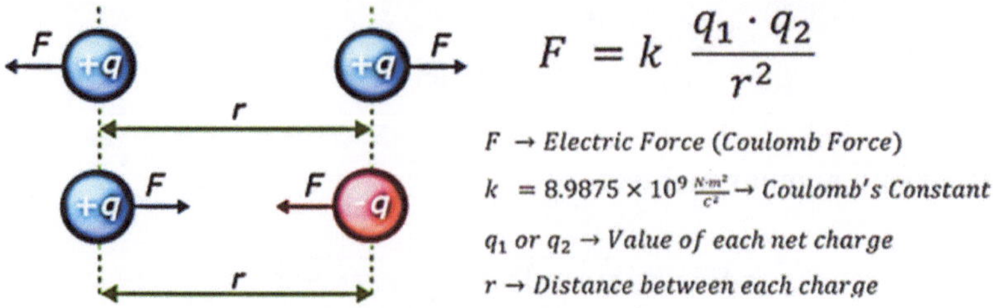

Fig. 3 A representation of electrostatic force

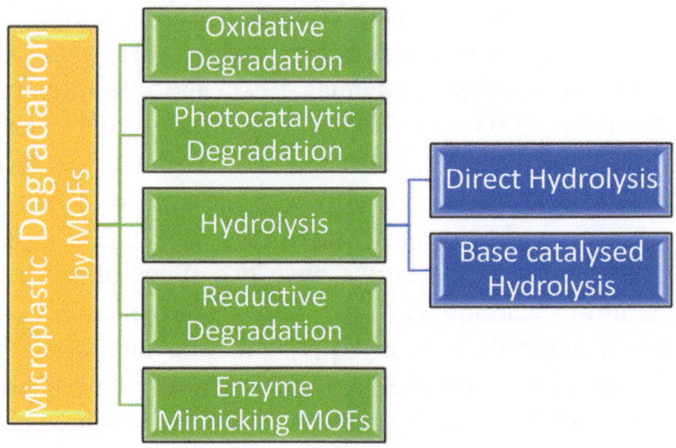

Fig. 4 Demonstration of different types of degradations facilitated by MOF

In quantum mechanics, a wave function of two identical electrons is based on the exchange effect. The exchange effect is applied to electrons and all other identical subatomic particles including bosons and fermions (Zhao et al., 2005). This effect is based on two Principles. First is the indistinguishability of these subatomic particles. It is the indistinguishability of their identity. There is no physical parameter that differs to describe the identity of paralleled spin subatomic particle. So, wave function or energy is written in vice versa form. And the second Principle is, all subatomic particles follow Pauli's exclusion Principle.

The wave function that describes two electrons is either unchanged or merely changes sign for the pair of identical subatomic particles. So, it should either be symmetric or antisymmetric to any variations. We consider the wave function of electrons is antisymmetric. For paralleled spin electrons, the wave function is antisymmetric. And when they are antiparallel it is symmetric. The electron cloud distribution is different in these two cases. The distribution difference causes different electrostatic forces in these two cases. The difference in electrostatic force is called the exchange energy that appears as an interaction between the two electrons of same atom or the different one of different microplastic moieties. The only thing they should be in close vicinity.

According to Pauli's exclusion Principle, electrons with the same spin are kept spatially apart. Pauli's Principle is applicable to all electrons similarly whether you are dealing with an atom or in a microplastic moiety. Therefore, conclusively, when two electrons come closer significantly to overlap each other, repulsive energy rises. A/R_n, n = 9 or 12, where R is the distance between microplastic moieties or non-bonded atoms. A is a constant and depends on the type of atom. With respect to time, electronic distribution around the nucleus changes or fluctuates. This is evident from Heisenberg's uncertainty Principle, where we cannot determine the momentum or velocity and position of electrons simultaneously.

$$\Delta x \cdot \Delta p \geq \frac{h}{4\pi}$$

where, **Δ**x is uncertainty in the position. **Δp** is uncertainty in momentum, (p = mv, v is velocity). These changes in electron position or fluctuation in the electron cloud. This is like the generation of instantaneous electric field that causes fluctuation of electron cloud to neighbor atom. So, this process continues, or constantly occurring. As electron clouds always fluctuate continuously. All the atoms in the surroundings adjust the spatial distribution of their electrons continuously. The net effect is a chain reaction, of electronic fluctuations that induce redistribution of electrons in other surrounding atoms. The electronic motion is correlated. And requires a quantum mechanical treatment to the process to quantize this force. Although, the magnitude of dispersion forces could be estimated by Hamaker rule (Dmitrieva & Plindov, 1983) described as retarded and non retarded. Retarded are those who are apart by the more distance than the wavelength of light. And non retarded, which are closer to the wavelength of light. Dispersion force decreases with distance R with 1/R (Zhao et al., 2022). Total force per unit area decreases as 1/R (Singh et al., 2021). Larger and heavier atoms show more dispersion forces. Straight-chain hydrocarbons show more dispersion than their branched counterparts. For halogens, it follows the trend RF < RCl < RBr < RI.

4.5 Solvation Effects

The presence of water or other solvents can influence electrostatic interactions. Solvent microplastic moieties can form solvation shells around charged or polar groups on the microplastics, affecting their overall charge distribution and making it more favorable for them to be adsorbed into the MOF foam.

4.6 Polarizability Energy

In any medium, when two microplastic moieties come adjacent, there is charge redistribution in each microplastic moiety. Redistribution of charge causes added attraction amongst microplastic molecules. Therefore, the energy associated with this charge redistribution is named polarization energy. When a microplastic moiety with polarizability is in an electric field, it would have Polarization energy,

$$\mathbf{E}_{Po} = -\frac{1}{2}\alpha\mathbf{E}^2$$

where, α—is polarizability, E is an electric field.

Polarization is an additive attraction energy. It is the summation of the polarizabilities of the atoms (Dmitrieva & Plindov, 1983). It could be estimated by the number of valence electrons as well as how tightly they are bound to the nucleus. The order of polarizability in amines, could be demonstrated as, $NH_3 < CH_3N < (CH_3)_2N < (CH_3)_3N$. It depends on the size and weight of a microplastic moiety. The more interacting surfaces provided more energy for polarizability. For halogens, it follows the trend presented as, F < Cl < Br < I. An increased polarizability of anion and polarizing power of cation is presented here for a few salts (Knight et al., 1985). [NaCl/ Na^+ Cl^-] < [$MgCl_2$/Mg^{2+} 2$(Cl)^-$] < [$AlCl_3$/Al^{3+} 3$(Cl)^-$] & [AlF_3/Al^{3+} 3$(F)^-$,] < [$AlCl_3$/Al^{3+} 3$(Cl)^-$] < [$AlBr_3$/Al^{2+} 3$(Br)^-$].

5 Catalytic Degradation by MOF

5.1 Depletion of Counterions

If the microplastics carry a net charge, the foam's electrostatic interactions can lead to the depletion of counterions in the surrounding solution, increasing the likelihood of microplastics being attracted to and absorbed by the MOF foam. It's important to note that the specific electrostatic interactions at play will depend on the charge distribution and functional groups present on both the MOF foam and the microplastics. The performance of a given MOF foam in adsorbing microplastics will be influenced by its structural properties, surface chemistry, and the characteristics of the microplastics in the environment. In summary, non-bonding electrostatic forces in MOF foam play a role in the absorption of microplastics by providing attractive interactions between charged or polar groups on the microplastics and the functional sites within the MOF foam structure, ultimately leading to the retention of microplastics within the foam.

Catalytic degradation of microplastics by Metal–Organic Frameworks (MOFs) can involve various mechanisms depending on the specific MOF and the reaction conditions (Fig. 4). Several types of catalytic degradation processes for microplastic degradation by MOFs include.

5.2 State of the Art for Catalytic Degradation by MOF

After capturing Microplastics, in metal–organic framework foam or membrane, It is necessary to degrade them, to avoid further spread of pollution (Wu et al., 2022). There have been many studies that detail the advanced approach of microplastic degradation by Metal–Organic Frameworks (Cao et al., 2023; Thakur et al., 2022). There are different types of degradation processes. It could be oxidative degradation or reductive degradation. Photocatalytic degradation (Ge et al., 2022) hydrolysis, catalyzed hydrolysis or it could be mixing of all the mechanisms for metal–organic frameworks.

In the oxidative degradation process, we mostly use transition metal which catalyzes oxidative reactions. MOFs with incorporated transition metal ions can catalyze oxidative reactions that break down the polymer chains in microplastics (Bule Možar et al., 2023). This mechanism generates reactive oxygen species (ROS), such as hydroxyl radicals (•OH) or superoxide radicals (O2•-), which attack the polymer bonds, leading to chain scission and the degradation of plastics into smaller fragments.

Photocatalytic Degradation is also very common method for microplastic degradation these days (Ge et al., 2022). Some MOFs exhibit photocatalytic properties, especially under exposure to UV or visible light. When irradiated with light, these MOFs can generate electron–hole pairs, leading to redox reactions that facilitate the degradation of microplastics by generating ROS and accelerating the degradation process (Razali et al., 2020).

Hydrolysis degradation process requires MOFs with hydrolytic activity that can catalyze the hydrolysis of ester bonds in many types of microplastics, such as polyethylene terephthalate (PET) and polyethylene (Golmohammadi et al., 2023). Microplastics are prone to absorb humidity, which is directly dependent on the microplastic's molecular structure. In this mechanism, water breaks polymeric chains into simple molecules (Golmohammadi et al., 2023). The hydrolysis reaction also involves breaking ester bonds, leading to the cleavage of polymer chains and the formation of smaller degradation products. Direct Acid Hydrolysis occurs in MOFs that can catalyze acid hydrolysis reactions where the MOF structure contains acidic groups. These acidic groups can enhance the hydrolysis of microplastics in the presence of an acidic solution, further breaking down the polymer chains (Sevim & Pan, 2018).

Base-catalyzed hydrolysis is similar to acid hydrolysis, MOFs can catalyze base-catalyzed hydrolysis reactions by providing basic sites within their structure. This can lead to the cleavage of chemical bonds within microplastics.

Biodegradation-biodegradation, or microbial biodegradation are the most harmless way of dealing with Microplastic pollution. Biological degradation by microorganisms utilizes carbon sources as their food (Bule Možar et al., 2023). For this purpose, there has been a study engineering green MOF that was very hydrophobic for efficient biodegradation (Chen et al., 2023). This process does not have any toxic byproducts, but the multiplication of microorganisms and transporting it to further pollutants provides a healthy cycle of ecological balance. Recent review articles and book chapters revealing biodegradation by enzyme complexes through photo radicals inducing oxidation and bacterial growth, is a milestone to have deep insight in the field (Devi & Kumar, 2023; Subair et. al., 2023; Tang et al., 2022). Other studies on biodegradation of microplastic include MOF membrane· (Gnanasekaran et al., 2021; https://doi.org/10.1016/j.seppur.2021.119655; Honarmandrad et al., 2023) with biodegradation process.

5.3 Combinations of Mechanisms

In some cases, MOFs can exhibit multiple catalytic mechanisms simultaneously, combining oxidative, hydrolytic, and other degradation pathways to enhance the overall degradation efficiency. The choice of MOF, reaction conditions, and the specific type of microplastic being targeted will determine the most effective catalytic degradation mechanism. Researchers are actively exploring various MOFs and optimizing reaction conditions to develop efficient and environmentally friendly methods for microplastic degradation.

The approach of MOF-mediated microplastic degradation combined with the confiscation of microplastics using MOF-foam is a multistep process aimed at efficiently removing and degrading microplastics in aquatic environments. Here are all the steps involved-

5.4 Steps in MOF-Mediated Microplastic Adsorption and Degradation

5.5 Mechanism of MOF Mediated Microplastic Degradation

First, we have to design MOF-Foam for Bulk- Production for that development of MOF-foam materials with high surface area and porosity needs to be tailored to effectively capture microplastics (Fig. 5). MOF-foam should be designed to float on water.

The next step is the deployment of MOF-foam. Deployment of MOF-foam involves placing MOF-foam in areas with known microplastic contamination, such as rivers, lakes, or oceans for maximum capture. Microplastic adsorption process takes place efficiently by MOF-foam when it is placed properly, in the right direction. Foam functions as a passive collector, absorbing and concentrating microplastics from the surrounding water due to its high adsorptive capacity, while MOF component acts as an active absorber to selectively bind certain chemical species. Periodic Collection is required for an efficient collection process. Retrieving MOF-foam periodically after a set duration is essential when it seems to be saturated with microplastics.

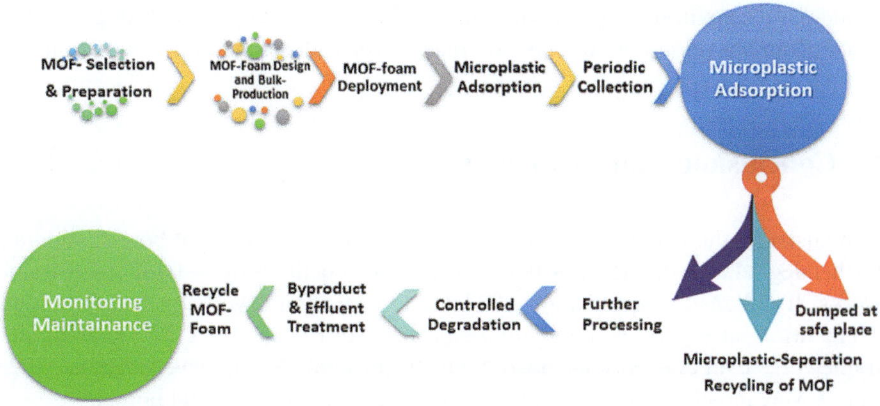

Fig. 5 A demonstration of schematics of the overall process from adsorption to degradation and recycling of microplastics by MOFs

Microplastic Isolation is the next step of the process, where it needs to separate the collected microplastics from the MOF-foam. For further processing, it may involve MOF-Mediated Microplastic Degradation. For that isolated microplastics are transferred to a controlled environment for degradation. MOF selection and preparation is required for specific MOFs engineered for microplastic degradation. Prepare the MOFs in bulk for use in the degradation process. Controlled degradation process involves initiating the degradation process using the chosen MOFs. This process may involve the following steps:

I. Exposure to specific triggers (e.g., heat, light, or chemical agents) that activate the MOFs.
II. The MOFs act as catalysts or mediators, promoting the breakdown of microplastics.
III. The degradation pathways are designed to convert microplastics into harmless byproducts or non-persistent compounds.

Byproduct Assessment is essential for continued processing. Then evaluation of the byproducts generated during the degradation process is done to ensure they are non-toxic and environmentally safe. In the process of effluent treatment, one should treat any effluents resulting from the degradation process to remove any remaining microplastic fragments or byproducts. After all these steps recycling MOF-Foam is crucial. For that, clean and recycle the MOF-foam for further use in the microplastic collection.

Monitoring and Maintenance is the last step of this whole process. Continuous monitoring of the MOF-foam to deployment sites is critical for microplastic levels and the condition of the MOF-foam. Maintenance and replacement of MOF-foam should be performed as needed to ensure optimal performance. This approach offers a comprehensive solution to the problem of microplastic pollution, combining the efficient collection of microplastics with their controlled degradation. It is essential to address the environmental impact and scalability of this approach through ongoing research and development to make it practical for widespread implementation.

6 Conclusions and Prospectives

Conclusively, the chapter provides a comprehensive overview of confiscation and further degradation of microplastics using different methods by metal–organic frameworks. It utilizes the application of Metal–Organic framework for adsorption and degradation sites. The chapter highlights the extra capabilities of metal–organic frameworks over conventional adsorbents, (Chen et al., 2023; Devi & Kumar, 2023; López-Magano et al., 2020) because of their high surface area and porosity.

The chapter discusses two-step processes, the first is confiscating microplastics—MOF-membranes and MOF-foam, as efficient filters for water systems. The second step process is the degradation of these captured microplastics within morph membranes and foam to completely get rid of plastic pollution. Further, it discusses non-bonding forces in detail, which are necessary to identify the right kind of MOFs suitable for the ecosystem. Non-bonding forces are the necessary forces for physisorption. Furthermore, it discusses the type of degradation possible for microplastic in the ecosystem.

In summary, the future of using MOFs for the adsorption and degradation of microplastics holds great potential. Continued research, technological innovation, and collaborative efforts across scientific, industrial, and regulatory sectors will be pivotal in realizing the promise of MOFs as a transformative tool in mitigating the pervasive impact of microplastic pollution on our environment.

References

Alkorta, I., Elguero, J., & Frontera, A. (2020). Not only hydrogen bonds: Other noncovalent interactions. *Crystals, 10*(3), 180. https://doi.org/10.3390/cryst10030180

Anand, U., Dey, S., Bontempi, E., Ducoli, S., Vethaak, A. D., Dey, A., & Federici, S. (2023). Biotechnological methods to remove microplastics: A review. *Environmental Chemistry Letters, 21*(3), 1787–1810. https://doi.org/10.1007/s10311-022-01552-4

Atzori, M., Serpe, A., Deplano, P., Schlueter, J. A., & Mercuri, M. L. (2015). Tailoring magnetic properties of molecular materials through non-covalent interactions. *Inorganic Chemistry Frontiers, 2*(2), 108–115.

Baskett, M., & Lahti, P. M. (2005). Crystallography and magnetism of 3-(N-Tert-Butyl-N-Aminoxyl) benzoic acid. *Polyhedron, 24*(16–17), 2645–2652.

Bauzá, A., Seth, S. K., & Frontera, A. (2018). Molecular electrostatic potential and "atoms-in-molecules" analyses of the interplay between π-hole and lone pair···π/X–H···π/metal···π interactions. *Journal of Computational Chemistry, 39*(9), 458–463. https://doi.org/10.1002/jcc.24869

Bradshaw, D., El-Hankari, S., & Lupica-Spagnolo, L. (2014). Supramolecular templating of hierarchically porous metal-organic frameworks. *Chemical Society Reviews, 43*(16), 5431–5443. https://doi.org/10.1039/C4CS00127C

Brunet, G., Safin, D. A., Robeyns, K., Facey, G. A., Korobkov, I., Filinchuk, Y., & Murugesu, M. (2017). Confinement effects of a crystalline sponge on ferrocene and ferrocene carboxaldehyde. *Chemical Communications, 53*(41), 5645–5648.

Bule Možar, K., Miloloža, M., Martinjak, V., Cvetnić, M., Kušić, H., Bolanča, T., Kučić Grgić, D., & Ukić, Š. (2023). Potential of advanced oxidation as pretreatment for microplastics biodegradation. *Separations, 10*(2), 132. https://doi.org/10.3390/separations10020132

Cacho-Bailo, F., Caro, G., Etxeberría-Benavides, M., Karvan, O., Téllez, C., & Coronas, J. (2016). MOF–polymer enhanced compatibility: Post-annealed zeolite imidazolate framework membranes inside polyimide hollow fibers. *Rsc Advances, 6*(7), 5881–5889.

Cao, Y., Sathish, C. I., Guan, X., Wang, S., Palanisami, T., Vinu, A., Yi, J. (2023). Advances in magnetic materials for microplastic separation and degradation. *Journal of Hazardous Materials*, 132537.

Céspedes, O., Ferreira, M. S., Sanvito, S., Kociak, M., & Coey, J. M. D. (2004). Contact induced magnetism in carbon nanotubes. *Journal of Physics: Condensed Matter, 16*(10), L155.

Chauhan, R., Madan, J., Kaushik, D., Sardana, S., Pandey, R. S., & Sharma, R. (2013). Inclusion complex of colchicine in hydroxypropyl-β-cyclodextrin tenders better solubility and improved pharmacokinetics. *Pharmaceutical Development and Technology, 18*(2), 313–322. https://doi.org/10.3109/10837450.2011.591801

Chauhan, P., Javed, S., Levendis, D. C., & Fernandes, M. (2022). Hydrophobicity directed guest-inclusion for structure-elucidation of enclatherated guests within a crystalline sponge by SC-XRD. *Journal of Molecular Structure, 1251*, 132054. https://doi.org/10.1016/j.molstruc.2021.132054

Chauhan, P., & Mishra, A. K. (2023). MOF: A futuristic material for dyes remediation. In A.K. Mishra (Ed.), *Nano-engineered Materials for Textile Waste Remediation*. Environmental Footprints and Eco-design of Products and Processes (pp. 129–151). Springer Nature: Singapore. https://doi.org/10.1007/978-981-19-7978-1_6.

Chen, Y., Huang, X., Zhang, S., Li, S., Cao, S., Pei, X., Zhou, J., Feng, X., & Wang, B. (2016). Shaping of metal-organic frameworks: from fluid to shaped bodies and robust foams. *Journal of the American Chemical Society, 138*(34), 10810–10813. https://doi.org/10.1021/jacs.6b06959

Chen, Y.-J., Chen, Y., Miao, C., Wang, Y.-R., Gao, G.-K., Yang, R.-X., Zhu, H.-J., Wang, J.-H., Li, S.-L., & Lan, Y.-Q. (2020). Metal-organic framework-based foams for efficient microplastics removal. *Journal of Materials Chemistry A, 8*(29), 14644–14652. https://doi.org/10.1039/D0TA04891G

Chen, X., Ma, H., Ji, X., Han, R., Pang, K., Yang, Z., Liu, Z., & Peng, S. (2023). Engineering green MOF-based superhydrophobic sponge for efficiently synchronous removal of microplastics and pesticides from high-salinity water. *Water Research, 243*, 120314. https://doi.org/10.1016/j.watres.2023.120314

Crystal—Magnetism. Encyclopedia Britannica. Retrieved May 17, 2020, from https://www.britannica.com/science/crystal.

Daze, K., & Hof, F. (2016). Molecular interaction and recognition. *Encyclopedia of Physical Organic Chemistry* 1–51. https://doi.org/10.1002/9781118468586.epoc3001.

Dedeler, A. P., A novel magnetic system with carbon nanotubes to remove microplastics from water.

Deumal, M., Cirujeda, J., Veciana, J., & Novoa, J. J. Structure–magnetism relationships in α-nitronyl nitroxide radicals. *Chemistry–A European Journal, 5*(5), 1631–1642.

Devi, J., & Kumar, A. (2023). Efficiency of MOFs in water treatment against the emerging water contaminants such as endocrine disruptors, pharmaceuticals, microplastics, pesticides, and other contaminants. In *Metal Organic Frameworks for Wastewater Contaminant Removal* (pp. 73–94). John Wiley & Sons, Ltd. https://doi.org/10.1002/9783527841523.ch4.

Di Ventra, M., & Pantelides, S. T. (2000). Hellmann-Feynman theorem and the definition of forces in quantum time-dependent and transport problems. *Physical Review B, 61*(23), 16207–16212. https://doi.org/10.1103/PhysRevB.61.16207

Dmitrieva, I. K., & Plindov, G. I. (1983). Dipole polarizability, radius and ionization potential for atomic systems. *Physica Scripta, 27*(6), 402.

Electrostatic Interaction—An overview|ScienceDirect Topics. Retrieved May 10, 2020, from https://www.sciencedirect.com/topics/chemistry/electrostatic-interaction.

Figgis, B. N., & Lewis, J. (1964). The magnetic properties of transition metal complexes. *Progress in Inorganic Chemistry, 6*, 37–239.

Ge, J., Zhang, Z., Ouyang, Z., Shang, M., Liu, P., Li, H., & Guo, X. (2022). Photocatalytic degradation of (micro)plastics using TiO2-based and other catalysts: Properties, influencing factor, and mechanism. *Environmental Research, 209*, 112729. https://doi.org/10.1016/j.envres.2022.112729

Gnanasekaran, G., Arthanareeswaran, G., & Mok, Y. S. (2021). A high-flux metal-organic framework membrane (PSF/MIL-100 (Fe)) for the removal of microplastics adsorbing dye contaminants from textile wastewater. *Separation and Purification Technology, 277*, 119655. https://doi.org/10.1016/j.seppur.2021.119655

Golmohammadi, M., Fatemeh Musavi, S., Habibi, M., Maleki, R., Golgoli, M., Zargar, M., Dumée, L. F., Baroutian, S., & Razmjou, A. (2023). Molecular mechanisms of microplastics degradation: A review. *Separation and Purification Technology, 309*, 122906. https://doi.org/10.1016/j.seppur.2022.122906

He, Y.-C., Yang, J., Kan, W.-Q., Zhang, H.-M., Liu, Y.-Y., & Ma, J.-F. (2014). A new microporous anionic metal-organic framework as a platform for highly selective adsorption and separation of organic dyes. *Journal of Materials Chemistry A, 3*(4), 1675–1681. https://doi.org/10.1039/C4TA05391E

A high-flux metal-organic framework membrane (PSF/MIL-100 (Fe)) for the removal of microplastics adsorbing dye contaminants from textile wastewater. *Separation and Purification Technology*, *277*, 119655 (2021). https://doi.org/10.1016/j.seppur.2021.119655.

Hobza, P. (2012). Calculations on noncovalent interactions and databases of benchmark interaction energies. *Accounts of Chemical Research, 45*(4), 663–672. https://doi.org/10.1021/ar200255p

Honarmandrad, Z., Kaykhaii, M., & Gębicki, J. (2023). Microplastics removal from aqueous environment by metal organic frameworks. *BMC Chemistry, 17*(1), 122. https://doi.org/10.1186/s13065-023-01032-y

1st International Conference on Noncovalent Interactions (ICNI). IUPAC|International Union of Pure and Applied Chemistry. Retrieved May 06, 2020, from https://iupac.org/event/1st-international-conference-on-noncovalent-interactions-icni/.

Kathuria, A., Abiad, M. G., & Auras, R. (2013). Deterioration of metal-organic framework crystal structure during fabrication of poly (l-lactic acid) mixed-matrix membranes. *Polymer International, 62*(8), 1144–1151.

Kim, H.-S., & Kee, H.-Y. (2016). Crystal structure and magnetism in α- RuCl 3: An ab initio study. *Physical Review B, 93*(15), 155143.

Kim, H., Doan, V. D., Cho, W. J., Valero, R., Aliakbar Tehrani, Z., Madridejos, J. M. L., & Kim, K. S. (2015). Intriguing electrostatic potential of CO: Negative bond-ends and positive bond-cylindrical-surface. *Scientific Reports, 5*(1), 1–9. https://doi.org/10.1038/srep16307

Kinetic Theory: Atomic and Molecular Explanation of Pressure and Temperature|Physics. Retrieved May 06, 2020, from https://courses.lumenlearning.com/physics/chapter/13-4-kinetic-theory-atomic-and-molecular-explanation-of-pressure-and-temperature/.

Knight, W. D., Clemenger, K., de Heer, W. A., & Saunders, W. A. (1985). Polarizability of alkali clusters. *Physical Review B, 31*(4), 2539.

Kumar Dey, T., Hou, J., Sillanpää, M., & Kumar Pramanik, B. (2023). Metal-organic framework membrane for waterborne micro/nanoplastics treatment. *Chemical Engineering Journal, 474*, 145715. https://doi.org/10.1016/j.cej.2023.145715

Lei, Z., Deng, Y., & Wang, C. (2018). Multiphase surface growth of hydrophobic ZIF-8 on melamine sponge for excellent oil/water separation and effective catalysis in a Knoevenagel reaction. *Journal of Materials Chemistry A, 6*(7), 3258–3263. https://doi.org/10.1039/C7TA10566E

Li, Q., Li, J., Fang, X., Liao, Z., Wang, D., Sun, X., Shen, J., Han, W., & Wang, L. (2018). Interfacial growth of metal-organic framework membranes on porous polymers via phase transformation. *Chemical Communications, 54*(29), 3590–3593. https://doi.org/10.1039/C7CC09905C

Li, T.-T., Fan, Y., Cen, X., Wang, Y., Shiu, B.-C., Ren, H.-T., Peng, H.-K., Jiang, Q., Lou, C.-W., & Lin, J.-H. (2020). Polypropylene/polyvinyl alcohol/metal-organic framework-based melt-blown electrospun composite membranes for highly efficient filtration of PM2.5. *Nanomaterials, 10*(10), 2025. https://doi.org/10.3390/nano10102025.

Lin, S., Song, Z., Che, G., Ren, A., Li, P., Liu, C., & Zhang, J. (2014). Adsorption behavior of metal-organic frameworks for methylene blue from aqueous solution. *Microporous and Mesoporous Materials, 193*, 27–34. https://doi.org/10.1016/j.micromeso.2014.03.004

Lin, J.-Y., Chen, P.-Y., Kwon, E., Oh, W. D., You, S., Huang, C.-W., Ghanbari, F., Wi-Afedzi, T., & Lin, K.-Y.A. (2021). One-step synthesized 3D-structured MOF foam for efficient and convenient catalytic reduction of nitrogen-containing aromatic compounds. *Journal of Water Process Engineering, 40*, 101933. https://doi.org/10.1016/j.jwpe.2021.101933

Liu, S., Zhai, L., Li, C., Li, Y., Guo, X., Zhao, Y., & Wu, C. (2014). Exploring and exploiting dynamic noncovalent chemistry for effective surface modification of nanoscale metal-organic frameworks. *ACS Applied Materials & Interfaces, 6*(8), 5404–5412. https://doi.org/10.1021/am500192b

Lodish, H., Berk, A., Zipursky, S. L., Matsudaira, P., Baltimore, D., & Darnell, J. (2000). Noncovalent bonds. In *Molecular Cell Biology* (4th ed.).

López-Magano, A., Jiménez-Almarza, A., Alemán, J., & Mas-Ballesté, R. (2020). Metal-Organic Frameworks (MOFs) and Covalent Organic Frameworks (COFs) applied to photocatalytic organic transformations. *Catalysts, 10*(7), 720. https://doi.org/10.3390/catal10070720

Mauro, A. (1962). Space charge regions in fixed charge membranes and the associated property of capacitance. *Biophysical Journal, 2*(2), 179–198. https://doi.org/10.1016/S0006-3495(62)86848-9

Mehmood, T., Mustafa, B., Mackenzie, K., Ali, W., Sabir, R. I., Anum, W., Gaurav, G. K., Riaz, U., Liu, X., & Peng, L. (2023). Recent developments in microplastic contaminated water treatment: Progress and prospects of carbon-based two-dimensional materials for membranes separation. *Chemosphere, 316*, 137704. https://doi.org/10.1016/j.chemosphere.2022.137704

Mohana, A. A., Rahman, M., Sarker, S. K., Haque, N., Gao, L., & Pramanik, B. K. (2022). Nano/microplastics: Fragmentation, interaction with co-existing pollutants and their removal from wastewater using membrane processes. *Chemosphere, 309*, 136682. https://doi.org/10.1016/j.chemosphere.2022.136682

Mokrousov, Y., Atodiresei, N., Bihlmayer, G., Heinze, S., & Blügel, S. (2007). The interplay of structure and spin-orbit strength in the magnetism of metal-benzene sandwiches: From single molecules to infinite wires. *Nanotechnology, 18*(49), 495402.

Mondal, S., Das, S., & Nandi, A. K. (2020). A review on recent advances in polymer and peptide hydrogels. *Soft Matter, 16*(6), 1404–1454. https://doi.org/10.1039/C9SM02127B

Mukherjee, A., Bose, S., & Das, P. (2024). *Efficient Microplastic Remediation through Best Possible Strategies: A Review*. CRC Press.

Nijem, N., Canepa, P., Kong, L., Wu, H., Li, J., Thonhauser, T., & Chabal, Y. J. (2012). Spectroscopic characterization of van Der Waals interactions in a metal organic framework with unsaturated metal centers: MOF-74–Mg. *Journal of Physics: Condensed Matter, 24*(42), 424203.

Ning, G.-H., Matsumura, K., Inokuma, Y., & Fujita, M. (2016). A saccharide-based crystalline sponge for hydrophilic guests. *Chemical Communications, 52*(43), 7013–7015.

Novoa, J. (1999). Electrostatic compression on non-covalent interactions: The case of π stacks involving ions. *New Journal of Chemistry, 23*(6), 577–579.

Novoa, J. J., & Deumal, M. (1997). Theoretical analysis of the packing and polimorphism of molecular crystals using quantum mechanical methods: The packing of the 2-hydro nitronyl nitroxide. *Molecular Crystals and Liquid Crystals Science and Technology. Section A. Molecular Crystals and Liquid Crystals, 305*(1), 143–156.

Pasanen, F., Fuller, R. O., & Maya, F. (2023). Fast and simultaneous removal of microplastics and plastic-derived endocrine disruptors using a magnetic ZIF-8 nanocomposite. *Chemical Engineering Journal, 455*, 140405.

Peng, Y., & Yang, W. (2020). 2D metal-organic framework materials for membrane-based separation. *Advanced Materials Interfaces, 7*(1), 1901514. https://doi.org/10.1002/admi.201901514

Peng, Y., Li, Y., Ban, Y., & Yang, W. (2017). Two-dimensional metal-organic framework nanosheets for membrane-based gas separation. *Angewandte Chemie International Edition, 56*(33), 9757–9761. https://doi.org/10.1002/anie.201703959

Pilkington, M., & Decurtins, S. (2003) A rational approach for the self-assembly of molecular building blocks in the field of molecule-based magnetism. *Crystal Design: Structure and Function*, 275–323.

Ploetz, E., Engelke, H., Lächelt, U., & Wuttke, S. (2020). The chemistry of reticular framework nanoparticles: MOF, ZIF, and COF materials. *Advanced Functional Materials, n/a*(n/a), 1909062. https://doi.org/10.1002/adfm.201909062.

Politzer, P., & Murray, J. S. (2019). An overview of strengths and directionalities of noncovalent interactions: σ-holes and π-holes. *Crystals, 9*(3), 165. https://doi.org/10.3390/cryst9030165

Ratera, I., & Veciana, J. (2011). Playing with organic radicals as building blocks for functional molecular materials. *Chemical Society Reviews, 41*(1), 303–349. https://doi.org/10.1039/C1CS15165G

Razali, N. A., Abdullah, W. R. W., & Zikir, N. M. (2020) Effect of thermo-photocatalytic process using zinc oxide on degradation of macro/micro-plastic in aqueous environment. *Journal of Sustainability Science and Management (Malaysia).*

Reiff, W. M., Schulz, C. E., Whangbo, M.-H., Seo, J. I., Lee, Y. S., Potratz, G. R., Spicer, C. W., & Girolami, G. S. (2009). Consequences of a linear two-coordinate geometry for the orbital magnetism and Jahn-Teller distortion behavior of the high spin iron (II) complex Fe [N (t-Bu) 2] 2. *Journal of the American Chemical Society, 131*(2), 404–405.

Rest, C., Kandanelli, R., & Fernández, G. (2015). Strategies to create hierarchical self-assembled structures via cooperative non-covalent interactions. *Chemical Society Reviews, 44*(8), 2543–2572.

Rzęsikowska, K., Krawczuk, A., & Kalinowska-Tłuścik, J. (2019). Electrostatic potential and non-covalent interactions analysis for the design of selective 5-HT7 ligands. *Journal of Molecular Graphics and Modelling, 91*, 130–139. https://doi.org/10.1016/j.jmgm.2019.06.007

Sanna, E., Escudero-Adán, E. C., Bauzá, A., Ballester, P., Frontera, A., Rotger, C., & Costa, A. (2015). A crystalline sponge based on dispersive forces suitable for X-ray structure determination of included molecular guests. *Chemical Science, 6*(10), 5466–5472. https://doi.org/10.1039/C5SC01838B

Scaramucci, A., Bousquet, E., Fechner, M., Mostovoy, M., & Spaldin, N. A. (2012). Linear magnetoelectric effect by orbital magnetism. *Physical Review Letters, 109*(19), 197203.

Scheiner, S. (2018). Comparison of various means of evaluating molecular electrostatic potentials for noncovalent interactions. *Journal of Computational Chemistry, 39*(9), 500–510. https://doi.org/10.1002/jcc.25085

Seber, G., Freitas, R. S., Oliveira, N. F., Jr., & Lahti, P. M. (2013). Crystal packing and magnetism in phenolic nitronylnitroxides: 2-(3′, 5′-dimethoxy-4′-hydroxyphenyl)-4, 4, 5, 5-tetramethyl-4, 5-dihydro-1 H-imidazole-1-oxyl. *Crystal Growth & Design, 13*(2), 893–900.

Sevim, K., & Pan, J. (2018). A model for hydrolytic degradation and erosion of biodegradable polymers. *Acta Biomaterialia, 66*, 192–199. https://doi.org/10.1016/j.actbio.2017.11.023

Sharma, R., Sastry, G. N. (2015). Deciphering the dynamics of non-covalent interactions affecting thermal stability of a protein: Molecular dynamics study on point mutant of thermus thermophilus isopropylmalate dehydrogenase. *PLoS One, 10*(12). https://doi.org/10.1371/journal.pone.0144294.

Shekhah, O., Liu, J., Fischer, R. A., & Wöll, C. (2011). MOF thin films: Existing and future applications. *Chemical Society Reviews, 40*(2), 1081–1106.

Shi, C., Zhang, S., Zhao, J., Ma, J., Wu, H., Sun, H., & Cheng, S. (2022). Experimental study on removal of microplastics from aqueous solution by magnetic force effect on the magnetic sepiolite. *Separation and Purification Technology, 288*, 120564. https://doi.org/10.1016/j.seppur.2022.120564

Silverman, J. N. (1984). Generalized Hellmann-Feynman and curvature theorems in classical and quantum mechanics. *International Journal of Quantum Chemistry, 25*(5), 915–928. https://doi.org/10.1002/qua.560250512

Singh, S., Kalyanasundaram, M., & Diwan, V. (2021). Removal of microplastics from wastewater: Available techniques and way forward. *Water Science and Technology, 84*(12), 3689–3704. https://doi.org/10.2166/wst.2021.472

Subair, A., Meera, G., Chellappan, S., Sajithkumar, K. J., Chinglenthoiba, C., Priya, K. L., Indu, M. S. (2023). Techniques for removal and degradation of microplastics. In C. Wang, S. Babel, E. Lichtfouse (Eds.), *Microplastic Occurrence, Fate, Impact, and Remediation* (pp. 127–153). Environmental Chemistry for a Sustainable World. Springer Nature Switzerland. https://doi.org/10.1007/978-3-031-36351-1_6.

Szilva, A., Costa, M., Bergman, A., Szunyogh, L., Nordström, L., & Eriksson, O. (2013). Interatomic exchange interactions for finite-temperature magnetism and nonequilibrium spin dynamics. *Physical Review Letters, 111*(12), 127204.

Tang, K. H. D., Lock, S. S. M., Yap, P.-S., Cheah, K. W., Chan, Y. H., Yiin, C. L., Ku, A. Z. E., Loy, A. C. M., Chin, B. L. F., & Chai, Y. H. (2022). Immobilized enzyme/microorganism complexes for degradation of microplastics: A review of recent advances, feasibility and future prospects. *Science of the Total Environment, 832*, 154868. https://doi.org/10.1016/j.scitotenv.2022.154868

Thakur, S., Mathur, S., Patel, S., & Paital, B. (2022). Microplastic accumulation and degradation in environment via biotechnological approaches. *Water, 14*(24), 4053. https://doi.org/10.3390/w14244053

Kinetic Molecular Theory|Boundless Chemistry. Retrieved May 06, 2020, from https://courses.lumenlearning.com/boundless-chemistry/chapter/kinetic-molecular-theory/.

Varadwaj, P. R., Varadwaj, A., & Marques, H. M. (2019). Halogen bonding: A halogen-centered noncovalent interaction yet to be understood. *Inorganics, 7*(3), 40. https://doi.org/10.3390/inorganics7030040

Wu, Y., Wang, X., Kirlikovali, K. O., Gong, X., Atilgan, A., Ma, K., Schweitzer, N. M., Gianneschi, N. C., Li, Z., Zhang, X., & Farha, O. K. (2022). Catalytic degradation of polyethylene terephthalate using a phase-transitional zirconium-based metal-organic framework. *Angewandte Chemie, 134*(24), e202117528. https://doi.org/10.1002/ange.202117528

Ye, S., Cheng, M., Zeng, G., Tan, X., Wu, H., Liang, J., Shen, M., Song, B., Liu, J., & Yang, H. (2020). Insights into catalytic removal and separation of attached metals from natural-aged microplastics by magnetic biochar activating oxidation process. *Water Research, 179*, 115876.

Yue, W., Tian, H., Hu, N., Geng, Y., & Wang, F. (2008). Crystal packing motifs of oligothiophenes end-capped with n-containing aryls. *Crystal Growth and Design, 8*(7), 2352–2358.

Zhao, H., Huang, X., Wang, L., Zhao, X., Yan, F., Yang, Y., Li, G., Gao, P., & Ji, P. (2022). Removal of polystyrene nanoplastics from aqueous solutions using a novel magnetic material: Adsorbability, mechanism, and reusability. *Chemical Engineering Journal, 430*, 133122. https://doi.org/10.1016/j.cej.2021.133122

Zhao, Y., Schultz, N. E., & Truhlar, D. G. (2005). *Exchange-correlation functional with broad accuracy for metallic and nonmetallic compounds, kinetics, and noncovalent interactions*. American Institute of Physics.

Zhu, S., Yuan, Q., Yin, T., You, J., Gu, Z., Xiong, S., & Hu, Y. (2018). Self-assembly of collagen-based biomaterials: preparation, characterizations and biomedical applications. *Journal of Materials Chemistry B, 6*(18), 2650–2676. https://doi.org/10.1039/C7TB02999C

Zohar, A. R., Levy, S. T. (2019). Attraction vs. repulsion—Learning about forces and energy in chemical bonding with the ELI-Chem simulation. *Chemistry Education Research and Practice, 20*(4), 667–684. https://doi.org/10.1039/C9RP00007K.

Chapter 13
Wastewater Treatment Plants and Microplastic Degradation

Vatika Soni, Komal Poonia, Sonu, Pankaj Raizada, Pardeep Singh, and Chaudhery Mustansar Hussain

Abstract Non-degradable microplastics are emerging pollutants and are discharged in huge amounts into the waterbodies with insufficient treatment. Due to their severe threat to biota, various treatment plants for treating wastewater are considered as the last line of defense which forms a barrier between harmful microplastics and the surroundings. As per the literature, various explored primary, secondary, as well as tertiary treatment phases in wastewater treatment plants remove around 65, 14, and 2% microplastic, respectively. Other than conventional treatment methods, herein we mainly explore the recently reported advanced microplastic degradation methods that achieved almost satisfying microplastic degradation activity. Out of these treatment techniques, advanced oxidation processes can only degrade microplastics partially into simpler molecules like CO_2 and H_2O under laboratory conditions, thus exhibiting a removal efficiency of >99%. A focal point of this review is the detailed analysis of the efficiency of advanced oxidation processes, with a specific emphasis on photocatalytic degradation. Photocatalytic degradation, as a subset of advanced oxidation processes, is dissected to underscore its potential in addressing the persistent nature of microplastic pollutants. The review manuscript further delves into an extensive exploration of various photocatalysts, elucidating their specific roles in the degradation process. The discussion encompasses the generation and utilization of different radicals during photocatalytic reactions, offering insights into the underlying mechanism governing microplastic breakdown.

Keywords Microplastic removal · Advanced oxidation processes · Photodegradation · Sustainable development goals

V. Soni · K. Poonia · Sonu · P. Raizada · P. Singh
School of Advanced Chemical Sciences, Shoolini University, Solan 173212, Himachal Pradesh, India

C. M. Hussain (✉)
Department of Chemistry and Environmental Science, New Jersey Institute of Technology, Newark, NJ 07102, USA
e-mail: chaudhery.m.hussain@njit.edu

A. K. Mishra et al. (eds.), *Microplastics*, https://doi.org/10.1007/978-981-97-6461-7_13

1 Introduction

Plastic products have been vital in civilization and have brought great convenience to our daily lives ever since the invention of the first synthesis resin (Lin et al., 2022). Usually, plastics are low-cost, and lightweight, with excellent durability and chemical stability. As per statistics data, plastics production may reach around 500 million tons in the coming 2025 year (Du et al., 2021a). Plastics as the persistent non-biodegradable pollutants are directly discharged into the water bodies for thousands of years through various routes with small recovery. Based on size, regular or irregularly shaped particles are classified as microplastics (<5 mm), mesoplastics (<10 nm), nanoplastics (<1000–1 nm), as well as macroplastics (>1 nm) (Klein, 2018). Microplastics with a diameter < 5 mm are generally formed by the decomposition of plastics present in a huge range in the atmosphere, marine, and terrestrial habitats (Ariza-Tarazona et al., 2019). On the basis of origin, microplastics are broadly categorized into primary and secondary. Primary microplastics are mainly synthesized for industrial and household applications, however, the UV irradiation and mechanical abrasion of large macroplastics form secondary. Microplastics adsorb various other toxic pollutants such as pharmaceuticals, heavy metals, metalloids, hydrocarbons, etc. due to their large surface area and small particle size (Corcoran, 2022). They may remain in the same position for a long time, thus affecting the sustainable use of land. Before exposure to fresh/marine water ecosystems or their environmental accumulation, microplastics from various human activities enter wastewater treatment plants. Large quantities of small microplastics are more harmful than plastics as they may enter the human body from various food chains. Also, various studies have reported that microplastics can seriously damage the reproduction as well as growth of organisms and even lead to death. Thus, the study of the effective wastewater treatment techniques is essential for the removal of microplastics in order to prevent their release in the natural environment (Chen et al., 2022).

Around the globe, various municipal wastewater treatment plants share a basic design with different stages including sedimentation, flocculation, and aeration etc. The removal/degradation efficacy may vary in contaminant water treatment plants due to the different morphological and physico-chemical properties of microplastics. Preliminary studies have verified that large microplastics in the treatment plants can be removed by either the primary treatment method or the secondary method. Primary treatment predominantly involves different processes like adsorption, coagulation, and ozonation to remove the bulk of the microplastics present in the contaminated water. For example; 92% and 74% microplastic removal efficiency was detected during the primary wastewater treatment. Clarification, as well as biochemical removal methods, are involved in secondary treatment, where microplastics have interacted with various microorganisms (Du et al., 2021a). Considering their size, microplastics with size >500 μm were mostly not present in the effluents after the treatment. The contaminated water is then passed to the tertiary treatment for their further removal through supplementary polishing steps. Tertiary steps include processes like reverse osmosis, micro/ultrafiltration, bioreactors, sand filtration, etc. The particles after the tertiary microplastic treatment may not go into the water ecosystem and might remain in the loading of the plant. Conclusively, primary, secondary, and tertiary contaminated water treatment plants can be successful in eliminating the large-sized microplastic particles, however, the small-sized particles are directly entering the municipal waters/freshwater rivers or may remain in the biosolid. Thus, we need more effective, advanced treatment plants to remove smaller-sized microparticles from the wastewater (Ahmed et al., 2022; Du et al., 2021a) (Fig. 1).

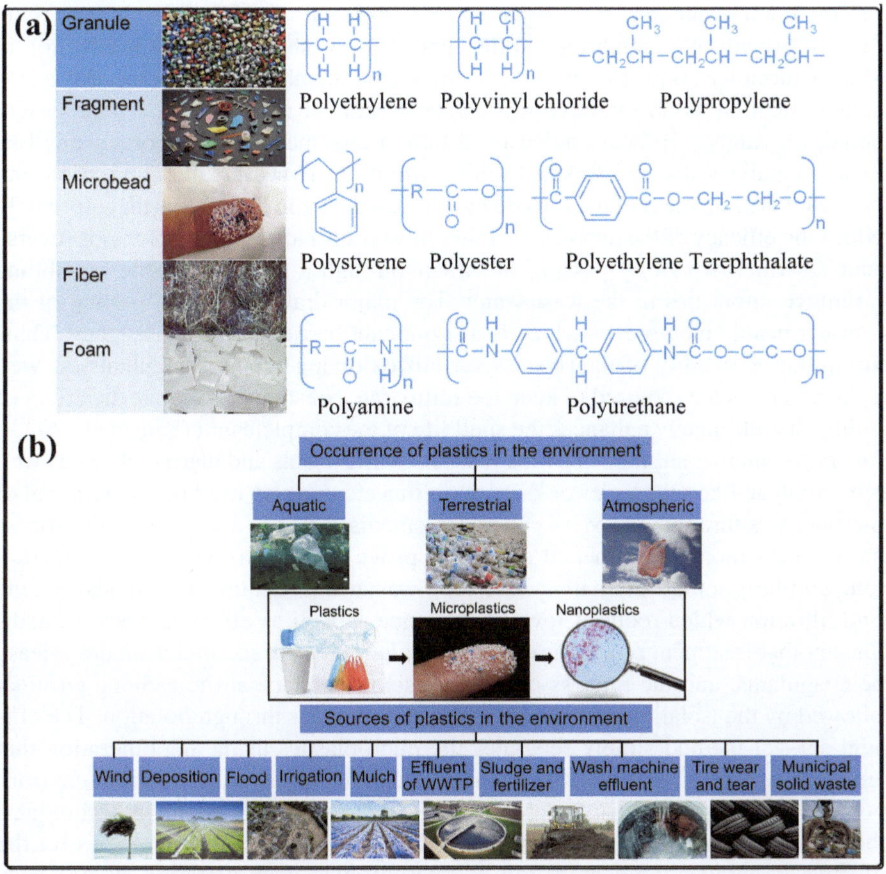

Fig. 1 **a, b** Commonly found microplastics and their sources in the environment. Reproduced from Ahmed et al. (2022) (License no. 5687731294741)

2 Current Trends of Microplastic Degradation Present in the Wastewater

Significant developments for wastewater treatment plants are required to enhance the efficacy of the microplastic contaminant (Wu et al., 2022). Concerning primary as well as secondary treatment methods, a few of the advancements are described. Membrane processes offer multifold advantages including utilization of less energy, ease of use, less complicated process, high stability, and ability to withstand large amounts of water (Golmohammadi et al., 2023). In membrane technology, dynamic membranes must have an adequate permeability for preserving the membrane-forming component and be strong to undergo the vital forces for a prolonged time. The fast growth of a membrane on the supportive mesh may lead to a reduction in effluent turbidity (Bacha et al., 2021). Similarly, membrane bioreactors have also gained attention as an effective wastewater treatment method. The whole system includes a catalysis process aided by different enzymes and microorganisms like organic catalysts that are linked by ultra-filtration. However, the congestion and managing the thickness of the formed biofilm are the main limitations that ultimately reduce the efficacy of the process. Another most used and effective process is reverse osmosis which works by passing the water through a semi-permeable membrane to remove impurities in the wastewater. The major drawback is the fouling of the semi-permeable membranes which is a significant barrier for its consistency. Thus, during the operability of the process various oxidizing agents, coagulants, as well as disinfectants are required to keep the reflux rate constant and reduce the effect of fouling that ultimately enhances the shelf life of the complete unit (Kim et al., 2022). Eukaryotic marine animals like blue mussels or lugworms and marine plants as well as microalgae like seagrasses or Zostera marina etc. are also used for the removal of microplastics through a process called bioremediation. It is a very difficult process that requires more experimental studies to prove the effectiveness of bioremediation. Similarly, other alternative advanced wastewater treatment methods include sand filtration which requires low maintenance as well as effective costs. In addition, another treatment process is electrocoagulation where sacrificial anodes release the coagulants, and the process of electrolysis takes place at the cathode terminal followed by the isolation of different toxic microplastics through flotation. The efficient sol–gel method simply separates the microplastics using agglomeration that can occur irrespective of the concentration, quantity, or any other external environmental factors. Apart from these abovementioned microplastic treatment techniques, photodegradation and photocatalytic degradation are the green technologies for the effective mineralization/degradation of microplastics. Few literature studies have reported that the degradation of microplastics in the wastewater occurs with the help of generated reactive oxygen species in the presence of ultraviolet and visible light energy.

3 Photocatalytic Degradation

The photodegradation process is a sustainable alternative approach for thwarting the pollution caused by microplastics in the environment (Xu et al., 2021). Microplastic photodegradation occurs naturally and is an uncontrollable process, thus it is very important to study the relationship between the aging properties of microplastics and the degree of their aging (Uheida et al., 2021). In a report by Song et al. (2017), the functional groups containing oxygen atoms and cracks were generated on the surface of microplastics as investigated by the aging process under lab accelerated degradation environment (Song et al., 2017). The photodecomposition of microplastics present in the water is extremely slow, Zhu and his coworkers in 2020 reported the aging of polystyrene microplastic under sunlight for 150 days through different generated radicals in the suspension (Zhu et al., 2020). Photocatalytic degradation using the semiconductors process is considered an advanced green treatment method for the degradation of smaller microplastics present in the wastewater within a short period. It is a key process for the degradation of polymers through free radical formation, addition of O_2, abstraction of hydrogen, crosslinking of chemical chains, and morphological changes (flakes and cracks). The photodegradation mechanism involves the bombardment of a photoactive semiconductor with photons (mainly UV and visible light) to generate reactive hydroxyl (•OH), singlet oxygen ($^1O_2$), hydrogen peroxide (H_2O_2), and superoxide (•O_2^-) radicals (Soni et al., 2021). The photocatalytic degradation mainly depends on the energy of the absorbed photon, its type, and the properties of the semiconductor material used (Soni et al., 2022). When the energy of the photon is higher than or equal to that of the band gap energy of the semiconductor, the electron in the valance band (VB) will be shifted to the conduction band (CB) producing positive holes (h^+) in the VB (Soni et al., 2023). Thus, the charge isolation on the surface of the semiconductor takes place which will further react with surface adsorbed O_2 and H_2O to generate reactive oxygen species to initiate photodegradation. These generated unstable reactive radical species distort the structure of microplastics directly by chain breakage and thus degrade/mineralize the microplastics adsorbed on the surface of a photocatalytic semiconductor to CO_2 and H_2O as shown in (Fig. 2a). Various visible or UV light active semiconductor materials are used for microplastic degradation like metal oxides, metal sulfides, metal-free, MOF based, etc. (Du et al., 2021a).

3.1 Photocatalytic Metal Oxides-Based Photocatalysts

To overcome the inherent limitations like wide band gap energy, and high charge recombination rate in bare metal oxides photocatalysts, various studies are focused on the modification of bare photoactive metal oxides by the deposition of plasmonic NPs, using dopants or through heterojunction formation with other potential semiconductors with apt band alignment (Du et al., 2021a). The overall aim is to modify the

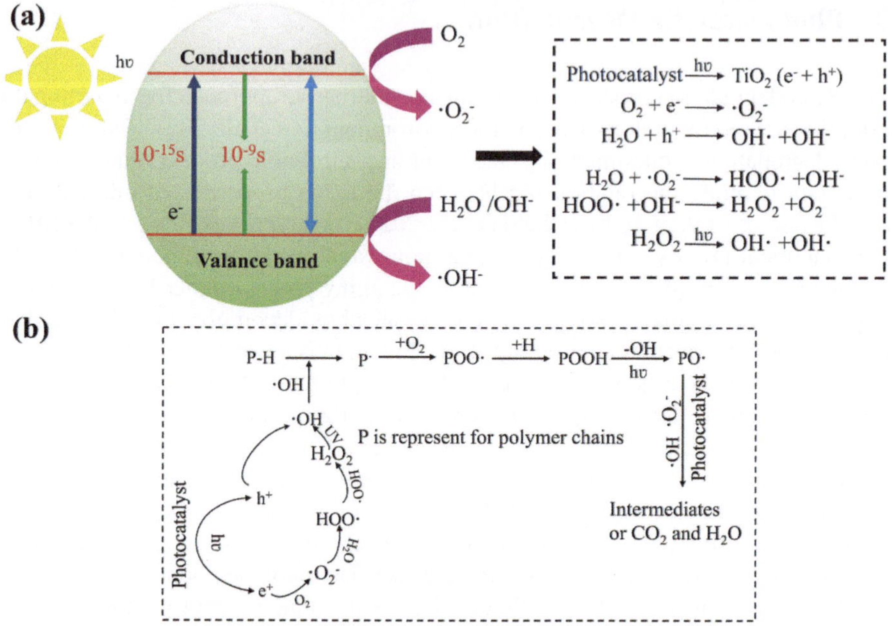

Fig. 2 **a** The proposed photocatalytic degradation mechanism of pollutants over semiconductor photocatalyst, **b** the plastic photo-degradation mechanism. Reproduced from Du et al., (2021a) (License no. 5687741095341)

surface condition of bare photocatalysts to enhance the microplastic photocatalytic degradation process by reducing charge recombination and enhancing visible light absorption. For instance, in a report, the addition of non-metal and metal dopants into the lattice of bare metal oxide semiconductor leads to increased microplastic degradation. Llorente-García et al. fabricated an N-doped TiO_2 coating via. the evaporation-induced self-assembly process to investigate the degradation rate of high-density polyethylene microplastic (Llorente-García et al., 2020). The breakdown of targeted polyethylene was performed at 400–800 nm visible light; and after 50 h the observed weight loss was calculated to be 4.65%. High-density polyethylene plastic is mainly degradable under UV light, and is highly resistant to visible light. From the experimental results, it was detected that when the size of polyethylene was halved, the final weight loss rate increased 21 times.

Similarly, the maximum weight reduction of 14.34%, polyethylene microplastics by Fe/Ag doped TiO_2 films was found which was relatively higher than the mono doped Ag-TiO_2 film (13.75%), and Fe-TiO_2-composite film (13.49%) (Asghar et al., 2010). The incorporation of both Fe and Ag dopants in TiO_2 mainly enhanced the isolation rate of charge carriers. SEM images demonstrated that on the exposure of microplastics to UV irradiations, different-sized cavities were observed on the surface of targeted polyethylene. The degradation of polyethylene-TiO_2/Fe/Ag under UV irradiation is initiated by the direct absorption of *hv,* where both Fe and Ag dopants serve as e^- and h^+ trapping sites. The charge isolation significantly enhanced the generation of $O_2^{\bullet-}$ and •OH radicals on the surface, which attack the polymer chains along with the matrix. In addition to these modification strategies, heterojunction formation has illustrated higher microplastic degradation efficiency owing to more spatial charge isolation as a result of band bending. Qin et al. studied the photodegradation of polyethylene terephthalate as well as polyethylene microplastics using various Ag_2O/Fe-MOF, Ag_2O, and Fe-MOF photoactive semiconductors (Qin et al., 2022). The potential reaction mechanism for microplastic degradation over Ag_2O/Fe-MOF is the development of a p–n junction between Ag_2O and Fe-MOF, which accelerates the movement of e^- and h^+ through the heterojunction by forming an internal electric field from Fe-MOF to Ag_2O. Under visible light, the photoexcited e^- on the surface of Ag_2O can be transported to the CB of Fe-MOF and utilized for producing $O_2^{\bullet-}$ radicals followed by H_2 production. Whereas, the h^+ are migrated from VB of Fe-MOF to VB of Ag_2O semiconductor for •OH radical generation. Furthermore, the generated electric field at the Ag_2O–Fe-MOF interface boosts the separation efficiency of the charge carrier in the bulk. The photoexcited radicals produced upon Ag_2O–Fe-MOF photocatalyst surface were utilized for decomposing the plastic molecules into short-chain organic molecules. The experimental results demonstrated that the net weight loss of polyethylene and polyethylene terephthalate using Ag_2O/Fe-MOF was significantly higher than bare Ag_2O, and Fe-MOF, respectively. Another experimental research by Feng et al. investigated the photocatalytic degradation of high-density polyethylene microplastic in the presence of NH_2-MIL-88B(Fe)/MoS_2 nanocomposite (Feng et al., 2022). The surface morphology of the high-density polyethylene samples was analyzed by SEM during the photolysis process in the presence of simulated sunlight. As the time of light exposure increased, several cracks and textures were observed on the high-density polyethylene surface. As-fabricated NH_2-MIL-88B(Fe)/MoS_2 heterojunction construction enhanced the e^-/h^+ pair isolation rate, thereby extending the lifetime of the radical and indirectly generating unstable •OH radicals. Under visible light irradiation, the photogenerated e^- combines with O_2 at the CB to generate $O_2^{\bullet-}$ and •OH radicals that play a significant role in breaking the structure of high-density polyethylene and catalysing the photodegradation reaction. As-fabricated chemically stable, environmentally friendly NH_2-MIL-88B(Fe)/MoS_2 nanocomposite showed good structural integrity, high-efficiency photocatalytic degradation and great recoverability for reuse of the catalyst.

Similarly, another BiOI-Fe_3O_4 micro-swimmer demonstrates the superior photodegradation of polystyrene latex microplastics in the presence of visible light (Khairudin et al., 2022). The n-p heterojunction at the interface of BiOI and Fe_3O_4 generated more photoexcited charge carriers and enhanced the ability of light absorption. The magnetically recyclable BiOI-Fe_3O_4 micro-swimmers adsorb and transport the targeted polystyrene latex till it forms a larger aggregate that settles down. When visible light is irradiated, the BiOI- Fe_3O_4 micro-swimmer binds with the surface generated $O_2^{\bullet-}$, and •OH radicals in the polystyrene latex suspension. These radicals help in the breakdown of the polystyrene structure, giving intermediate products which are photo-catalytically oxidize to simple CO_2 and H_2O.

3.2 MOF-Based Semiconductor Photocatalysts

Metal–organic frameworks (MOFs) are a promising class of photocatalysts made up of metal clusters along with organic linkers and serve as a platform for the transportation of photogenerated electrons via ligand-to-metal charge transfer (LMCT) (Poonia et al., 2023). Wherein, the organic coordinating ligands photoexcited e^- to the metal clusters in MOF. A TiO_2/MIL-100 (Fe) photocatalyst was synthesized via solvothermal/microwave method and post-annealing method, resulting in better photocatalytic performance for polyethylene terephthalate (PET) microplastic degradation. The extended absorption of the catalyst, higher surface area, and lower bandgap up to 2.21–2.65 eV (Rojas-Guerrero et al., 2023). Moreover, the BET surface area of MIL-100 (Fe) increased with increasing MIL-10 (Fe) (128.2 m^2/g) loading in TiO2/MIL-100 (Fe) (198.8 m^2/g) composites. Acidic pH was the ideal range for heterogeneous photocatalytic degradation, owing to prominent Coulomb attraction between catalyst and plastic resulting in a degraded polymer matrix. Similarly, at a pH range of 3–10, ZIF-67 exhibited an 88.3% adsorption rate of polystyrene (PS) microplastics, which was ascribed to a greater electrostatic attraction between the negatively charged PS microplastics and the positively charged ZIF-67 (Wan et al., 2022). Hydrogen bonding as well as π-π stacking of ZIF-67 and PS enhanced their elimination in an aqueous solution (Fig. 3a). However, the repulsive force between anionic-charged ZIF-67 and negatively charged PS microplastics decreased their adsorption under extremely alkaline circumstances (pH 11–12). Additionally, because of the hydrophilicity of MOF complexes, they are more likely to participate in the degradation process. For example, compared to ZIF-8, Zr-MOF had a larger pore capacity and therefore offered a free surface for frequent free radical movement (Nikhar et al., 2023). Under UV light, Zr-MOF experiences a charge transfer mechanism that changes Zr^{4+} into Zr^{3+}. These Zr^{3+} oxidize into Zr^{4+} by producing an $\bullet O^{2-}$ radical through their reaction with molecular oxygen. The efficiency of catalysts is ascribed to the LUMO–HOMO mechanism, which establishes the band gap between the electron-donating capacity of pollutants and the capacity to receive electrons from the catalyst. Because of its intricate molecular structure, Zr-MOF performed better than polyvinyl propylene (PVP) in polystyrene sulfonate (PSS) degradation (95%) while ZIF-8 performed better in PVP (65%) because of its high electron density imidazole-ring, which facilitates excellent charge transfer. The effective reaction between microplastics and generated radicals successfully breaks down plastic molecules into short-chain organic compounds (Golmohammadi et al., 2023). For instance, in a nanoflower hybrid, MoS_2 dopant in NH2-MIL-88B(Fe) created a heterojunction, which increased catalytic performance by the adsorption step. The original High-density polyethylene (HDPE) surface was smooth, but light irradiation from 6 to 48 h made it rough. With an increase in photodegradation time, the HDPE broke down into tiny particles, creating holes in its interior chamber that resembled honeycombs. Under light circumstances, the quantity of small particles rose as the particles aged and their size shrank. The fraction of HDPE with particle sizes less than 3 μm rose from 20 to 95.6%, with around 80% of the initial HDPE falling into this range (Feng et al., 2022). Similarly, as per the bandgap alignments (Fig. 3b), Fe-MOF demonstrated a lower valance band (n-type semiconductor) and a greater conduction band compared to Ag_2O (p-type semiconductor). When implanted in porous Fe-MOF, this resulted in the formation of a strong heterojunction Ag_2O/Fe-MOF corresponding to a weight loss (27.5 mg) of polyethylene glycol (PEG) and the total organic content up to 141.6 mg l^{-1} whereas Ag_2O alone was only 47.9 mg l^{-1} (Qin et al., 2022). This may be attributed to p-n heterojunction structure after the integration of Ag_2O and Fe-MOF, which made electron–hole separation easier. Furthermore, structural defects in the Ag_2O/Fe-MOF provided more open frameworks and active binding sites in the MOF.

3.3 Photocatalytic Bismuth-Based Semiconductor Photocatalysts

Since most bismuth-based photocatalysts have a bandgap of less than 3.0 eV, which allows them to be used in visible light. The electrical structure of the band produces hybrid O 2p and Bi 6s orbitals, which enhance the mobility of charge carriers created during photosynthesis. (Mengting et al., 2022). Bi-based photocatalysts are more conducive to hole creation and transportation, promoting photocatalytic activity. Similarly, mobile electrons (e^-) produced in the conduction band (CB) and holes (h^+) in the valence band (VB) of BiOCl particles adsorbed oxygen molecules that produce $\bullet O^{2-}$ species and hydroxyl ions can trap photogenerated holes to produce hydroxyl radicals (•OH). Energy provided with prolonged light exposure (90 h) formed radicals (•OH and $\bullet O^{2-}$) on BiOCl films which caused cavity formation on the surface of PS via chain scission, and oxidation processes (Sarwan et al., 2020). When oxygen breaks a chain, carbonyl or carboxyl groups are produced, which start breakdown processes. Polymeric chains are attacked by radicals with a carbon core, which breaks lengthy chains and forms volatile organic chemicals such as ethanol, formaldehyde, acetaldehyde, and ethene. Further oxidation of these intermediates can produce CO_2 and H_2O. Recently, Du et al. introduced potassium hexafluorophosphate (KPF6) to modify BiOBr nanosheets, to improve microplastic degradation performance. The in-situ generation of H_2O_2 using KPF6/BiOBr materials also decomposed discarded polyethylene bags, resulting in a 2.33 times higher mass loss of plastics than pure BiOBr (Du et al., 2021b). Even after 8 h of degradation, polyethylene plastic remained a smooth membrane with a rough surface. In contrast, 20 wt % KPF6/BiOBr solution containing in-situ H_2O_2 showed rougher surfaces and more surface fractures. This might be because of increased •OH oxidation of the surface chemical structure, which starts the breakdown of polymers and embrittlement of PE. In order to find possible byproducts of polyethylene degradation with in-situ H_2O_2 generation, the LC/MS approach was employed. These include h+, $\bullet O^{2-}$, and •OH radicals, which depolymerize polyethylene polymeric chains, generating hexane (P1), carbon chain free radicals (P2), alkoxy radicals (P3), and

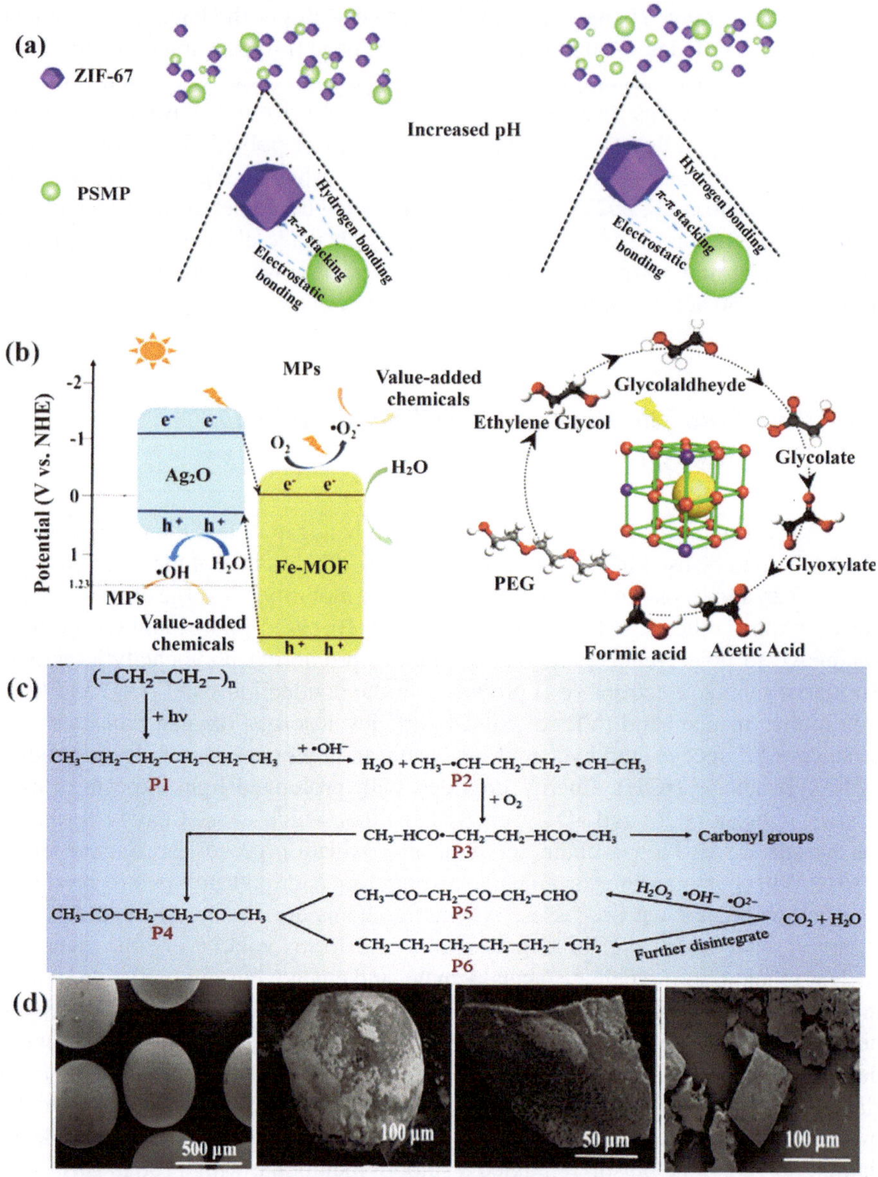

Fig. 3 **a** The diagrammatic illustration of the influence of increasing pH on the adsorption of PS microplastics by ZIF-67. Reproduced from Wan et al. (2022) with permission CC by 4.0. **b** A potential photocatalytic mechanism for MPs conversion in Ag_2O/Fe-MOF shows potential for PEG conversion Reproduced from Qin et al. (2022) with permission CC BY-NC-ND 4.0 DEED. **c** Possible process of PE degradation and intermediated formed using KPF6/BiOBr photocatalyst. Reproduced from Du et al., (2021b) License number 5692890097891. **d** The SEM image displaying the photodegraded PE by BiOCl-1 at different time intervals (1, 5, and 10 h). Reproduced from Jiang et al. (2021) License number 5693500488631

carbonyl groups (P5) (Fig. 3c). Advanced water treatment systems are increasingly using micromotors/microswimmers for environmental sensing and pollutant removal (Jurado-Sánchez & Wang, 2018). Bi_2WO_6 micromotors similar to perovskites are synthesized by self-assembling nanoparticles into three-dimensional microspheres using the Ostwald ripening technique. Fe_3O_4 nanoparticles were applied to visible light-driven $BiVO_4$ micromotors in the form of stars to enhance their magnetic separation abilities. In a time duration of a week, micromotors were able to reduce the bulk of polylactic acid by 3% and effectively absorb 70% of the distributed fragments in a 5 cm channel (Khairudin et al., 2022). The FTIR analysis of polystyrene latex/ethanol solution before and after the photocatalytic degradation process showed that new peaks were formed in the hydroxyl (1740–1720 cm^{-1}) and carbonyl (1658–1648 cm^{-1}) region. Multiple peaks indicated alkene C= C bonds and conjugated C=C were formed through β-scission, whereas intramolecular H transfer verified the breakdown of polystyrene latex microplastics. The production of C–OH and C=O bonds, respectively, was the outcome of the macromolecular structure interaction with the water molecule.

Surface functional groups, in particular the surface OH– group, have a major role in catalyst degradation. Examples of this include the enhancement of •OH formation and the advantage of surface OH– discrete BiOCl in photoinduced oxygen vacancy and defects will make the surface porous ultimately increasing the adsorption. This statement was in agreement with XPS analysis which depicted a noticeable C–O peak in the O 1s spectrum and revealed that mannitol-treated BiOCl effectively attacked and oxidized C on the microplastic surface (Jiang et al., 2021). The Bi–O and OH peaks confirmed the adsorption of the catalyst on the microplastic, simultaneously SEM images showed (Fig. 3d) that mannitol-treated BiOCl broke down the PE-S microplastic, which was formerly a smooth spherical into a rough and porous structure. The microplastic particles broke down into smaller pieces and hemispheres, with a less smooth surface after 5 h whereas 10 h degrading period further broke down into smaller pieces.

3.4 Other Photocatalysts

When exposed to light energy, hydroxyl groups or surface-bound water molecules, produce superoxide anion radicals ($\bullet O^{2-}$) and hydroxyl radicals ($^{\bullet}OH$). These radicals demineralize into CO_2, H_2O, and CH_4 after depolymerizing microplastic polymer into monomers (Lastovina & Budnyk, 2021). In a recent study, the production of –OH, C=O, and C–H groups was shown by in-situ DRIFTS and mass spectrometer investigation of the degradation process, demonstrating that polystyrene underwent photodegradation. Proton nuclear magnetic resonance is also utilized for high precision in evaluating the low molecular intermediate products, in addition to assessing the mass loss and shape (Ouyang et al., 2021). Synergistic H_2 evolution and microplastic degradation were instigated by a very effective MXene/$ZnxCd_{1\text{-}x}S$ photocatalyst. When 3 g of waste polyethylene terephthalate bottles were broken down into ethylene glycol and terephthalic acid (Cao et al., 2022). Only terephthalic acid and ethylene glycol were the primary components before the degradation process, but following the reaction, organic compounds such as methanol, formate, and acetic acid were discovered by proton nuclear magnetic resonance analysis (Fig. 4a). A similar report stated that under visible light illumination, the hydrothermally aided synthesis of $NiAl_2O_4$ spinel is advantageous for the degradation of polypropylene (PP). The catalytic activity was compared for the hydrothermal technique and the co-precipitation method for the synthesis of $NiAl_2O_4$ spinel. However, because of the heating rate, reaction duration, and annealing, hydrothermally produced $NiAl_2O_4$ spinel encourages greater PP-LDPE degradation as compared to co-precipitation (Venkataramana et al., 2021). Evidently, the –CH group of polymer diminishes around 2912 cm^{-1} as a result of oxidative breakdown. PP-LDPE had a starting weight of 0.04 g, however, because of the photooxidation process with synthesized $NiAl_2O_4$ spinel effectively decreased the weight of PP-LDPE. Moreover, photocatalysts have also been used for PMS activation; since PMS is activated by the generation of the electron/hole equivalent reaction intermediate of catalyst. Wan et al. stated that in electron transport via the Type II mechanism of CdS/CeO_2, oxygen molecules may convert to superoxide radicals because they cannot directly oxidize H_2O to form •OH whereas O_2 and PMS may be reduced to $•O^{2-}$ and $SO_4^{•-}$ by the ensuing e^- respectively (Fig. 4b). Wherein PET-12 plastic and $•O^{2-}$ may be directly oxidized by the h+ in valence band of CdS and provide a negligible quantity of $^1O_2$ (Wan et al., 2023). The direct activation of PMS by Ce^{3+} and Ce^{4+} transition ions resulted in the generation of $SO_4^{•-}$, which oxidizes PET-12 plastic by immediately breaking down the polymer into intermediates with a rate of 93.92% weight loss upon adding 3 mM PMS. The results were in best alignment with EPR, stating that the most important contributions radicals were $SO4^{•-} > •OH > •O^{2-} > h^+ > {}^1O_2$. Reactor development, reaction system, and catalyst material design all affect photocatalytic degradation efficiency. Nonetheless, Exploring the activity of photocatalysts beyond UV and visible light, modification for photocatalysts is often towards increasing the spectrum absorption range. MoS_2 generated photogenerated electrons and holes when exposed to NIR lamp light. Due to the greater LUMO of polydopamine, electrons can move under boron-nitride transport to the CB of MoS_2, inhibiting complexation until the Fermi energy level achieves equilibrium. Owing to their great potential, hole interactions with water molecules are challenging to create •OH (Xu et al., 2023). For fiber-based microplastics (FMPs), hole oxidation can be employed, and for organic materials, photoelectrons can create superoxide radicals that oxidize them. TOC values and weight loss of degrading solutions and FMPs were noted in the study indicating within 48 h, weight loss rates for PLA-FMPs and PET-FMPs were 92 and 35%, respectively. Consistent with the results of the TOC test, the hydrolysis of polyester-based polymers produced mostly aldehydes. In addition, the hydrolysis solution can produce a variety of organic compounds (Fig. 4c). Hydrolysis-susceptible plastics may be degraded by high concentrations of hydrogen (acid) in an aquatic environment or organic water molecules additionally the generation of radicals is accelerated. The study of heterojunction between g–C_3N_4 and WO_3 used EPR to detect active radicals, revealed a clear hydroxyl radical signal at pH 7

(Wang et al., 2023). However, detecting $\cdot O^{2-}$ is crucial in oxidative stress and can lead to the formation of other reactive oxygen species (ROS), which is not evident at pH > 14 (Fig. 4d). H-NMR was in strong agreement with the radical generation process and confirmed PET breakdown products showed that terephthalic acid and ethylene glycol decomposed first, with formate, acetic acid, methanol, and ethanol forming later.

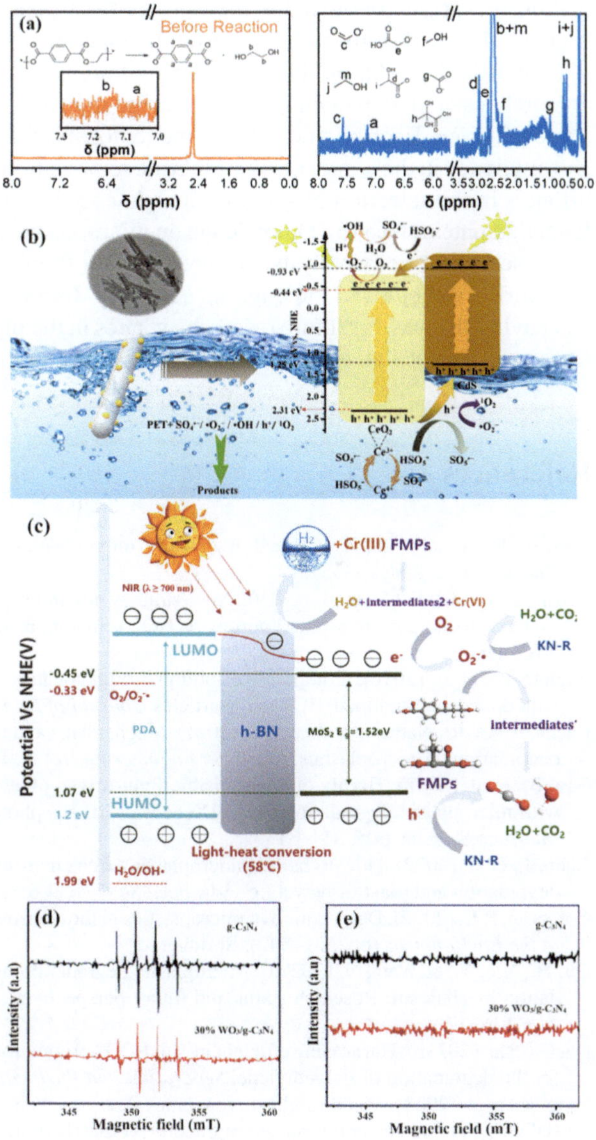

Fig. 4 **a** 1 H NMR spectroscopy of MXene/ $ZnxCd_{1-x}S$ before and after the microplastic degradation reaction. Reproduced from Cao et al. (2022) License number 5693500872635. **b** Schematic diagram of the electron transfer process and catalytic mechanism. Reproduced from Wan et al. (2023) License number 5692891229708. **c** diagrammatic illustration of charge transfer mechanism of the MoS_2/BN system for fiber-based microplastics. Reproduced from Xu et al. (2023) License number 5692900005985. The EPR spectra of g–C_3N_4 and 30% WO_3/g–C_3N_4 show peaks at **d** pH = 7 and **e** pH > 14. Reproduced from Wang et al. (2023) License number 5693501367582

4 Conclusion

This book chapter provides a nuanced exploration of the existing landscape of microplastic pollution, introducing the scale and implications of this environmental challenge. The study systematically reviews conventional and advanced methodologies employed for microplastic remediation, emphasizing the need for innovative approaches to enhance efficiency. The book chapter delves into an exhaustive examination of various photoactive metal oxides, carbon, bismuth, and metal–organic framework-based semiconductor, providing a comparative analysis of their roles in microplastic degradation. By synthesizing findings from diverse studies, it offers a comprehensive understanding of the specific radical pathways involved in photocatalytic reactions. Each photocatalyst is meticulously evaluated for its unique properties and efficiency in the context of microplastic remediation, aiding researchers and practitioners in the selection of suitable candidates for distinct environmental scenarios. It amalgamates a wealth of knowledge on microplastic pollution, advanced oxidation processes, and photocatalysts. By presenting a thorough analysis of the literature, this work contributes to the ongoing scholarly discourse, offering valuable insights and paving the way for future research avenues in the pursuit of sustainable solutions to address the escalating crisis of microplastic pollution.

References

Ahmed, R., et al. (2022). Critical review of microplastics removal from the environment. *Chemosphere, 293*, 133557.

Ariza-Tarazona, M. C., et al. (2019). New strategy for microplastic degradation: Green photocatalysis using a protein-based porous N-TiO2 semiconductor. *Ceramics International, 45*(7), 9618–9624.

Asghar, W., et al. (2010). Comparative solid phase photocatalytic degradation of polythene films with doped and undoped TiO2 nanoparticles. *Journal of Nanomaterials, 2011*, 20.

Bacha, A.-U.-R., Nabi, I., & Zhang, L. (2021). Mechanisms and the engineering approaches for the degradation of microplastics. *Acs Es&t Engineering, 1*(11), 1481–1501.

Cao, B., et al. (2022). Highly-efficient visible-light-driven photocatalytic H2 evolution integrated with microplastic degradation over MXene/ZnxCd1-xS photocatalyst. *Journal of Colloid and Interface Science, 605*, 311–319.

Chen, J., et al. (2022). How to build a microplastics-free environment: Strategies for microplastics degradation and plastics recycling. *Advanced Science, 9*(6), 2103764.

Corcoran, P. L. (2022). Degradation of microplastics in the environment. *Handbook of Microplastics in the Environment* (pp. 531–542). Springer.

Du, H., Xie, Y., & Wang, J. (2021a). Microplastic degradation methods and corresponding degradation mechanism: Research status and future perspectives. *Journal of Hazardous Materials, 418*, 126377.

Du, C., et al. (2021b). Harnessing efficient in-situ H2O2 production via a KPF6/BiOBr photocatalyst for the degradation of polyethylene. *Separation and Purification Technology, 279*, 119734.

Feng, X., et al. (2022). Visible-light-driven removal of tetracycline hydrochloride and microplastics (HDPE) by nano flower hybrid heterojunction NH2-MIL-88B (Fe)/MoS2 via enhanced electron-transfer. *Separation and Purification Technology, 302*, 122138.

Golmohammadi, M., et al. (2023). Molecular mechanisms of microplastics degradation: A review. *Separation and Purification Technology, 309*, 122906.

Jiang, R., et al. (2021). Microplastic degradation by hydroxy-rich bismuth oxychloride. *Journal of Hazardous Materials, 405*, 124247.

Jurado-Sánchez, B., & Wang, J. (2018). Micromotors for environmental applications: A review. *Environmental Science: Nano, 5*(7), 1530–1544.

Khairudin, K., Bakar, N. F. A., & Osman, M. S. (2022). Magnetically recyclable flake-like BiOI-Fe3O4 microswimmers for fast and efficient degradation of microplastics. *Journal of Environmental Chemical Engineering, 10*(5), 108275.

Kim, S., et al. (2022). Advanced oxidation processes for microplastics degradation: A recent trend. *Chemical Engineering Journal Advances, 9*, 100213.

Klein, S., et al. (2018). Analysis, occurrence, and degradation of microplastics in the aqueous environment. *Freshwater Microplastics: Emerging Environmental Contaminants?* 51–67

Lastovina, T. A., & Budnyk, A. P. (2021). A review of methods for extraction, removal, and stimulated degradation of microplastics. *Journal of Water Process Engineering, 43*, 102209.

Lin, Z., et al. (2022). Current progress on plastic/microplastic degradation: Fact influences and mechanism. *Environmental Pollution, 304*, 119159.

Llorente-García, B. E., et al. (2020). First insights into photocatalytic degradation of HDPE and LDPE microplastics by a mesoporous N-TiO2 coating: Effect of size and shape of microplastics. *Coatings, 10*(7), 658.

Mengting, Z., et al. (2022). Advances in BiO X-based ternary photocatalysts for water technology and energy storage applications: Research trends, challenges, solutions, and ways forward. *Reviews in Environmental Science and Bio/technology, 21*(2), 331–370.

Nikhar, S., Chakraborty, M., & Kumar, P. (2023). Relative capability analysis on carboxylate and imidazole based MOFs for PVP and PSS photo catalytic degradation in water. *Inorganic Chemistry Communications, 156*, 111132.

Ouyang, Z., et al. (2021). Recent advances in photocatalytic degradation of plastics and plastic-derived chemicals. *Journal of Materials Chemistry A, 9*(23), 13402–13441.

Poonia, K., et al. (2023). Recent advances in Metal Organic Framework (MOF)-based hierarchical composites for water treatment by adsorptional photocatalysis: A review. *Environmental Research* 115349.

Qin, J., et al. (2022). In-situ formation of Ag2O in metal-organic framework for light-driven upcycling of microplastics coupled with hydrogen production. *Applied Catalysis b: Environmental, 319*, 121940.

Rojas-Guerrero, C., et al. (2023). Solar photocatalytic degradation of polyethylene terephthalate nanoplastics: Evaluation of the applicability of the TiO2/MIL-100 (Fe) composite material. *Journal of Environmental Chemical Engineering, 11*(5), 110415.

Sarwan, B., et al. (2020). Visible light photocatalytic deterioration of polystyrene plastic using supported BiOCl nanoflower and nanodisk. *European Polymer Journal, 134*, 109793.

Song, Y. K., et al. (2017). Combined effects of UV exposure duration and mechanical abrasion on microplastic fragmentation by polymer type. *Environmental Science & Technology, 51*(8), 4368–4376.

Soni, V., et al. (2021). Sustainable and green trends in using plant extracts for the synthesis of biogenic metal nanoparticles toward environmental and pharmaceutical advances: A review. *Environmental Research, 202*, 111622.

Soni, V., et al. (2022). Current perspective in metal oxide based photocatalysts for virus disinfection: A review. *Journal of Environmental Management, 308*, 114617.

Soni, V., et al. (2023). Fabricating cattle dung-derived nitrogen-doped biochar supported oxygen-deficient ZnO and Cu2O-based novel step-scheme photocatalytic system for aqueous Doxycycline hydrochloride mitigation and Cr (VI) reduction. *Journal of Environmental Chemical Engineering, 11*(5), 110856.

Uheida, A., et al. (2021). Visible light photocatalytic degradation of polypropylene microplastics in a continuous water flow system. *Journal of Hazardous Materials, 406*, 124299.

Venkataramana, C., et al. (2021). Photocatalytic degradation of polyethylene plastics by NiAl2O4 spinels-synthesis and characterization. *Chemosphere, 265*, 129021.

Wan, H., et al. (2022). Removal of polystyrene microplastics from aqueous solution using the metal–organic framework material of ZIF-67. *Toxics, 10*(2), 70.

Wan, Y., et al. (2023). Enhanced degradation of polyethylene terephthalate plastics by CdS/CeO2 heterojunction photocatalyst activated peroxymonosulfate. *Journal of Hazardous Materials, 452*, 131375.

Wang, X., et al. (2023). Preparation of heterojunction C3N4/WO3 photocatalyst for degradation of microplastics in water. *Chemosphere* 139206.

Wu, X., et al. (2022). Critical effect of biodegradation on long-term microplastic weathering in sediment environments: A systematic review. *Journal of Hazardous Materials, 437*, 129287.

Xu, Q., et al. (2021). Coagulation removal and photocatalytic degradation of microplastics in urban waters. *Chemical Engineering Journal, 416*, 129123.

Xu, N., et al. (2023). Near-infrared-dominated photothermal nanofluids for degradation and H2-producing photochemical reforming of fiber-based microplastics. *Ceramics International, 49*(15), 24730–24743.

Zhu, K., et al. (2020). Long-term phototransformation of microplastics under simulated sunlight irradiation in aquatic environments: Roles of reactive oxygen species. *Water Research, 173*, 115564.